THE ENCHANTING ARITHMETIC

IV EDITION

- R. SRIDHARAN

No.	III edition	IV edition
1.	Total number of pages is less by nearly 70 pages than that of proposed IV edition.	Total number of pages is more than by nearly 70 pages with extra information that that of III edition.
2.	Less information about number types and classification of numbers are given.	More information about number types and classification of numbers are given elaborately.
3.	Features of number one and nine are not discussed separately.	Features of number one and nine are discussed separately.
4.	Comparison of numbers is not tabulated.	Comparison of numbers with their arithmetic properties is tabulated.
5.	Diagrammatic representation of square numbers and Hogben numbers not given.	Diagrammatic representation of square numbers and Hogben numbers in spiral form is given.
6.	Prime numbers are not discussed separately.	Prime numbers are discussed separately.
7.	Magic circle is not explained.	Magic circle is introduced.
8.	Solved examples with exercises are not given.	Solved examples with exercises are given. This section will be useful in preparing competitive exams.
9.	Explanation about the formation of square numbers is not given.	The formation of square numbers is elaborately given.
10.	Relationship among the power numbers is not given.	Pictorial relationship among the power numbers is given.

Dedicated

To

My wife Jayasri

PREFACE

Mathematics is an important subject without which we cannot imagine the modern world of science and technology. In mathematics, arithmetic is the backbone of mathematical sciences. Here we deal with some basic concepts of arithmetic and their applications which you will also feel amazing and make more interesting to read a lot.

This book is written in a simple language with examples, and definitions for easy understanding of the readers from various backgrounds. I sincerely hope this book will provide a better understanding of basic concepts in arithmetic. Any suggestions and constructive criticism towards this book can be sent to sri_chemist@yahoo.co.in and sri1chemist@gmail.com will be gratefully acknowledged. Further contact can be made through 97895 28940.

Author

S.No.	Contents	Page no.
1.	Basic arithmetic	7
2.	Addition	33
3.	Subtraction	43
4.	Multiplication	45
5.	Powers	77
6.	Division	111
7.	Mirror image and Palindrome numbers	140
8.	Arithmetic of zero and infinity	151
9.	Sequences and series	157
10.	Polygonal numbers	163
11.	Mathematics can be fun	211
	Annexure Annexure 1:- Glossary of numbers Annexure 2:- Comparison of mathematical properties Annexure 3:- Comparison of numbers Solved examples with exercise Subject index References	229 254 256 271 275 282

1. BASIC ARITHMETICS

Arithmetic is an important subject. The concepts and ideologies discussed in arithmetic are to be borne in mind from layman to the learned man for their betterment of life.

The word arithmetic means the 'art of calculation', but with the passage of time, the subject of arithmetic has been transformed into theory of numbers.

The set of positive integers 1, 2, 3, 4, 5 … naturally thought of first in the process of counting, is one component of what are called as real numbers or natural numbers and denoted as 'N'. Real numbers are positive numbers. It starts from 1 onwards. Real numbers form the alphabets of arithmetic. The notation used is

$$N = \{1, 2, 3, 4, 5, 6...\}$$

One important factor about natural number is that two separate natural numbers cannot have same value. That is either one number is greater than the other number or lesser than the other number. The mathematical symbol to show this comparison is '>' which stands for 'greater than' and the opposite symbol '<' stands for lesser than. For example 2>1 and 3<4.

Note: - Counting of the objects by 1, 2, 3, 4, 5 … is called as cardinal numbers. Counting of the objects by first, second, third, fourth, fifth… is called as ordinal numbers.

India contributed the "zero" and the set of whole numbers include zero in the real number system. Thus the whole number set denoted by 'W' is {0, 1, 2, 3, 4...}

The addition operation of natural numbers has closure property that is when two natural numbers are added again we will get a natural number or positive number. But the subtraction operation of natural number like the larger number from smaller number will give negative number and is not a natural number. So the need of negative numbers came into existence. The set of negative integers is denoted as {-1, -2, -3, -4, -5...}

So to the whole number set, was added the set of negative numbers to make up

$$Z = \{...-5, -4, -3, -2, -1, 0, 1, 2, 3, 4, 5...\}$$

as the set of positive and negative integers including zero. The positive number and the negative number are like a mirror image of natural number. That is negative numbers are opposite of positive numbers.

Then the need arose to include the fractions, both proper and improper fractions to form the set Q of rational numbers. This is definable as a number capable of having thrown in the form of p/q where p and q are integers and q≠0.

Even after this the system was found to be inadequate when one thought of numbers like √2, √3, √7 and even ∏ (used in the calculation of circumference or area of a circle) are not numbers that can be put in the form p/q where p and q are integers and q≠0, a definable classification of rational numbers. Logically that which is not rational, got designated as irrational numbers.

This completes the formulation of real numbers – either as rational or irrational. That is real numbers comprise both the rational and irrational numbers. The complex numbers consist of real number and an imaginary number.

The classification of numbers can be diagrammatically represented as follows.

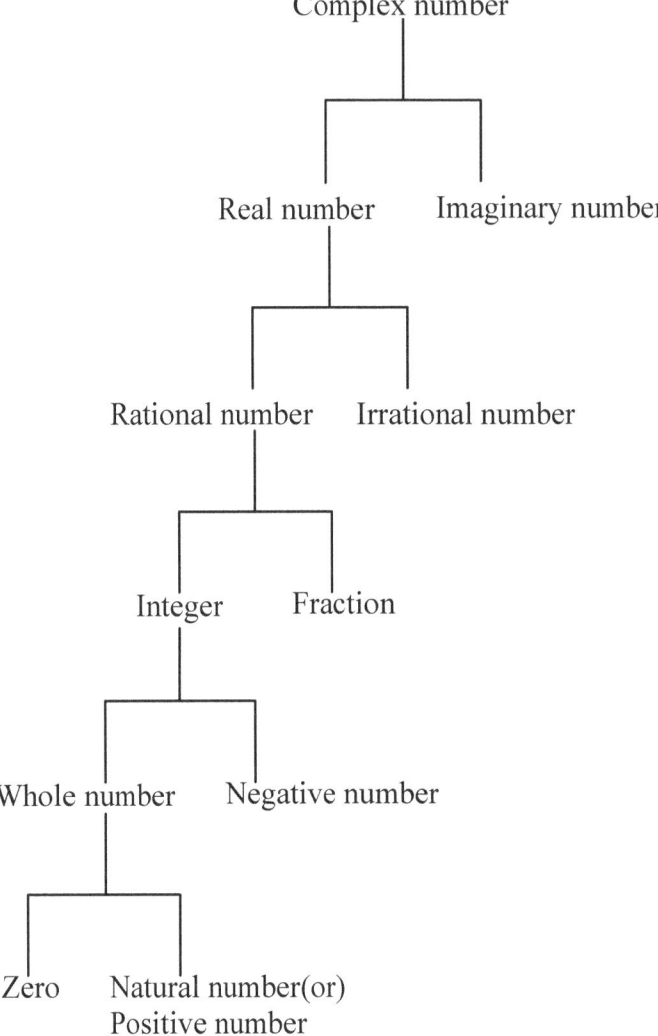

In this treatise, mostly simplified methods of algebraical operations, among numbers are dealt with and hence often we talk about integers. (The term integers and numbers refer the same context only).

Generally based upon the digits in a number, it is classified into single digit number, two digit number, three digit number and so on. The common name for indicating a number having more than two digits is multidigit number.

Even number or Paired number: -

If a real number is divisible by 2 or multiples of 2 means it is an even number. We can also call even number as paired numbers. For example 14 is divisible by 2 and so it is an even number.

The even number series is 0, 2, 4, 6, 8, 10, 12…

Even numbers end with the digits 0, 2, 4, 6 or 8.

All even numbers are composite numbers but all composite numbers are not even numbers. For example 27 is an odd number as well as a composite number but 23 is only an odd number but not a composite number.

All the real numbers as well as negative numbers when multiplied by 2 will give even numbers.

The even numbers which are power numbers are called as even power numbers. For example 64 is an even power number.

Super even number: -

If all the digits in a number are even then it is called as a super even number. For example 6248 is a super even number and 6148 is not a super even number.

Obviously super even numbers start from two digit number onwards.

The sequence of super even number is 20, 22, 24, 26, 28, 40, 42, 44, 46, 48, 60…

Odd number or unpaired number: -

If a real number is indivisible by 2 means it is an odd number. We can also call odd number as unpaired numbers. For example 15 is indivisible by 2 and so it is an odd number. The odd number series is 1, 3, 5, 7, 9, 11 …

Odd numbers end with the digits 1, 3, 5, 7 or 9.

All prime numbers are odd numbers but all odd numbers are not prime numbers. For example 13 is an odd number as well as a prime number but 21 is an odd number but not a prime number.

For an odd number 'n', "n^2-1" is divisible by 8. For example if n=13 then the resultant of $13^2-1=168$ is divisible by 8.

The odd numbers are opposite of even numbers.

Super odd number: -

If all the digits in a number are odd then it is called as a super odd number. For example 519 is a super odd number and 619 is not a super odd number.

The super odd numbers start from two digit number onwards.

The sequence of super odd number is 11, 13, 15, 17, 19, 31, 33, 35, 37, 39, 51…

Super odd numbers are opposite of super even numbers.

Alternating number: -

A number is called as an alternating number, if in its representation odd and even digits come alternately. For example 254387 is an alternating number because it contains the odd and even digits alternatively.

Obviously alternating numbers start from two digit number onwards only.

The alternating number sequence is 10, 12, 14, 16, 18, 21, 23, 25, 27, 29, 30, 32, 34, 36, 38, 41, 43, 45, 47, 49, 50, 52, 54, 56, 58, 61, 63, 65, 67, 69, 70, 72, 74, 76, 78, 81, 83, 85, 87, 89, 90, 92, 94, 96, 98, 101…

The alternating number which is odd is called as odd alternating number and the alternating number which is even is called as even alternating number. For example 101 is an odd alternating number and 21456 is an even alternating number.

Obviously the super odd number and super even number will not become alternating number.

Consecutive number or Following number: -

It is the following number which present successively in the given number sequence. For example in the case of real number sequence, 6 is the following number of 5 and in the case of square number sequence 36 is the following number of 25 and so on.

If a, b and c are any three consecutive numbers or following numbers in real number sequence then $b^2 = ac+1$. For example 2, 3 and 4 then $3^2 = (2 \times 4) +1$ in real number sequence.

In real number sequence, consecutive numbers will not become even numbers, odd numbers, palindrome numbers and power numbers. Except 2 and 3, prime numbers will not come consecutively. But composite numbers will come as consecutive numbers. For example (14, 15) (20, 21) and (99, 100) are some of the consecutive composite numbers.

Precursor number: -

Precursor number is the number which occurs before the given number in the sequence. For example 15 is the precursor number of the 16 in the real number sequence.

Prime number: -

Prime number is a number which does not have a factor or divisor apart from 1 and the number itself. The prime number series is 2, 3, 5, 7, 11, 13, 17, 19, 23, 29, 31, 37, 41, 47, 53, 59, 61, 67, 71, 73, 79...

Prime numbers are discussed in chapter 6.

Divisible number: -

A real number which has a factor/s apart from 1 and the number itself is called as divisible number. For example 21 is a divisible number because it has two factors 3 and 7 apart from 1 and the number itself.

A divisible number may be either an odd number or an even number. The combination of both the odd and even divisible numbers in real number series is called as composite numbers.

Composite number: -

All the real numbers except the prime numbers are called as composite numbers. It contains both odd divisible number and even divisible number and so called as composite numbers. The composite numbers are opposite to prime numbers. The composite number series is 4, 6, 8, 9, 10, 12, 14, 15, 16, 18, 20...

A formula for finding out the composite number is $(n^4 + 4n)$ is a composite number provided that n>1. There are 8769 composite numbers below 10,000.

Pronic number or Oblong number or Heteromecic number: -

It is a number which is the product of two consecutive integers, that is, $n (n + 1)$ where 'n' is a real number. This can also be represented as (n^2+n). The pronic number series is 0, 2, 6, 12, 20, 30, 42, 56, 72, 90, 110, 132, 156, 182, 210, 240, 272, 306, 342, 380, 420, 462, 506, 552, 600, 650, 702, 756, 812, 870, 930, 992, 1056, 1122, 1190, 1260, 1332, 1406, 1482, 1560, 1640, 1722, 1806, 1892, 1980, 2070, 2162, 2256, 2352, 2450, 2550, 2652, 2756, 2862, 2970, 3080, 3192, 3306, 3422, 3540, 3660, 3782, 3906, 4032, 4160, 4290, 4422, 4556, 4692, 4830...

Obviously all pronic numbers are composite numbers.

The pronic numbers are also called as rectangular numbers.

The nth pronic number is twice the nth triangular number. For example the fifth pronic number is 20 and is equal to twice the fifth triangular number that is 2×10.

Pronic numbers will form triangular numbers.

It is to be noted that all pronic numbers are even numbers. If the pronic number contains all the digits as even digits then it is a super even pronic number. For example 240 is a super even pronic number and 210 is not a super even pronic number.

The nth pronic number is 'n' is less than the nth square number. For example the fifth pronic number is 20 and is equal to '5', less than the corresponding fifth square number that is 25-5=20.

Reflux number or Mirror image number: -

This is the number obtained by the mirror image of the original number. For example if the number 176 is placed before a mirror the image obtained is 671 and is its reflux number.

If the reflux number is same as that of the original number itself means it is called as palindrome number. For example 1221, whose reflux number is palindromic of the original number and so it is a palindrome number. Note that all palindrome numbers are reflux numbers but all reflux numbers are not palindrome numbers.

The palindrome numbers are discussed in chapter 7.

Transpose number: -

The numbers which are derived by interchanging the digits of the original number are called as transpose number. For example 21 and its transpose number is 12.

Obviously for a single digit number there is only one transpose number available and is the original number itself. For 213 the transposed numbers are 231, 132, 123, 312 and 321.

Cyclic numbers: -

The numbers formed by constructing new numbers from the same digits by moving them cyclically are called as cyclic numbers. For example for 2387, the cyclic numbers are 3872, 8723 and 7238.

For an 'n' digit number, 'n' cyclic numbers are possible. That is, for a five digits number five cyclic numbers are possible including the original number itself.

All cyclic numbers are transpose numbers but all transpose numbers will not become cyclic numbers.

Obviously cyclic number starts from two digit number onwards only.

For a two digit number except for palindromic two digit number, two cyclic numbers are possible. For a palindromic two digit number, only one cyclic number available and is the original number itself.

For a three digit number except for palindromic three digit number, three cyclic numbers are possible. For a palindromic three digit number, only one cyclic number available and is the original number itself. The same is applicable for higher digit numbers.

This is applicable for higher multidigit numbers.

The result of summation of cyclic numbers with 'n' digits is found by multiplying the sum of digits of the original number and 'n' 1s. For example the summation of cyclic numbers for 123 is (1+2+3) ×111=666. That is 123+231+312=666. Likewise for a 4 digit number we have to multiply with 1111 and so on.

A curious relationship in the case of the cyclic number of 1,552 is 1,552 + 5,521 + 5,215 + 2,155 = 14,443. Now remove the unit decimal of the cyclic number and then add. Continue the process till single digit number is obtained.
155 + 552 + 521 + 215 = 1,443

$$15 + 55 + 52 + 21 = 143$$
$$1 + 5 + 5 + 2 = 13$$

Probable number: -

The number of probable numbers that can be written using the given digits is called as the probable number. For example the number of probable three digit numbers that can be formed by using 'm' digits is m^3. That is by using 1 and 2 the number of probable three digit numbers are 2^3 (here m=2). The numbers are 111, 222, 112, 221, 211, 122, 121 and 212.

For repunit repdigit numbers there is no probable number.

Repdigit numbers and Repunit numbers: -

Numbers formed with repeated digit is called as repdigit numbers. The sequence of repdigit numbers is 00, 11, 22, 33, 44, 55, 66, 77, 88, 99, 111, 222, 333, 444, 555, 666, 777, 888, 999, 1111, 2222, 3333, 4444, 5555, 6666, 7777, 8888, 9999, 11111, 22222, 33333, 44444, 55555, 66666, 77777, 88888, 99999…

These are denoted by $_nR$ where 'n' is the number which is repeating and 'R' is the number of repetitions. For example $_53$ means 555.

The numbers formed with only 1's are called as repunit numbers.

Other natural numbers which are not having repeated digits are called as non repdigit numbers.

Repunit numbers and repdigit numbers will not become power numbers. That is 111, 2222, 33, 888 and so on like numbers are not power numbers.

All the Repunit numbers and the repdigit numbers are palindrome numbers, but the reverse is not true. Also all repdigit numbers are multiples of repunits.

The repdigit numbers which are even are called as even repdigit numbers and the repdigit numbers which are odd are called as odd repdigit numbers. For example 222 is an even repdigit number and 555 is an odd repdigit number. Also all the odd repdigit numbers are super odd repdigit numbers and all the even repdigit numbers are super even repdigit numbers.

All the repdigit numbers are composite numbers.

Repunit numbers and repdigit numbers will not become alternate numbers.

Strobogrammatic number: -

A strobogrammatic number is a number that in a given base and with the given set of glyphs or styles, appears the same whether viewed normally or upside down. In base 10, given a set of glyphs where 0, 1 and 8 are symmetrical around the horizontal axis, and 6 and 9 are the same as each other upside down. The strobogrammatic numbers sequence is 0, 1, 8, 11, 69, 88, 96, 101, 111, 181, 609, 619, 689, 808, 818, 888, 906, 916, 986, 1001…

1881 and 1961 were the most recent strobogrammatic years; the next strobogrammatic year will be 6009.

Although amateur mathematicians are quite interested in this concept, professional mathematicians generally are not. Like the concept of repunits, repdigit and palindromic numbers, the concept of strobogrammatic numbers is base-dependent. Unlike palindromicity it is also font dependent. The concept of strobogrammatic numbers is not neatly expressible algebraically, the way that the concepts of repunits is, or even the concept of palindromic numbers.

In Roman numeral or Devanagari, the numbers listed above are not strobogrammatic at all.

Complementary number: -

It is the number complement to the given decimal number with respect to the base 10 (or 100, 1000, etc, depending on how many digits you have). It is found out by the subtraction of all the digits from nine but the last or unit digit from ten. For example the complementary number of 123 is 1000-123 and is 877. Another example is the complementary number of 758120 is 241880 (Note: - The last digit zero in the number is subtracted from 10, keeping zero 1 carried over. Again subtract the second digit from 9 and the resultant is added with the carryover).

Here is another example where we have to skip over some zeros:
Example: What is the complement of 1700?
- Skip over the two zeros
- The "10" complement of the 7 is 3,
- The "9" complement of 1 is 8,

So the answer is 8300.

A check for complementary number is, the sum of complementary number and the original number will be 10 or 100 or 1000, etc, depending on how many digits you have.

Upside down number: -

A number 'n' is called as upside-down number if its i-th leftmost digit and its i-th rightmost digit are complements, i.e., their sum is 10. For example, 1289 is an upside down number because $1 + 9 = 2 + 8 = 10$. If the number has odd digits, the middle digit should always be 5 for an upside down number. For example 42586 is an upside-down number because $4 + 6 = 2 + 8 = 5 + 5 = 10$.

Obviously from the definition it follows that upside-down numbers are zeroless.

The sequence of upside-down numbers is 5, 19, 28, 37, 46, 55, 64, 73, 82, 91, 159, 258, 357, 456, 555, 654, 753, 852, 951, 1199, 1289, 1379…

The numbers 5, 55, 555… are repdigit numbers of five as well as upside down numbers.

The upside down number which is odd is called as odd upside down number and the upside down number which is even is called as even upside down number. For example 1289 is an odd upside down number and 654 is an even upside number.

Further if the odd upside down number contains all the digits as odd, then it is a super odd upside down number and if the upside number contains all digits as even, then it is a super even upside down number. For example 1379 is a super odd upside down number and 46 is a super even upside down number.

The upside down number which is prime number is called as prime upside down number and the upside down number which is composite is called as composite upside down number. For example 73 is a prime upside down number and 82 is a composite upside number.

Climbing number or Ascending number: -

The digits in the number 1789 are in ascending order. So it is called as climbing number or ascending number. The next climbing number after 1789 is 2345. The sequence of ascending number is 12, 13, 14, 15, 16, 17, 18, 19, 23, 24, 25, 26, 27, 28, 29, 34, 35, 36, 37, 38, 39, 45, 46, 47, 48, 49, 56, 57, 58, 59, 67, 68, 69, 78, 79, 89…

There are two types of ascending numbers namely regular ascending number and irregular ascending number. If the digits in the ascending number are consecutive then it is a consecutive digit ascending number or regular ascending number. Other ascending numbers are called as normal ascending number or irregular ascending number. For example 12345 is a consecutive digit ascending number or regular ascending number and 379 is an irregular ascending number.

Obviously single digit numbers are not ascending numbers. That is ascending number series starts from two digit number onwards.

Ascending numbers will not become palindrome numbers.

Ascending numbers will not become repunit and repdigit numbers.

Ascending numbers will not become undulating numbers.

Ascending numbers will not become seesaw numbers.

The ascending number which is also odd number is called as odd ascending number and the ascending number which is also even number is called as even ascending number. For example 1789 is an odd ascending number and 2358 is an even ascending number.

Further if the ascending number contains all the digits as even, then it is a super even ascending number and if the ascending number contains all the digits as odd then it is a super odd ascending number. For example 248 is a super even ascending number and 157 is a super odd ascending number.

The ascending number which is also prime number is called as prime ascending number and the ascending number which is also a composite number is called as composite ascending number. For example 89 is a prime ascending number and 1789 is a composite ascending number.

The consecutive ascending number consisting of all the digits in the given numbering system is called as pandigital ascending number. That is 123456789 is the pandigital ascending number in decenary numbering system.

The power number which is ascending in nature is called as power ascending number. For example 256 is a power ascending number.

The pronic number which is ascending in nature is called as ascending pronic number. For example 156 is an ascending pronic number.

Descending number: -

If the digits in a number are in a descending order means then it is called as descending number. For example 8731 is a descending number. The sequence of descending number is 10, 20, 21, 30, 31, 32, 40, 41, 42, 43, 50, 51, 52, 53, 54, 60, 61, 62, 63, 64, 65, 70, 71, 72, 73, 74, 75, 76, 80, 81, 82, 83, 84, 85, 86, 87, 90, 91, 92, 93, 94, 95, 96, 97, 98…

There are two types of descending numbers namely regular descending number and irregular descending number. If the digit in the descending number is consecutive then it is a consecutive digit descending number or regular descending number. Other descending numbers are normal descending numbers or irregular descending number. For example 54321 is a consecutive descending number and 821 is an irregular descending number.

Obviously single digit numbers are not descending numbers.

The reflux number of the descending number is ascending number and vice versa.

Descending numbers will not become palindrome numbers.

Descending numbers will not become repunit numbers and repdigit numbers.

Descending numbers will not become undulating numbers.

Descending numbers are opposite of ascending numbers.

The descending number which is also odd number is called as odd descending number and the descending number which is also even number is called as even descending number. For example 8731 is an odd descending number and 8532 is an even descending number.

Further if the descending number contains all the digits as even, then it is a super even descending number and if the descending number contains all the digits as odd then it is a super odd ascending number. For example 820 is a super even descending number and 531 is a super odd descending number.

The descending number which is also prime number is called as prime descending number and the descending number which is also composite number is called as composite descending number. For example 71 is a prime descending number and 9871 is composite descending number.

The consecutive descending number consisting of all the digits in the given numbering system is called as pandigital descending number. That is 987654321 is the pandigital descending number in decenary numbering system.

The power number which is descending in nature is called as power descending number. For example 81 is a power descending number.

The pronic number which is descending in nature is called as descending pronic number. For example 210 is a descending pronic number.

Seesaw numbers: -

If the left hand side of the digits in a number with respect to the middle digit/s is increasing/decreasing and again the right hand side of the number is decreasing/increasing then the number is called as seesaw number. For example 786, 7553, 613... are some of the seesaw numbers.

Obviously the seesaw numbers start from three digit numbers onwards only.

Seesaw numbers are of regular or smooth seesaw number and irregular or normal seesaw number. For example 27863 is a normal seesaw number and 75357 is a smooth seesaw number.

Based on the digital value of the middle digit/s the seesaw numbers can be classified into positive seesaw number and negative seesaw number.

Positive seesaw number: - If the left hand side of the digits in a number with respect to the middle digit/s is increasing and again the right hand side of the number is decreasing, then the number is called as positive seesaw number. For example 137, 785, 694, 1357642...are some of the positive seesaw number.

Positive seesaw number which is palindromic in nature is called as positive palindromic seesaw number. For example 23521 is a positive palindromic seesaw number.

Positive seesaw number which is even is called as even positive seesaw number and the positive seesaw number which is odd is called as odd positive seesaw number. For example 694 is an even positive seesaw number and 785 is an odd positive seesaw number.

Further if the positive seesaw number contains all the digits as even, then it is a super even positive seesaw number and if the positive seesaw number contains all the digits as odd then it is a super odd positive seesaw number. For example 842 is a super even positive seesaw number and 751 is a super odd positive seesaw number.

Positive seesaw number which is prime number is called as prime positive seesaw number and the positive seesaw number which is composite number is called as composite positive seesaw number. For example 137 is a prime positive seesaw number and 785 is a composite positive seesaw number.

Positive seesaw number which is smooth type is positive smooth seesaw number or simply can be called as peak number. For example 35653 is a positive smooth seesaw number or peak number.

Negative seesaw number: - If the left hand side of the digits in a number with respect to the middle digit/s is decreasing and again the right hand side of the number is increasing then the number is called as negative seesaw number. For example 612, 613, 902, 923, 1304…are some of the negative seesaw number.

Negative seesaw number which is even is called as even negative seesaw number and the negative seesaw number which is odd is called as odd negative seesaw number. For example 1304 is an even negative seesaw number and 923 is an odd negative seesaw number.

Further if the negative seesaw number contains all the digits as even, then it is a super even negative seesaw number and if the negative seesaw number contains all the digits as odd then it is a super odd negative seesaw number. For example 824 is a super even negative seesaw number and 517 is a super odd negative seesaw number.

Negative seesaw number which is prime number is called as prime negative seesaw number and the negative seesaw number which is composite number is called as composite negative seesaw number. For example 613 is a prime negative seesaw number and 612 is a composite negative seesaw number.

Negative seesaw number which is palindromic in nature is called as negative palindromic seesaw number. For example 532126 is a negative palindromic seesaw number.

Negative seesaw number which is smooth type is negative smooth seesaw number or simply can be called as valley number. For example 71017 is a negative smooth seesaw number or valley number.

Seesaw numbers will not become repunit numbers. In other words the repunit numbers and the repdigit numbers can be called as balanced seesaw numbers.

Undulating number: -

Undulating numbers are numbers of the form abc…abc… in base 10. It is like the ups and downs of unceasing sea waves. This property is significant starting from 3-digit numbers onwards only and so we will not consider numbers below 100. The sequence of undulating numbers is 101, 102, 103, 104, 105, 106, 107, 108, 109, 121, 123, 124, 125, 126, 127, 128, 129, 130, 131, 132, 134, 135, 136, 137, 138, 139, 140, 141, 142, 143, 145, 146, 147, 148, 149, 150, 151, 152, 153, 154, 156, 157…

Based upon the number of repetitions, the undulating number can be termed as first order undulating number, second order undulating number, third order undulating number and so on. For example 157 is a first order undulating number, 157157 is a second order undulating number and so on.

The undulating numbers are of two types namely normal or irregular undulating number and smooth or regular undulating number. In normal undulating number the digits are present randomly with ups and downs. For example 906907906 is a normal undulating number. In smooth undulating number the digits are repeating uniformly. For example 74747474 is a smooth undulating number.

If the smooth undulating number has two repeating digits of the form abababab… then it is called as two digit smooth undulating number and if it has three repeating digits of the form abcabcabc… then it is called as three digit smooth undulating number and so on. For example 2323 is a second order two digit smooth undulating number. The smooth undulating numbers are represented by ^nR where 'n' is the number and 'R' is the number of repetitions or undulations. For example the smooth undulating number 2323 can be represented as $^{23}2$. All the two digit smooth undulating numbers are divisible by 101. Similarly all the three digit smooth undulating numbers are divisible by 1001 and so on.

The undulating numbers which are palindromic in nature are called as palindromic undulating numbers. For example 101, 111, 121, 131, 141…are some of the palindromic undulating numbers.

The undulating numbers which are prime numbers are called as prime undulating numbers and the undulating numbers which are composite are called as composite undulating numbers. For example 101, 131, 151… are prime undulating numbers and 111, 121, 141… are composite undulating numbers. In the case of smoothly undulating numbers, 12121 is a smoothly undulating composite number and 72727 is a smoothly undulating prime number.

The undulating numbers which are odd are called as odd undulating numbers and the undulating numbers which are even are called as even undulating numbers. For example 101 is an odd undulating number and 202 is an even undulating number.

Further if the undulating number contains all the digits as even, then it is a super even undulating number and if the undulating number contains all the digits as odd then it is a super odd undulating number. For example 82480 is a super even undulating number and 1517 is a super odd undulating number.

The undulating number which are power numbers are called as power undulating number, for example 121 is a power undulating number.

Any number having the abcabc… numeral pattern, that is three digit undulating numbers (replace each letter with a digit. E.g. 123123) is divisible by 77.

Repunit numbers and repdigit numbers will not become undulating numbers.

The alternating number which is undulating in nature is called as undulating alternating number. For example 52543 is an undulating alternating number.

Pandigital number: -

A pandigital number is an integer that in a given base has among its significant digits each digit used in the base at least once. That is a pandigital number contains all the base digits in a numbering system at least once. For example, 1223334444555567890 is a pandigital number in base 10.

The pandigital base 10 number sequence is given by 1023456789, 1023456798, 1023456879, 1023456897, 1023456978, 1023456987, 1023457689…

Other natural numbers which are not pandigital in nature is called as non pandigital numbers.

If the pandigital number contains the digits in the ascending order is called as an ascending pandigital number and if the pandigital number contains the digits in the descending order is termed as descending pandigital number. It is obvious that both the ascending and descending pandigital numbers will have all the digits of the numbering system in a regular fashion. So both the ascending and descending pandigital numbers are called as smooth or regular pandigital numbers. Other pandigital numbers are called as normal or irregular pandigital numbers. For example 9876543210 is a regular pandigital number and 1023456789 is an irregular pandigital number.

The pandigital number which is odd is called as odd pandigital number and the pandigital number which is even is called as even pandigital number. For example

1023456978 is an even pandigital number and 1023457689 is an odd pandigital number.

It is obvious that there is no super odd pandigital number and super even pandigital number possible.

Also repunit and repdigit numbers will not form pandigital numbers.

The following table lists the smallest pandigital numbers of a few selected bases:

Base	Smallest pandigital number	Equivalent value in base 10
2	10	2
3	102	11
4	1023	75
10	1023456789	1023456789
16	1023456789ABCDEF	1162849785405935
Roman numerals	MCDXLIV	1444

It is obvious that if the base value is high, more pandigital numbers without redundant digits or repeated digits are possible and if the base value is small fewer pandigital numbers without redundant digits are possible.

Pandigital numbers are useful in fiction and in advertising. The Social Security Number 987-65-4321 is a zeroless pandigital number reserved for use in advertising. Some credit card companies use pandigital numbers with redundant digits as fictitious credit card numbers.

Some more examples of pandigital numbers are

123456789 is the first as well as smallest zeroless pandigital number.

987654321 is the largest zeroless pandigital number without redundant digits.

1023456789 is the first pandigital number.

1234567890 is the first pandigital number with the digits in order.

9876543210 is the largest pandigital number without redundant digits.

The smallest pandigital palindromic number in base 10 is 1023456789876543201.

9814072356 $(=99,066^2)$ is the largest pandigital square number.

12345678987654321 is a pandigital number with all the digits except zero in both ascending and descending orders. It is the square of 111111111 and is also a palindrome number.

A repeating set of all the digits 0-9 placed successively is obtained by the division 13,717,421/111,111,111 = 0.1234567890123456789...

No base 10 pandigital number can be a prime number if it doesn't have redundant digits. That is a pandigital number can be a prime number if and only if it has repeated digit/s. The sum of the digits 0 to 9 is 45, passing the divisibility rule for both 3 and 9. The first base 10 pandigital prime is 10123457689.

Special multiplication with pandigital numbers:

In the following tabular column, if the zeroless pandigital number is multiplied by two gives the pandigital number.

Pan-digital number	2× a zero-less pan-digital number	Uses each of 9 digits 1-9 once
1,037,246,958	= 2×518,623,479	= 9×57,624,831
1,046,389,752	= 2×523,194,876	= 9×58,132,764
1,286,375,904	= 2×643,187,952	= 9×71,465,328
1,307,624,958	= 2×653,812,479	= 9×72,645,831
1,370,258,694	= 2×685,129,347	= 9×76,125,483
1,462,938,570	= 2×731,469,285	= 9×81,274,365

The pandigital number 381,654,729 is a special number in which

the first one digit is divisible by 1
the first two digits is divisible by 2
the first three digits is divisible by 3
the first four digits is divisible by 4
the first five digits is divisible by 5
the first six digits is divisible by 6
the first seven digits is divisible by 7
the first eight digits is divisible by 8
and the whole number is divisible by 9.

Multiplication of the following ascending pandigital number gives the product with special last digits of descending pandigital number.

$$\underline{123456789} \times 989010989 = 122,100,120,\underline{987,654,321}$$

Concatenation number: -

The number formed by placing the same number right to the original number is called as concatenation number. For example the concatenation numbers of 21540 is 2154021540, 215402154021540…

The concatenation number for a single digit number should be a repunit number or repdigit number. The concatenation number for a two digit and multiple digit number is an undulating number provided the number is not a repunit or repdigit number. For example the concatenation number of 123 is 123123 is an undulating number.

It is to be noted that all concatenation numbers are undulating numbers but the reverse is not true. For example 201 is an undulating number but not a concatenation number.

Based upon the number of repetitions the concatenation number is classified into first order concatenation number, second order concatenation number, third order concatenation number and so on. For example 123 is a first order concatenation number, 123123 is a second order concatenation number, 123123123 is a third order concatenation number and so on.

It is to be noted that concatenation numbers can be formed from all type of natural numbers.

Obviously the concatenation number of a pandigital number is called as a pandigital concatenation number. For example 123456789123456789 is a pandigital concatenation number.

The concatenation number of a palindrome number is also a palindrome number.

The concatenation number of a power number will not be a power number.

The concatenation number of a prime number will be a composite number. The concatenation number of a composite number is again a composite number.

The concatenation number of an odd number is an odd number and the concatenation number of an even number is an even number. Further the concatenation number of a super odd number will be a super odd concatenation number and the concatenation number of a super even number will be a super even concatenation number.

The concatenation number of an ascending number or a descending number will be an undulating number. For example the concatenation number of 541 will give 541541 is an undulating number.

The concatenation number of repunit number as well as repdigit number will also be a repunit number and repdigit respectively. That is all repunit and repdigit numbers are concatenation numbers but the reverse is not true.

Digitally balanced number: -

The number having equal number of zeros, ones, twos, threes and so on upto nines in decimal system is called as a digitally balanced number in decenary system. For example 33112244665577008899 is a digitally balanced number.

3125467890, 11224466557700889933, 11332244665577008899... are some of the examples of digitally balanced numbers.

It is to be noted that all digitally balanced numbers are pandigital numbers but the reverse is not true.

The digitally balanced numbers are present in all the numbering systems. In the case of binary system, the binary number having equal number of ones and zeros is called as a digitally balanced number in binary system and in the case of ternary system, the ternary number having equal number of ones, twos and zeros is called as a digitally balanced number in ternary system and so on.

Peak number: -

The number which is having the digits ascending with respect to the middle digit/s and descending afterwards can be called as a peak number. For example 123474321 is a peak number.

A palindromic peak number is the peak number which is palindromic in nature. For example 135531 is a palindromic peak number.

A prime number which is having the peak number property is called as peak prime number. For example 346586543 is a peak prime number.

Valley number: -

The number which is having the digits descending with respect to the middle digit/s and ascending afterwards can be called as a valley number. For example 743212346 is a valley number.

A palindromic valley number is the valley number which is palindromic in nature. For example 5310135 is a palindromic valley number.

A prime number which is having the valley number property is called as valley prime number. For example 987646789 is a valley prime number.

Valley numbers are opposite of peak numbers.

Number digit or Digit sum: -

The unit digit derived after the sum of the digits of a number is called as the number digit or digit sum. This can also be called as digital root. Obviously the number digit for a single digit number is the same number itself. For example the number digit of 4519 is 4+5+1+9=19 and again 1+9=10 to give 1+0=1. That is 1 is the number digit of 4519.

A quicker way of finding the number digit is by the method of casting out or cancelling of nines. For example 2351068, then 3+6=9; 1+8=9 and so the left out digits are 5, 2 and 0 and the addition of which will give the number digit, that is 7. Another example is 991 and if we remove 9 then the left out number 1 is the digit sum.

Number digit of a number will not be zero. Also the number digit of a number is always a positive number.

The number digit of a negative number like -732 is -3. Since the number digit should not be a negative number add 9 to the result to get 6 and is the number digit of the given negative number.

Every number has its unique number digit. But the number digit may be associated with more than one number. For example 3254 has the number digit 5 while the number digit 5 is also possessed by numbers 32, 41, 536, 2327 and so on.

Similarly a unit digit is derived after the multiplication of the digits of the number like the addition. For example in the number 362, 3×6×2=36 again 3×6=18 and further 1×8=8 and so 8 is the unit digit derived after the multiplication of the digits of the number. Like the number digit obtained after addition process, the unit digit obtained from multiplication process is unique for a number but one drawback of this process is, if the number contains zero as one of the digit/s or if the product of the digits contains zero as one of the digit/s then the unit digit derived after the multiplication of the digits of the number will become zero. For example in the number 4519, 4×5×1×9=180 again 1×8×0=0 and so the unit digit derived after the multiplication of the digits of the number is zero.

VARIOUS TYPES OF REPRESENTATION OF NUMBERS

In ancient days numbers were represented by placing a line to represent one, two lines for two, three lines for three and four for four. Five is represented by crossing the four lines and the higher numbers are represented by repeating the process. But the method is difficult for representing higher and bigger numbers.

So the digits were developed in various parts of the world to represent the numbers. But it is still found difficult to represent higher numbers. For example the Roman number system which was widely used in the middle Ages, finds it difficult to represent bigger numbers. Except the Indo Arabic numbers which are currently in use, other number representations in various parts and languages of the world are devoid of zero.

The following table list out some of the different types of mentioning of numbers in various parts of the world and their corresponding equivalents in Indo Arabic numbers.

Name	0	1	2	3	4	5	6	7	8	9	10	100	1000
Bengali	-	১	২	৩	৪	৫	৬	৭	৮	৯	-	-	-
Guajarati	-	૧	૨	૩	૪	૫	૬	૭	૮	૯	-	-	-
Gurumukhi	-	੧	੨	੩	੪	੫	੬	੭	੮	੯	-	-	-
Hindi or Devanagiri	-	१	२	३	४	५	६	७	८	९	-	-	-
Japanese	零	一	二	三	四	五	六	七	八	九	十	百	千
Kannada	-	೧	೨	೩	೪	೫	೬	೭	೮	೯	-	-	-
Malayalam	-	൧	൨	൩	൪	൫	൬	൭	൮	൯	൰	-	-
Roman	-	I	II	III	IV	V	VI	VII	VIII	IX	X	C	M
Tamil	-	௧	௨	௩	௪	௫	௬	௭	௮	௯	௰	௱	௲
Telugu	-	౧	౨	౩	౪	౫	౬	౭	౮	౯	-	-	-
Thai	-	๑	๒	๓	๔	๕	๖	๗	๘	๙	-	-	-

NUMBERING SYSTEMS

There are two types of numbering systems called as non positional numbering system and positional numbering system.

Non positional number system is based upon the additive approach method. For example in the Roman number system I, V, X, C and M symbols are used. These symbols are simply added to find out the value of a particular number. The number 23 is represented as XXIII in Roman number system. The major disadvantage of this type of number system is, it is difficult to perform the arithmetic operations.

The positional number system uses only few symbols called as digits. Such symbols specify different values depending upon the position where they occupy in the number. For example in the case of decimal number system, in the number 456 the most significant digit (MSD) of the number is 4 and the least significant digit (LSD) of this number is 6.

Based upon the integers represented by the repeating terms or digits various numbering systems are followed. For example if the numbering system involves the digits from 0 to 9 it is called as decenary system and if the numbering system involves 0 and 1 means it is called as binary system and so on.

There are four commonly used positional number system that is decimal number system, binary number system, octal number system and hexadecimal number system.

Decenary system: -

It has 0, 1, 2, 3, 4, 5, 6, 7, 8 and 9 numbers. It is based on the powers of 10. It is the usual way of representing numbers. For example $121_{10} = 121$ and the number can be represented as the powers of 10 like $[(1 \times 10^2) + (2 \times 10^1) + (1 \times 10^0) = 121_{10}]$. But generally we are not representing the subscript '10' since decenary system is the common method of numbering system.

Here 10 is called as radix of the decenary number system. Radix or base is the first higher number obtained by using the digits of the numbering system. That is in decenary system 10 is the first higher number using the digits of decenary system.

Place value: -

Place value of a digit is the value of the digit in the place where it is occupying the number. For example the place value of 7 in the number 607354 is 7000.

Place value of a digit in the number is calculated from the right side to the left side of the number that is from unit digit onwards.

Western method of assigning place value is one, ten, thousand, ten thousand, hundred thousand, million, ten million, hundred million, billion and so on.

Indian method of assigning place value is one, ten, thousand, ten thousand, lakh, ten lakh, crore, ten crore, hundred crore and so on.

Face value: -

Face value of a digit is the value of the digit seen as such in the number. For example the face value of 7 in the number 607354 is 7.

Zero has no face value but it has the place value. In the above case the face value of 0 is 0 and the place value of 0 is 10,000.

Infinity has no place value but it has the face value.

Absolute value: -

It is the magnitude of a number irrespective of the sign involved in representing the number. For example absolute value of -607354 is 607354. The absolute value of 607354 is 607354.

Binary system: -

It has 0 and 1 numbers. Like in decenary system the binary system can also be represented as the powers of 2. Conversion of decenary number system to binary number system is as follows. For example the number 121 can be converted into binary number or binary digit (bit) as follows.

121/2 = 60 remainder is 1

60/2 = 30 remainder is 0

30/2 = 15 remainder is 0

$$15/2 = 7 \quad \text{remainder is } 1$$
$$7/2 = 3 \quad \text{remainder is } 1$$
$$3/2 = 1 \quad \text{remainder is } 1$$

Now the binary number equivalent of 121 is 1111001_2.

Conversion of binary number to decenary number is as follows.

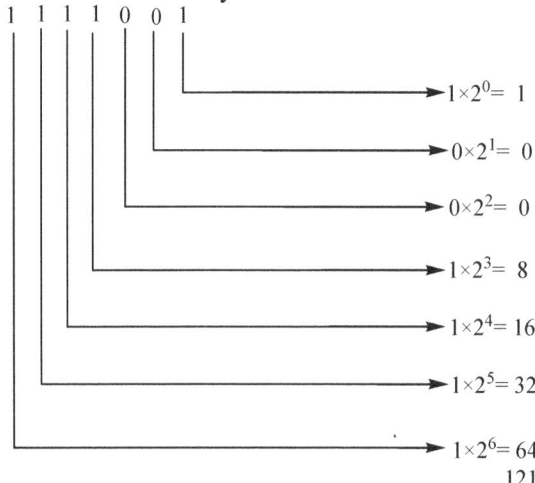

Now the decenary equivalent of 1111001_2 is 121.

Evil number: -The number 'n' is evil number if it has an even number of 1's in its binary expansion. The sequence of evil numbers is 3, 5, 6, 9, 10, 12, 15, 17, 18, 20, 23, 24, 27, 29, 30, 33, 34, 36, 39, 40, 43, 45, 46, 48…

There are 4999 evil numbers below 10,000.

Odious number: -The number 'n' is odious if it has an odd number of 1's in its binary expansion. The sequence of odious number is 1, 2, 4, 7, 8, 11, 13, 14, 16, 19, 21, 22, 25, 26, 28, 31, 32, 35, 37, 38, 41, 42, 44, 47, 49, 50…

There are 5000 odious numbers below 10,000.

Digitally balanced number: -If a number in its binary numeral system representation has the same number of ones and zeros, then it is called as a digitally balanced number. For example $177_{10} = 10110001_2$. Here the number 177 is a digitally balanced number because it has equal number of zeros and ones in its binary number. Some of the digitally balanced numbers in binary system are3 (10_2), 9 (1001_2), 10 (1010_2), 12 (1100_2), 35 (100011_2), 37 (100101_2), 38 (100110_2), 41 (101001_2), 42 (101010_2), 44 (101100_2), 49 (110001_2), 50 (110010_2)…

If a number in its binary numeral system representation has different number of ones and zeros, then it is called as a digitally imbalanced number.

For example $4_{10} = 100_2$. Here the number is a digitally imbalanced number because it has unequal number of zeros and ones in its binary number.

The binary number which is palindromic in nature is called as palindromic binary number. For example 1100011_2 is a palindromic binary number.

The binary number which is undulating in nature is called as undulating binary number. For example 1001001001_2 is an undulating binary number.

The binary number which is repunit in nature is called as repunit binary number. For example 111111_2 is a repunit binary number.

Pernicious number: -A number is called as pernicious number if it contains a prime number of ones in its binary representation. For example $21 = 10101_2$ is pernicious number since it contains 3 ones and 3 is a prime number. Some of the pernicious numbers are 3, 6, 7, 9, 10, 11, 13, 14, 17…

Ternary system: -

It has 0, 1 and 2 numbers. This system of representation is based on powers of 3. Here the radix is 3. For example the ternary equivalent of 121_{10} is 11111_3.

Pental number system: -

It has 0, 1, 2, 3 and 4 numbers. For example 121 is converted into pental system as follows.

$121/5 = 24$ remainder is 1

$24/5 = 4$ remainder is 4

Now the pental number equivalent of 121 is 441_5.

Conversion of pental number into decenary number is as follows.

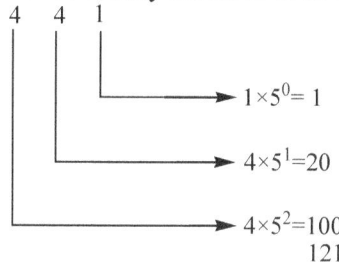

$1 \times 5^0 = 1$

$4 \times 5^1 = 20$

$4 \times 5^2 = 100$

121

Now the decenary equivalent of 441_5 is 121.

Octal number system: -

It has 0, 1, 2, 3, 4, 5, 6 and 7 numbers. For example 121 is converted into octal number as follows.

$121/8 = 15$ remainder is 1

$15/8 = 1$ remainder is 7

Now the octal number equivalent of 121 is 171_8.

Conversion of octal number into decenary number is as follows.

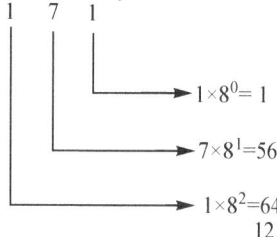

$1 \times 8^0 = 1$

$7 \times 8^1 = 56$

$1 \times 8^2 = 64$

121

Now the decenary equivalent of 171_8 is 121.

It is easy to convert from octal to binary system and vice versa since three binary digits make one octal digit.

The following table illustrates the equivalent of binary digit and octal digit.

Binary digit	Octal digit
000	0
001	1
010	2
011	3
100	4
101	5
110	6
111	7

For example, to convert from octal to binary digit, replace all octal digits by their binary equivalents. For example $(347)_8 = (011\ 100\ 111)_2 = (11100111)_2$

To convert binary to octal digit, partition the binary digits in groups of three, starting from the right and then replace each group by its octal digit.

For example $(10111011)_2 = (010\ 111\ 011)_2 = (273)_8$.

Dozenal system or duodecimal system: -

This system can also be called as duodecimal system. The radix of the system is 12. It has 1, 2, 3, 4, 5, 6, 7, 8, 9, A and B digits.

Hexadecimal number system: -

It has 0, 1, 2, 3, 4, 5, 6, 7, 8, 9, A, B, C, D, E and F digits. This numbering system is used in computer microprocessor work. Since numbers (0 to 9) and alphabets (A to F) are used to represent the digits in hexadecimal number system, it is also called as the alphanumeric number system.

For example 121 is converted into hexadecimal system as follows.

$121/16 = 7$ remainder is 9

Now the hexadecimal number equivalent of 121 is 79_{16}.

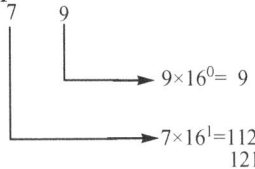

Now the decenary equivalent of 79_{16} is 121.

Note: - Similar method can be used to convert any decenary number to a number of required radix (base) value and vice versa.

A group of 4 bits can be represented by an equivalent hexadecimal digit as follows. $(1101\ 0010\ 1010\ 1100)_2 = (D2AC)_{16}$.

Features of numbering systems: -

The concepts and ideologies which are applicable to one numbering system need not be applicable to other numbering systems.

For example (i) 67 is palindromic in bases 5 (232_5) and 6 (151_6) while in base 10 this number is not a palindromic number.

$$\text{Some more examples are} \quad 57_{10} = 111_7.$$
$$86_{10} = 222_6.$$

$$2000_{10} = 5555_7.$$
$$342_{10} = 666_7.$$

In some cases resemblances among the numbering systems are also observed. For example 7447 is a palindrome in base 2 and in base 10. The copalindrome numbers are the numbers which are palindrome in nature in atleast two numbering systems. So the number 7447 is called as copalindrome number or multibase number.

The numbers which are palindromic in nature in both decimal and binary number system are $3_{10} = 11_2$

$$5_{10} = 101_2$$
$$7_{10} = 111_2$$

$$9_{10} = 1001_2$$
$$33_{10} = 100001_2$$
$$99_{10} = 1100011_2$$

$$313_{10} = 100111001_2$$
$$585_{10} = 1001001001_2$$
$$717_{10} = 1011001101_2$$

$$7447_{10} = 1110100010111_2$$
$$9009_{10} = 10001100110001_2$$
$$15351_{10} = 11101111110111_2$$

$$32223_{10} = 111110111011111_2$$
$$39993_{10} = 1001110000111001_2$$
$$53235_{10} = 1100111111110011_2$$

$$53835_{10} = 1101001001001011_2$$
$$73737_{10} = 10010000000001001_2$$
$$585585_{10} = 10001110111101110001_2$$

Further equivalent multibase palindromic numbers
$$11001111100011_2 = 1001001_3$$

Further multibase palindromic numbers or copalindromic numbers are

7997 is a palindrome in base 4 and in base 10.

1441 is a palindrome in base 6 and in base 10.

7667 is a palindrome in base 6 and in base 10.

The numbers which are palindromic in nature in both decimal and octal number system are $121_{10} = 171_8$

$$292_{10} = 444_8$$
$$333_{10} = 515_8$$

$$373_{10} = 565_8$$
$$414_{10} = 636_8 \text{ and so on.}$$

The higher multibase palindromic numbers are

6886 is a palindrome in base 9 and in base 10.

585 is a palindrome in bases 2, 8 and in base 10.

719848917 is a palindrome in bases 2, 8 and in base 10.

121 is palindromic in bases 3, 7, 8, and 10

373 is palindromic when expressed in bases 4, 8, 9, and 10.

786435 is multi-palindromic in bases 2, 4, 7, and 8.

The numbers which are palindromic in nature in both decimal and octal number system are $353_{10} = 161_{16}$

$$626_{10} = 272_{16}$$
$$787_{10} = 313_{16}$$

$$979_{10} = 3D3_{16}$$
$$1991_{10} = 7C7_{16} \text{ and so on.}$$

(ii) $922 = 1234_9$ is an ascending number in base 9 but a descending number in base 10. Some more examples are $1534_{10} = 4321_7$.
$$310_{10} = 1234_6.$$
$$1865_{10} = 12345_6.$$

(iii) 31 is a repunit number in base 5 (111), and base 2 (11111).
Further $127_{10} = 1111111_2$
$255_{10} = 11111111_2$
$9841_{10} = 111111111_3$

$312_{10} = 2222_5$
$7812_{10} = 222222_5$
$93_{10} = 333_5$

$93_{10} = 333_5$
$468_{10} = 3333_5$
$86_{10} = 222_6$

$259_{10} = 1111_6$
$172_{10} = 444_6$
$1036_{10} = 4444_6$

$215_{10} = 555_6$
$1295_{10} = 5555_6$
$7775_{10} = 55555_6$

$57_{10} = 111_7$ (that is: $57 = 7^2 + 7^1 + 7^0$)
$400_{10} = 1111_7$
$114_{10} = 222_7$

$1200_{10} = 3333_7$
$8403_{10} = 33333_7$
$73_{10} = 111_8$

$146_{10} = 222_8$
$438_{10} = 666_8$
$9362_{10} = 22222_8$

$1170_{10} = 2222_8$
$30_{10} = 33_9$
$273_{10} = 333_9$

$1755_{10} = 3333_8$
$640_{10} = 2222_9$
$7381_{10} = 11111_9$

(iv) 51 is an undulating number in base 4.
59 is an undulating number in base 4.
61 is an undulating number in base 6.

65 is an undulating number in base 8.

290 is an undulating number in base 12.

(v) 147 is digitally balanced in base 2 and base 4, because in such bases it contains all the possible digits an equal number of times.

149 is digitally balanced in base 2.

(vi) Some curious facts with various numbering systems

$$2040_{10} = 2040_5 + 2040_7 + 2040_8.$$
$$2101_{10} = 2101_5 + 2101_7 + 2101_8.$$
$$4202_{10} = 4202_5 + 4202_7 + 4202_8.$$

$$3121_{10} = 3121_5 + 3121_7 + 3121_8.$$
$$4141_{10} = 4141_5 + 4141_7 + 4141_8.$$

$$1304_{10} = 1304_6 + 1304_9.$$
$$2545_{10} = 2545_6 + 2545_9.$$

A comparison of binary, ternary, pental, octal, decimal, dozenal and hexadecimal system is as follows.

Numbering system	Binary	Ternary	Pental	Octal	Decimal	Dozenal	Hexadecimal
Radix value	2	3	5	8	10	12 or C	16 or G
Equivalent for 0_{10}	0000	0000	0	0	0	0	0
Equivalent for 1_{10}	0001	0001	1	1	1	1	1
Equivalent for 2_{10}	0010	0002	2	2	2	2	2
Equivalent for 3_{10}	0011	0010	3	3	3	3	3
Equivalent for 4_{10}	0100	0011	4	4	4	4	4
Equivalent for 5_{10}	0101	0012	10	5	5	5	5
Equivalent for 6_{10}	0110	0100	11	6	6	6	6
Equivalent for 7_{10}	0111	0101	12	7	7	7	7
Equivalent for 8_{10}	1000	0102	13	10	8	8	8
Equivalent for 9_{10}	1001	0110	14	11	9	9	9
Equivalent for 10_{10}	1010	0111	20	12	10	A	A
Equivalent for 11_{10}	1011	0112	21	13	11	B	B
Equivalent for 12_{10}	1100	1000	22	14	12	10	C
Equivalent for 13_{10}	1101	1001	23	15	13	11	D
Equivalent for 14_{10}	1110	1002	24	16	14	12	E
Equivalent for 15_{10}	1111	1010	30	17	15	13	F

ARITHMETIC OPERATIONS

The four important basic operations of arithmetic are addition, subtraction, multiplication and division. Multiplication can also be called as consecutive addition and similarly division can also be called as consecutive subtraction. Apart from these basic arithmetic operations or properties the following other arithmetic operations can also be studied and compared with other mathematical operations.

Additive property: - If A+A=2A then the property is known as additive property. For example 9+3=12. This is called as arithmetic addition.

Subtractive property: - If A-A=0 then the property is known as subtractive property. For example 9-3=6. This is called as arithmetic subtraction.

Multiplicative property: - If A×A=A^2 then the property is known as multiplicative property. For example 9×3=27. This is called as arithmetic multiplication.

Divisive property: - If A/A=1 then the property is known as divisive property. For example 9/3=3. This is called as arithmetic division.

Determinative property: - If A, B are two criteria and if A=B or A≠B then the property is known as determinative property. For example $\sqrt{9} = 3$ and $\sqrt{9} \neq 2$. This is called as arithmetic powers.

Reflux property: - If A' is obtained from A by mirror then the property is known as reflux property. Note that A' may be equal to A or may not be equal to A. For example the reflux number of 176 is 671.

Palindrome property: - If A' is equal to A by mirror then the property is known as palindrome property. For example in the case of 121 the reflux number of 121 is the same number itself and so it is a palindrome number. Note that a system obeying palindromic property will also obey reflux property but the converse is not true.

Symmetric property: - If A*B = B*A then the property is known as symmetric property. For example $3^2 = 9$; $9 = 3^2$. This is arithmetic powers. It is to be noted that reflux property is obtained by keeping in front of the mirror and symmetric property is obtained by the reversal of operations.

Transitive property: - If A=B and B=C then A=C is known as transitive property. For example $3^2 = 9$; $9 = 5 + 4$ then $3^2 = 5 + 4$

Identity property: - If A*I = A (Note: - * means an arithmetic operation) then the property is known as identity property. Here I is called as identity element. For example $5 + 0 = 5$ Here 0 is the identity element.

Inverse property: - If A*A^{-1} = 0 (Note: - * means an arithmetic operation) then the property is known as inverse property. Here A^{-1} is called as inverse element. For example $2 - 2 = 0$ Here -2 is the inverse element.

Commutative property: - If A*B = B*A (Note: - * means an arithmetic operation) then the property is known as commutative property. For example 3×2 = 6 = 2×3. This is the property of arithmetic multiplication.

Anti commutative property: - If A*B = $-$ (B*A) (Note: - * means an arithmetic operation) then the property is known as anti commutative property. For example $5 - 2 = - (2 - 5)$. This is arithmetic subtraction.

Left Associative property: - If A*(B*C) = (A*B)*C (Note: - * means an arithmetic operation) then the property is known as left associative property. For example 3 × (2×5) = (3×2) ×5 = 30. This is arithmetic multiplication.

Right associative property: - If (A*B)*C = A*(B*C) (Note: - * means an arithmetic operation) then the property is known as right associative property. For example (3×2) ×5 = 3 × (2×5) = 30. This is arithmetic multiplication.

Left cancellation property: - If A*B = A*C (Note: - * means an arithmetic operation) then B = C is known as left cancellation property.
For example 5×9 = 5×3×3 then canceling the common terms we get 9 = 3×3.

Right cancellation property: - If A*B = C*B (Note: - * means an arithmetic operation) then A = C is known as right cancellation property.
For example $2^2×5 = 4×5$ then canceling the common terms we get $2^2 = 4$.

Left distributive property: - If A (B*C) = AB * AC (Note: - * means an arithmetic operation) then the property is known as left distributive property. For example 3(10 − 4) = 3(10) − 3(4) = 18.

Right distributive property: - If (A*B) C = AC * BC (Note: - * means an arithmetic operation) then the property is known as right distributive property. For example (10 − 4)3 = (10)3 − (4)3 = 18.

Closure property: - If A*B (Note: - * means an arithmetic operation) is also the element of same order or in the same number system is known as closure property. For example real number system 3 + 2 = 5

Upset reflux property: - If $\overset{\overline{\text{ıııııııı}}}{\underset{\wedge}{\wedge}}$ by mirror then the property is known as upset reflux property. For example $\overset{8008}{\underset{\overline{\text{ıııııııı}}}{}}$ 8008 where the symbol ıııııııı represents a mirror.

Upset property or Reciprocal property: - If A*(1/A) = 1 (Note: - * means an arithmetic operation) then the property is known as upset property. The number 1 divided by the rational number 'A' is called as the reciprocal number.
For example 2 × (1/2) = 1. This is arithmetic multiplication.

Opposite property: - The opposite of a number 'a' is '-a'. If 'a' is a positive number then its opposite number will be a negative number and vice versa. If a=0, then its opposite number is again zero and zero is the only number that is its own opposite. The opposite property of multiplication is division and the opposite property of power number is finding out the power roots.

Note: -Opposite property should not be confused with reflux property. Opposite property deals with the positive and negative signs of a real number while the reflux property deals with the mirror image of the number.

The opposite of the opposite number is equal to the original number.
That is – (-a) = a. Also the sum of any number and its opposite number is zero.
That is a + (-a) = 0.

Negative property: - The number obtained by changing the direction or sign of a positive number keeping the magnitude same is called as negative property. The number obtained is called as negative number. For example the negative number of 3 is -3. But the reverse is not true. That is the negative property for -3 is 3 which is a positive number.

All the negative numbers will become opposite numbers but all the opposite numbers will not become negative numbers.

A comparison of these mathematical properties is listed as an annexure.

Branches of arithmetic

Algebra: - It is the arithmetical study of powers.

Binary arithmetic: -It is the arithmetical study of the numbering system with the base value 2.

Complex numbers: -It is the study of the number which consists of a real part and an imaginary part, represented as 'a+ib' where 'a' and 'b' are real parts and 'i' is an imaginary part with the value $i=\sqrt{-1}$.

Determinants: -It is the arithmetical study of square matrices.

Matrix: -It is the arithmetical study of array of numbers in rows and columns.

Mods: - It is the arithmetical study of systematic repetitive processes.

Probability: -It is the arithmetical study of probable events that can be possible during an operation.

Sets: - It is the arithmetical study of well defined collection of objects.

Statistics: - It is the arithmetical approach of arrangement and classification of the data collected.

Vectors: -It is the arithmetical study of properties involving both magnitude and direction.

Polygonal numbers: -

When the real numbers are expressed as dots based on the geometrical shapes formed they are classified as triangular numbers, square numbers, pentagonal numbers, and so on. The general terminology of these numbers is polygonal numbers and is discussed in chapter 10.

2. ADDITION

Addition can also be called as summation and it is denoted by the symbol "+" and is pronounced as plus. For example in 5+3=8. Here 5 is called as addendum, 3 is called as adduct and 8 is called as the sum or resultant.

Check for addition: -
The number digit of the result is equal to the sum of the number digits of the addendum and adduct involved in the addition.

For example 3989   number digit 2
<u>(+) 2403</u>   number digit 0 or 9
6392   number digit 2

Now the sum of the number digits of addendum and adduct is equal to 11 whose number digit is 2.

The addition of numbers which are having number digits either of 3, 6 or 9 among themselves will give the result either of 3, 6 or 9 as their number digits. For example 6 + 6 will give the resultant with number digit 3 while 6 + 9 will give the result with number digit 6.

It is to be noted that the number digit calculation can be simplified by 'casting out of nine' method.

Commutative property: -
Addition is commutative.
For example 5+3=8 and 3+5=8

Addition with odd and even numbers: -

Addition	Odd number	Even number
Odd number	Even number	Odd number
Even number	Odd number	Even number

For example, odd number + even number = odd number
$$5 + 4 = 9$$
In a broader sense, even + even + even +...+ even = even
even + even + even +...+ odd = odd
odd + odd + odd+...+odd = even if 'n' number of "even" odd numbers are present.
odd + odd + odd+...+odd = odd if 'n' number of "odd" odd numbers are present.

But this is not applicable in the case of addition of super odd numbers and super even numbers. That is if we add a super odd number with another super odd number, the resultant will be an even number only and not a super even number.

For example super odd number + super odd number = even number
$$315 + 739 = 1054$$

Addition with positive and negative numbers: -

Addition	Positive number	Negative number
Positive number	+	±
Negative number	±	-

For example Negative number + Positive number = Negative or positive number
(-4) + 2 = -2 and another example is 4 + (-2) = 2.

Addition of prime and composite numbers: -

Addition	Prime number	Composite number
Prime number	Composite	Composite
Composite number	Composite	Prime or composite

For example Prime + Composite = Composite
79 + 9 = 88

Addition of repdigit and non repdigit numbers: -

Addition	Repdigit number	Non repdigit number
Repdigit number	Repdigit number	Non repdigit number
Non repdigit number	Non repdigit number	Repdigit or non repdigit number

For example Non repdigit number+repdigit number=Repdigit or non repdigit number
123 + 321 = 444 and another example is 323 + 1224 = 1547

Addition of pandigital and non pandigital numbers: -

Addition	Pandigital number	Non pandigital number
Pandigital number	Pandigital number	Pandigital or non pandigital number
Non pandigital number	Pandigital or non pandigital number	Pandigital or non pandigital number

For example Non pandigital number + Non pandigital number =
Pandigital or Non pandigital number
13,576,498 + 200,000,000 = 213,576,498 and 576,498 + 123,214 = 699,712

Addition table: -

Addition	1	2	3	4	5	6	7	8	9	10
1	2	3	4	5	6	7	8	9	10	11
2	3	4	5	6	7	8	9	10	11	12
3	4	5	6	7	8	9	10	11	12	13
4	5	6	7	8	9	10	11	12	13	14
5	6	7	8	9	10	11	12	13	14	15
6	7	8	9	10	11	12	13	14	15	16
7	8	9	10	11	12	13	14	15	16	17
8	9	10	11	12	13	14	15	16	17	18
9	10	11	12	13	14	15	16	17	18	19
10	11	12	13	14	15	16	17	18	19	20

For example from the table we can see that 8+3=11

Addition rule: -

The first or unit digit of a number is to be added with first or unit digit of the other number/s followed by 10th or second digit of the number with the 10th or second digit of the other number/s and then 100th or third digit of the number with the corresponding 100th or third digit of the other number/s and so on from right to left. For example 5132

132

$$\frac{30}{5294}$$

Note: - (a) In the number 5132, '2' is the first or unit digit of the number, '3' is the 10th digit of the number, '1' is the 100th digit of the number and '5' is the 1000th digit of the number.

(b) The rule of addition is applicable for the addition of two and more numbers simultaneously.

(c) If the addition of first or unit digit of a number with first or unit digit of the other number/s is 10 and above, the unit digit of the addition is placed in the resultant and the tenth digit is to be 'carried over' and added with the 10th or second digit of the other number/s and so on.

(d) The addition rule is applicable for decimal number addition also.

For example 383.246   315.3*
$$\frac{123.169}{506.415}\qquad \frac{269.123}{584.423}$$

*It is to be noted that 315.3 can be considered as 315.300 and by applying zeros to the right hand side of the decimal fraction number does not alter the value of that number.

One simple method of addition of consecutive natural numbers can be done by Gauss method of 'addition by pairing'. For example the sum of consecutive natural numbers from 1 to 100 is done as follows.

$$1 + 2 + 3 + 4 + ... + 96 + 97 + 98 + 99 + 100$$

If we pair the numbers present in the extreme end numbers we will get fifty pairs with sum 101. If we multiply the number of pairs by this extreme end numbers sum we will get the sum of consecutive natural numbers.

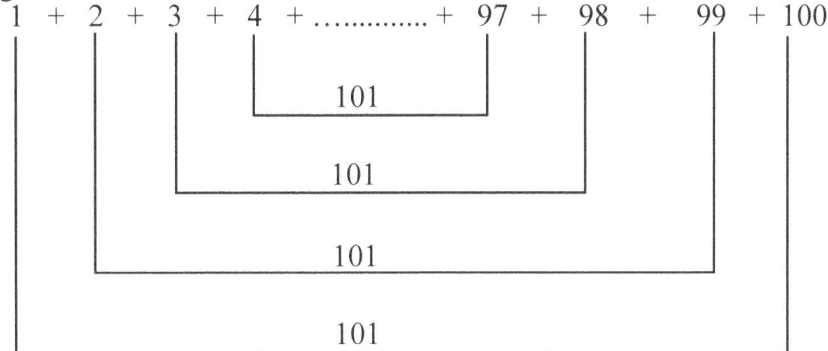

That is the sum of the above consecutive natural numbers will be $101 \times 50 = 5,050$.

If the sum of any three numbers is zero then the sum of their cubes is equal to three times their product. For example 3, 4 and -7

Sum of the above numbers $3 + 4 + (-7) = 0$

Sum of their cubes is $27 + 64 + (-343) = -252$

Product of the above numbers $3 \times 4 \times (-7) = -84$

Three times the product is $3 \times (-84) = -252$

Fun with addition: -

123456789×1	123456789×2	123456789×3	123456789×4
987654321×1	987654321×2	987654321×3	987654321×4
$+1$	$+2$	$+3$	$+4$
$1,111,111,111$	$2,222,222,222$	$3,333,333,333$	$4,444,444,444$

Likewise we can extend the series up to 9.

Some more fun with additions

$1 + 11 = 12 \times 1 = 12$

$2 + 22 = 12 \times 2 = 24$

$3 + 33 = 12 \times 3 = 36$

………………………

$9 + 99 = 12 \times 9 = 108$

Similarly

$1 + 11 + 111 = 123 \times 1 = 123$

$2 + 22 + 222 = 123 \times 2 = 246$

$3 + 33 + 333 = 123 \times 3 = 369$

………………………………

$9 + 99 + 999 = 123 \times 9 = 1107$

Likewise

$1 + 11 + 111 + 1111 = 1234 \times 1 = 1234$

$2 + 22 + 222 + 2222 = 1234 \times 2 = 2468$

$3 + 33 + 333 + 3333 = 1234 \times 3 = 3702$

………………………………………

$9 + 99 + 999 + 9999 = 1234 \times 9 = 11106$ and so on.

Further fun with additions are

$3340 = 3333 + 3 + 4 + 0$

$3341 = 3333 + 3 + 4 + 1.$

$3342 = 3333 + 3 + 4 + 2.$

$3343 = 3333 + 3 + 4 + 3.$

$3344 = 3333 + 3 + 4 + 4.$

$3345 = 3333 + 3 + 4 + 5.$

$3346 = 3333 + 3 + 4 + 6.$

$3347 = 3333 + 3 + 4 + 7.$

$3348 = 3333 + 3 + 4 + 8.$

$3349 = 3333 + 3 + 4 + 9.$

$4510 = 4444 + 55 + 11 + 0.$

$4511 = 4444 + 55 + 11 + 1.$

$4512 = 4444 + 55 + 11 + 2.$

$4513 = 4444 + 55 + 11 + 3.$

$4514 = 4444 + 55 + 11 + 4.$

$4515 = 4444 + 55 + 11 + 5.$

$4516 = 4444 + 55 + 11 + 6.$

$4517 = 4444 + 55 + 11 + 7.$

$4518 = 4444 + 55 + 11 + 8.$

$4519 = 4444 + 55 + 11 + 9.$

$6680 = 6666 + 6 + 8 + 0.$

$6681 = 6666 + 6 + 8 + 1.$

The following additions uses the digits from 0 to 9 only once

$$879 + 426 = 1305$$
$$879 + 624 = 1503$$
$$859 + 347 = 1206$$

$$859 + 743 = 1602$$
$$789 + 264 = 1053$$
$$789 + 246 = 1035$$

$$756 + 342 = 1098$$
$$765 + 324 = 1089$$
$$657 + 432 = 1089$$

$$675 + 423 = 1098$$
$$589 + 473 = 1062$$
$$589 + 437 = 1026$$

If we arrange the whole numbers in a triangle shape, the addition of each numbers in a row gives interesting pattern as follows.

Whole number triangle	Resultant
0	0
1 0 1	2
2 1 0 1 2	6
3 2 1 0 1 2 3	12
4 3 2 1 0 1 2 3 4	20
5 4 3 2 1 0 1 2 3 4 5	30
…………………………	….

The addition resultant of each row is a multiple of two. That is the difference between each consecutive row is 2, 4, 6, 8, 10 and the resultant of the next row in this triangle is 42 with the difference of 12 with the earlier row.

The number which is three times or third multiple's sum of the digits is equal to the sum of the digits itself is 27. The number is 27 and its digits sum of the digits is 2+7=9 and three times the sum of the digits is 9×3=27. This can be extended to all the multiples of nine.

(a) 2621 = 2222 + 66 + 222 + 111.

(b) 2623 = 2222 + 66 + 2 + 333.

(c) 2997 = 222 + 999 + 999 + 777.

(d) 2999 = 2 + 999 + 999 + 999.

(e) 3254 = 33 + 2222 + 555 + 444.

(f) 3259 = 33 + 2222 + 5 + 999.

(g) 2261 = 2222 + 22 + 6 + 11.

(h) 1543 = 1111 + 55 + 44 + 333.

(i) 2359 = 2222 + 33 + 5 + 99.

(j) 1454 = 11 + 444 + 555 + 444.

(k) 2354 = 2222 + 33 + 55 + 44.

(l) 2263 = 2222 + 2 + 6 + 33.

(m) 2532 = 2222 + 55 + 33 + 222.

(n) 1276=1111+22+77+66.

(o) 1187=111+111+888+77.

(p) 1099=1+0+999+99.

(q) 1185=11+1111+8+55.

(r) 1098=11+0+999+88.

(s) 198 = 11 + 99 + 88.

(t) 9653 = 99 + 666 + 5555 + 3333.

(u) 4243 = 444 + 22 + 444 + 3333.

(v) 4332 = 444 + 3333 + 333 + 222.

(w) 4335 = 444 + 3333 + 3 + 555.

(x) 4793 = 4444 + 7 + 9 + 333.

(y) 5054 = 555 + 0 + 55 + 4444.

(z) 5143 = 555 + 111 + 4444 + 33.

(aa) 5693 = 5555 + 6 + 99 + 33.

(ab) 5786 = 5555 + 77 + 88 + 66.

(ac) 6779 = 6666 + 7 + 7 + 99.

(ad) 6864 = 6666 + 88 + 66 + 44.

(ae) 6953 = 66 + 999 + 5555 + 333.

(af) 7496 = 777 + 44 + 9 + 6666.

(ag) 7586 = 777 + 55 + 88 + 6666.

(ah) 7672 = 777 + 6666 + 7 + 222.

(ai) 7679 = 7 + 6666 + 7 + 999.

(aj) 7851 = 7777 + 8 + 55 + 11.
(ak) 7941 = 7777 + 9 + 44 + 111.
(al) 7942 = 7777 + 99 + 44 + 22

(am) 7946 = 7777 + 99 + 4 + 66.
(an) 8486 = 888 + 44 + 888 + 6666.
(ao) 8664 = 888 + 6666 + 666 + 444.

(ap) 8753 = 88 + 7777 + 555 + 333.
(aq) 8758 = 88 + 7777 + 5 + 888.
(as) 5872 = 5555 + 88 + 7 + 222.

(at) 4339 = 4 + 3333 + 3 + 999.
(au) 6221 = 666 + 2222 + 2222 + 1111.
(av) 6223 = 666 + 2222 + 2 + 3333.

(aw) 6225 = 666 + 2 + 2 + 5555.
(ax) 6593 = 6 + 5555 + 999 + 33.
(ay) 4336 = 4 + 3333 + 333 + 666.

(az) 3985 = 3333 + 9 + 88 + 555.
(ba) 3521 = 3333 + 55 + 22 + 111.
(bb) 9658 = 99 + 666 + 5 + 8888.

(bc) 9386 = 99 + 333 + 8888 + 66.
(bd) 9563 = 9 + 5555 + 666 + 3333.
(be) 9568 = 9 + 5 + 666 + 8888.

Two digit number miracle: -

Take any two digit number except the palindrome numbers, write it backwards and subtract the smaller number from the bigger number.

Suppose if it comes as single digit number put zero before the number and consider it as a double digit number. Now write the result backwards and add this to the result itself and surprisingly we will get the summation always as 99 irrespective of the two digit number taken.

For example 23 and if it is written backwards it will be 32.

Subtract the smaller 32-23=9 and put zero before the resultant and now if the resultant is written backwards 90.

Now add 9+90=99.

Three digit number miracles: -

Case (a) Write three digit numbers from 1 to 9 without repeating the digits. Such type of three digit numbers like the following three, three digit numbers have the special property. The third row of three digit number is obtained by multiplying 3 with the first row of three digit number. The middle row of three digit number is obtained by subtracting the first row of three digit number from the third row of three digit number.

192	219	273	327
384	438	546	654
576	657	819	981

Case (b) Take any three digit number except the palindrome numbers, write it backwards and subtract the smaller number from the bigger number.

Now write the result backwards and add this to the result itself and surprisingly the summation will always be 1089, irrespective of the three digit number taken.

For example 398 and if it is written backwards it will be 893.

Subtract the smaller, 893-398=495 and if it is written backwards 594. It should be noted that the middle digit is always 9 and the first digit is complementary with the last digit.

Now add 495+594=1089.

The number 495 is called as Kaprekar constant for three digits number containing a set of three digits that are not identical.

Likewise we can find the Kaprekar constant for four digits number containing a set of four digits that are not identical. 6174 is reached as one repeatedly subtracts the highest and lowest numbers that can be constructed from a set of four digits that are not all identical. Thus, starting with 1234, we have

$$4321 - 1234 = 3087, \text{ then}$$
$$8730 - 0378 = 8352, \text{ and}$$
$$8532 - 2358 = 6174.$$

Repeating from this point onwards leaves the same number (that is $7641 - 1467 = 6174$). In general, when the operation converges it does so in atmost seven iterations. The number is called as Kaprekar constant for four digit number containing a set of four digits that are not identical.

However, in base 10 a single such constant only exists for numbers of 3 or 4 digits; for more digits, the numbers enter into one of several cycles.

One more interesting fact is that; write any five digit number with one condition of without repeating any digit. Write the palindrome of the number. Then subtract the smaller number from the bigger number. Now the resultant will be a divisor of 99 irrespective of five digit number taken. For example 25,417 and write it backwards. Now subtract the bigger number from smaller number, will give 46,035 can be divided by 99.

Case (c) Lori's Rule: Start with any three-digit number.

To get the hundreds digit of the next number in the sequence, take the starting number's hundreds digit and double it. If the double is more than 9, then add the double's digits together to get a one-digit number.

Do the same thing to the tens and units digits of the starting number to obtain the tens and units digits of the new number.

Repeat Steps 2 and 3 as often as necessary to find the special happening.

Example: 567, 135, 261, 432, 864, 738 and finally we get the starting number 567.

Identity property: -

The number zero is called as additive identity. It can also be called as subtractive identity. That is n+0=n where n ≠ 0. For example 33 + 0 = 33 = 33 - 0.

Columnwise addition of consecutive numbers: -

Take a sequence of consecutive number or a sequence of multiples of numbers say 5,6,7,8,9,10,11,12,13,14,15 and 16. They can be put in sequences as
(5, 6, 7), (8, 9, 10), (11, 12, 13) and (14, 15, 16).

Now write 5, 6, 7 as such and below them 8, 9, 10 in the reverse order, then below them 11, 12, 13 as such and then below them 14, 15, 16 in the reverse order and do the column wise addition as follows.

$$
\begin{array}{ccc}
5 & 6 & 7 \\
10 & 9 & 8 \\
11 & 12 & 13 \\
\underline{16} & \underline{15} & \underline{14} \\
42 & 42 & 42
\end{array}
$$

Surprisingly the column totals are the same. If the series is stopped with 3 rows only then the column totals come in a sequence skipping by 1.

$$
\begin{array}{ccc}
5 & 6 & 7 \\
10 & 9 & 8 \\
\underline{11} & \underline{12} & \underline{13} \\
26 & 27 & 28
\end{array}
$$

Note: The common difference in the column totals is one.

It is seen that when the count of rows is even the column totals are same. When the count of rows is odd then the column totals form a sequence.

$$
\begin{array}{ccc}
5 & 6 & 7 \\
\underline{10} & \underline{9} & \underline{8} \\
15 & 15 & 15
\end{array}
$$

If we interchange any two adjacent numbers in a row then the column totals form the same sequence of consecutive numbers 41, 42 and 43 but not in the same order. When two numbers interchanged in a row are not adjacent, then the column totals form the sequence 40, 42 and 44 with a column gap of two.

$$
\begin{array}{ccc}
6 & 5 & 7 \\
10 & 9 & 8 \\
11 & 12 & 13 \\
\underline{16} & \underline{15} & \underline{14} \\
43 & 41 & 42
\end{array}
\qquad
\begin{array}{ccc}
5 & 6 & 7 \\
9 & 10 & 8 \\
11 & 12 & 13 \\
\underline{16} & \underline{15} & \underline{14} \\
41 & 43 & 42
\end{array}
\qquad
\begin{array}{ccc}
5 & 6 & 7 \\
8 & 9 & 10 \\
11 & 12 & 13 \\
\underline{16} & \underline{15} & \underline{14} \\
40 & 42 & 44
\end{array}
\qquad
\begin{array}{ccc}
7 & 6 & 5 \\
10 & 9 & 8 \\
11 & 12 & 13 \\
\underline{16} & \underline{15} & \underline{14} \\
44 & 42 & 40
\end{array}
$$

Note: - When the numbers interchanged are adjacent then the column totals have the common difference of one and when the numbers interchanged are alternate then the column totals have the common difference of two.

Polite number: -

In number theory, a polite number is a positive integer that can be written as the sum of two or more consecutive positive integers. Other positive integers are called as impolite numbers. For example 6 = 1 + 2 + 3, 18 = 3 + 4 + 5 + 6, 41= 20 + 21 are all polite numbers.

Since the polite numbers are formed by the consecutive numbers, these polite numbers can also be called as staircase numbers. That is the polite number 15 can be arranged as a staircase number using 1, 2, 3, 4 and 5.

The sequence of polite numbers is 3, 5, 6, 7, 9, 10, 11, 12, 13, 14, 15, 17...

If the polite number representation starts with 1, the number so represented is triangular number. All positive integers are polite numbers except powers of two. The impolite numbers are exactly the powers of two. The sequence of impolite numbers is 2, 4, 8, 16, 32, 64, 128, 256, 512…

If the polite number representation starts with a number other than 1, the number is called as trapezium number because if we represent the consecutive addition in dots it will form a trapezium. For example in the above case 18 (= 3 + 4 + 5 + 6) is a trapezium number and is geometrically represented as follows.

So all trapezium numbers are polite numbers but all polite numbers are not trapezium numbers. The sequence of trapezium numbers is 18, 62…

Partition number: -

The partition number is defined as the number of ways a given number can be written as a sum of positive integers. For example the number '4' can be written as a sum of positive integers in five different ways. That is 4, 3+1, 2+1+1, 2 +2 and 1+1+1+1. So 5 is the partition number of 4. The partition number sequence is 1, 2, 3, 5, 7, 11, 15, 22, 30, 42, 56, 77, 101, 135, 176, 231, 297, 385, 490, 627, 792, 1002, 1255, 1575, 1958, 2436, 3010, 3718, 4565, 5604, 6842, 8349, 10143, 12310, 14883, 17977, 21637, 26015, 31185, 37338, 44583, 53174, 63261, 75175, 89134, 105558, 124754, 147273, 173525… This is the partition number sequence of the natural number 1, 2, 3, 4, 5, 6, 7, 8, 9, 10, 11, 12… respectively.

Palindromic partition number is a partition number which is palindromic in nature. For example 11, 22… are some of the palindromic partition numbers.

Prime partition number is a partition number which is also a prime number and composite partition number is a partition number which is also a composite number. For example 7 is a prime partition number and 30 is a composite partition number.

Even partition number is a partition number which is also an even number and odd partition number is a partition number which is also an odd number. For example in the above case, 7 is an odd partition number and 30 is an even partition number.

Repunit partition number is a partition number which is repunit in nature. For example the partition numbers like 11, 22… are repunit partition numbers.

About addition: -

Addition is the only unique and basic arithmetic operation in mathematics upon which other operations are derived based on the requirements. That is subtraction is nothing but negative addition, multiplication is consecutive addition and division is consecutive subtraction or consecutive negative addition.

Powers are special consecutive additions in which same number is taken for the consecutive addition.

3. SUBTRACTION

Subtraction is the reverse process of addition. It is denoted by the symbol "-" and is pronounced as minus. For example 11-9=2. Here 11 is called as subtrahend, 9 is called as minuend and 2 is called as the difference or result.

Subtraction can also be called as negative addition. That is in the above case the equation can be written as 11 + (-9) = 2.

Digital root rule of subtraction: -

(a)The difference of the digital roots of subtrahend and minuend should be equal to the digital root of the result.

For example 653 Digital root is 5
         (-) 186 Digital root is 6
             467 Digital root is 8

The difference in digital root between subtrahend and minuend is -1. Since the digital root should not be negative, take complement by subtracting from 9. We get 8 and is the digital root of the resultant.

(b) The subtraction of numbers which are having number digits either of 3, 6 or 9 among themselves will give the resultant either of 3, 6 or 9 as their number digits. For example 222(number digit 6) – 201(number digit 3) = 21(number digit 3).

Subtraction rule: -

The first decimal number is to be subtracted with first decimal number followed by 10th or second decimal number with the 10th or second decimal number and then 100th or third decimal number with the corresponding 100th or third decimal number and so on from right to left.

For example 4320
         (-) 1120
             3200

If any of the decimal number in the subtrahend is smaller than its corresponding decimal number in the minuend then we have to borrow 10 from its immediate left neighbor decimal number and do the subtraction.

For example 2320
         (-) 1129
             1191

If the subtrahend itself is smaller than the minuend then we will get negative number.

For example 234
         (-) 916
             -682

Unlike addition, the subtraction rule is applicable for the subtraction of two numbers only. To subtract more than two numbers, subtract the first two, and then subtract the third number from the difference of first two and so on.

Subtraction must be performed in the order that the numbers are listed. For example $(100 - 50) - 5 \neq 100 – (50 - 5)$

Non Commutative property: -

Subtraction is not commutative.

$11 - 9 \neq 9 - 11$

Subtraction with odd and even numbers: -

Subtraction	Odd number	Even number
Odd number	Even number	Odd number
Even number	Odd number	Even number

For example even number – odd number = odd number

$$12 - 3 = 9$$

Note:-For both addition and subtraction, the operations to odd number and even numbers are same.

Special subtractions: -

(a)   987654321   sum of the decimals in this number is 45
 (-) 123456789   sum of the decimals in this number is 45
    864197532   sum of the decimals in this number is 45

Similarly 87654321   sum of the decimals in this number is 36
   (-)12345678   sum of the decimals in this number is 36
    75308643   sum of the decimals in this number is 36

(b) 12345678 – 12345 = 12,333,333.

Subtraction table: -

Subtraction	1	2	3	4	5	6	7	8	9	10
1	0	1	2	3	4	5	6	7	8	9
2	-1	0	1	2	3	4	5	6	7	8
3	-2	-1	0	1	2	3	4	5	6	7
4	-3	-2	-1	0	1	2	3	4	5	6
5	-4	-3	-2	-1	0	1	2	3	4	5
6	-5	-4	-3	-2	-1	0	1	2	3	4
7	-6	-5	-4	-3	-2	-1	0	1	2	3
8	-7	-6	-5	-4	-3	-2	-1	0	1	2
9	-8	-7	-6	-5	-4	-3	-2	-1	0	1
10	-9	-8	-7	-6	-5	-4	-3	-2	-1	0

From the table we can see that 3-8 = -5.

Subtraction with positive and negative numbers: -

Subtraction	Positive number	Negative number
Positive number	±	±
Negative number	±	±

From the table we can find that negative number – negative number = ±
For example (-4)-(-2) = -2 and (-4)-(-6) = 2.

   Note: -If addition is compared with credit and subtraction to debit account then the above rules can be explained easily.
For example (-3) + (-7) = (-10). Here two debts are added.
        (+3) – (-7) = (+10). Here subtracting or removing the debts, which mean crediting in the account.

Check for subtraction: -

   To check the subtraction results, add the difference obtained with the minuend. For example 324-119=205. Now add the difference and minuend we will get back the subtrahend, 205+119=324.

4. MULTIPLICATION

Multiplication is denoted by the symbol "×" and is pronounced as into. For example 3×5=15. Here the number 3 is called as multiplicand, 5 is the multiplier and 15 is called as the product or result. Both the multiplicand and multiplier can be collectively called as factors. Here 3 and 5 are factors of 15.

Multiplication is consecutive addition or compound addition. That is multiplication is nothing but adding a number with the same number or with different number several times. For example in the above case, the equation can be written as 5+5+5=15. Instead of saying 'five plus five plus five', we can abbreviate it by saying 'five three times' or 'three fives' are fifteen. If the multiplication is carried with the same numbers then it will give a special type of numbers called as powers and if the multiplication is carried with the different numbers then it will give the normal multiplication products. In this chapter the normal multiplication is discussed and the powers are discussed in chapter 5.

The other symbols for multiplication are *, • and ()(). That is the above equation can also be written as 3*5=15; 3 • 5=15 and (3)(5)=15.

Note: -Lesser the difference between the multiplicand and multiplier higher the value of product. For example 9×3=27 but 9×9=81.

Multiplication rule: -

The first decimal number of the multiplier is to be multiplied with each of the decimal numbers of the multiplicand individually from right to left followed by 10th or second decimal number of the multiplier with each of the decimal numbers of the multiplicand but placing the resultant below the 10th decimal place from right to left and then 100th or third decimal number of the multiplier with each of the decimal numbers of the multiplicand but placing the resultant below the 10th decimal place from right to left and so on. Then finally add to get the product.

For example 532×234

$$
\begin{array}{r}
2128 \\
1596 \\
1064 \\
\hline
124488
\end{array}
$$

Note: - The order in which the numbers are multiplied in the case of multiplication with more numbers does not affect the result, just like the addition process.

For example 313 × 224 × 2370 = 224 × 313 × 2370 = 2370 × 224 × 313.

Digital root rule of multiplication: -

(a)The number digit of the product of two numbers is equal to the product of the number digits of the numbers involved in the multiplication.

For example 6347   number digit 2

$$
\begin{array}{r}
\underline{×206} \\
1307482
\end{array}
$$
    number digit 8

1307482   number digit 7

Now the product of the number digits of multiplicand and multiplier is 16(=2×8) whose number digit is 7.

(b) The multiplication of numbers which are having number digits either of 3, 6 or 9 among themselves will give the result either of 3, 6 or 9 as their number digits. Also it is to be noted that the number digit of any number multiplied by 9 will always be 9 only. For example 27(number digit 9) × 21(number digit 3) = 567(number digit 9).

Note: -This rule is applicable for digital root rule of subtraction also.

(c) The pattern of digital roots for 4 table is 4, 8, 3, 7, 2, 6, 1, 5 and 9. This pattern repeats itself if we continue the table on for higher numbers. Similarly we can find the patterns of digital roots for 2, 3, 5, 6, 7 and 8 tables which repeat themselves uniquely. But the digital root for 9 table is one and the same that is 9 only.

Binary method of multiplying two numbers: -
Step 1: Divide the multiplicand by 2, omitting fractions and multiply the multiplier by 2 and continue the operations till the multiplicand becomes 1.
For example 41×57
          20×114
          10×228
          5×456
          2×912
          1×1824
Step 2: Delete all the even numbers of the multiplicands and the corresponding multipliers on the right hand side and retain the remaining ones.
          41×57
          5×456
          1×1824
Step 3: Add the multipliers of all the numbers on the right hand side of step 2 and the result will be the required product.
That is 57+456+1824=2337 and this is the product of 41×57.
   Logic of the method: Consider the numbers of the multipliers in step 2
      $57=2^0×57$
     $456=2^3×57$
    $1824=2^5×57$ and now
    $2337=(2^0+2^3+2^5)×57=(1+8+32)×57=41×57.$

Some easy methods of multiplication: -
   An easy method of multiplying two numbers which give the totaling of the unit digits to ten and the tenth digits of both the numbers is same, is as follows. For example 43×47 (Here the total of the units is 10 and the tenth digit is same, that is 4).
Step 1: -Multiply the unit digits and write the product in the right hand side as 21.
Step 2: -Multiply the tenth digit with its following digit and write the product in the left hand side and now the resultant of 43 × 47 = 2021.
Another example 61 × 69
Step 1: - Right hand side product is 1 × 9 = 09 (In this method of multiplication 100 is taken as the base so we should always have two digits on the right hand side.)
Step 2: - Left hand side product is 6 × 7 = 42 and now the resultant is 4209.
   A simple method of multiplication with 10 and multiples of 10, which is 10^n is to write the multiplicand followed by 'n' number of zeros.
For example 284×1000=284000.
   A Simple method of multiplication when both multiplicand and multiplier are between 100 and 109: - When both the numbers are between 100 and 109, the product will be 1 followed by inserting the results of addition of unit digits and multiplication of unit digits. If the resultants of addition of unit digits and multiplication of unit digits is a single digit number means, put 0 before the single digit number and then insert. For example 102×103, the product will be 1 followed by 05 and 06. That is 10506. Another example 106×104=11024.

A simple method of multiplication when both multiplicand and multiplier is between 90 and 99: - Find the difference of multiplicand with 100 and the difference of multiplier with 100. Write the product of the differences in the right hand side followed by writing the subtraction resultant of multiplicand and the difference of multiplier with 100 in the left hand side. For example in the case of 96×91, the product of the difference of multiplicand with 100 and the difference of multiplier with 100 is 36. The subtraction resultant of multiplicand and the difference of multiplier with 100 is 87. Now the product of 96×91=8736.

The multiplication of a single digit number by 11 is equal to writing the same number two times adjacently. Similarly the multiplication of a two digit number by 101 is equal to writing the same number two times adjacently and likewise for a three digit number 1001 and so on.

For example 9 × 11=99

$$23×101=2,323$$
$$321×1001=321,321$$
$$5241×10001=52415241$$

....

One way of multiplying by 11 with the multidigit multiplicand is, to write the last digit of the number followed by the addition of the last digit number with the penultimate digit and so on like an arithmetic addition from right to left side.

For example 5826×11=(5+1)(5+8+1)(8+2)(2+6)6

↓

(Remainder is one and is carried over)
=64086.

One way of multiplying by 12 is to double the digit from right side and add to its neighbor.

For example 74238×12

Note: -Put zero before the multiplicand and start the process as directed.

074238×12= (0×2+7+1)(7×2+4+1)(4×2+2)(2×2+3+1)(3×2+8+1)(8×2)
=890856.

An easy method of multiplying a number by 5^n is put 'n' zeros to the right of the multiplicand and then divide the number so formed by 2^n. For example $9754×125=9754×5^3=9754000/2^3=9754000/8=1219250$.

Easy method of multiplication of some numbers: -

Multiplication by 5: Multiply by 10 and divide by 2.

Multiplication by 9: Multiply by 10 and subtract the original number.

Multiplication by 12: Multiply by 10 and add twice the original number.

Multiplication by 13: Multiply by 10 and add thrice the original number.

Multiplication by 18: Multiply by 20 and subtract twice the original number.

Multiplication by 19: Multiply by 20 and subtract the original number.

Multiplication by 98: Multiply by 100 and subtract twice the original number.

Multiplication by 99: Multiply by 100 and subtract the original number.

Commutative property: -

Multiplication is commutative.

3×5 = 15 = 5×3.

Identity property: -

The number one is called as multiplication identity. It can also be called as divisive identity. That is any number either multiplied or divided by 1 will give the same number.

That is $n \times 1 = n = n/1$ provided $n \neq 0$. For example $33 \times 1 = 33 = 33/1$.

Multiplication with odd and even numbers: -

Multiplication	Odd number	Even number
Odd number	Odd number	Even number
Even number	Even number	Even number

From the table we can tell that even number × odd number = even number

$$2 \times 11 = 22$$

In a broader sense, odd × odd × odd ×…× odd = odd (all odd numbers)

even × even × even ×…×even = even (all even numbers)

even × even × even ×…×odd = even (at least one even number).

Multiplication with positive and negative numbers: -

Multiplication	Positive number	Negative number
Positive number	+	-
Negative number	-	+

From the table we can tell that $(-3) \times (-6) = 18$.

Multiplication table: -

Multiplication	1	2	3	4	5	6	7	8	9	10
1	1	2	3	4	5	6	7	8	9	10
2	2	4	6	8	10	12	14	16	18	20
3	3	6	9	12	15	18	21	24	27	30
4	4	8	12	16	20	24	28	32	36	40
5	5	10	15	20	25	30	35	40	45	50
6	6	12	18	24	30	36	42	48	54	60
7	7	14	21	28	35	42	49	56	63	70
8	8	16	24	32	40	48	56	64	72	80
9	9	18	27	36	45	54	63	72	81	90
10	10	20	30	40	50	60	70	80	90	100

From the table we can tell that $6 \times 7 = 42$.

Some curious arithmetic operations: -

(a) $19 \times 0.95 = 18.05$
   $19 - 0.95 = 18.05$

(b) $1.2 \times 6 = 7.2$
   $1.2 + 6 = 7.2$

(c) $6.25 \div 5 = 1.25$
   $6.25 - 5 = 1.25$

(d) $1\frac{1}{2} \times 3 = 9/2$ $1\frac{1}{3} \times 4 = 16/3$ $1\frac{1}{4} \times 5 = 25/4$ ………
   $1\frac{1}{2} + 3 = 9/2$ $1\frac{1}{3} + 4 = 16/3$ $1\frac{1}{4} + 5 = 25/4$ ………

Likewise we can extend the above type of curious multiplication and addition operation.

(e) 1, 2 and 3 are the only three consecutive numbers whose summation and product are same. That is $1 \times 2 \times 3 = 6 = 1 + 2 + 3$

and similarly $(-1) \times (-2) \times (-3) = (-6) = (-1) + (-2) + (-3)$

(f) $2 \times 2 = 4$ and $2 + 2 = 4$.
Similarly $4 \div 2 = 2$ and $4 - 2 = 2$.

(g) $19 = (1 \times 9) + (1 + 9)$
$29 = (2 \times 9) + (2 + 9)$
$39 = (3 \times 9) + (3 + 9)$

$49 = (4 \times 9) + (4 + 9)$
$59 = (5 \times 9) + (5 + 9)$
$69 = (6 \times 9) + (6 + 9)$

$79 = (7 \times 9) + (7 + 9)$
$89 = (8 \times 9) + (8 + 9)$
$99 = (9 \times 9) + (9 + 9)$

This works for the numbers 19, 29, 39, 49, 59, 69, 79, 89 and 99 only.

(h) Difference between addition sum and multiplication product of a real number 'n' with the same number is

$$product = \frac{n}{2}(sum) \text{ where } n \triangleright 0$$

For example if n = 11, then its addition sum is $11 + 11 = 22$ and the multiplication product is $11 \times 11 = 121$

Now the multiplication product = (n/2) × addition sum

$$121 = 5.5 \times 22$$

FEATURES OF NUMBER 1

The number 1 is called as unit number.

The number 1 is the multiplication identity. It can also be called as division identity.

1 is an odd number.

$1^n = 1$, where 'n' is a positive number. That is 1 is a special power number.

The reciprocal number of 1 is 1.

The negative number as well as opposite number of 1 is -1.

The number 1 is neither a prime number nor a composite number. A prime number means it should have two trivial divisors, that is 1 and the number itself and a composite number means it should have at least one more divisor apart from the trivial divisors. Since the number 1 has only one divisor it is neither a prime number nor a composite number.

The number one can be considered as an automorphic number, Armstrong number, and polygonal number.

All positive numbers are derived from one only. That is 5 means it contains five ones and so on. This is called the 'oneness'. These ones put together and gives 1, 2 (1+1), 3 (1+1+1)... That is 'one' alone exists and all others are its own manifestations.

Repunit numbers: -

The numbers formed with only 1's are called as repunit numbers. That is 11, 111, 1111... are called as repunit numbers. These are denoted by 1_n where 'n' is the number of repetitions. For example 1_5 means 11111.

Repunit numbers will not become power numbers.

When the repunit number 11 is multiplied by the multiples of 11, the difference between any two successive products is constant, that is 121.

$$11 \times 11 = 121 \qquad 11 \times 66 = 726$$
$$11 \times 22 = 242 \qquad 11 \times 77 = 847$$
$$11 \times 33 = 363 \qquad 11 \times 88 = 968$$

$$11 \times 44 = 484 \qquad 11 \times 99 = 1,089$$
$$11 \times 55 = 605 \qquad 11 \times 110 = 1,210$$
....

For example $1210 - 1089 = 121$.

Also when the repunit number 11 is multiplied by the two digit consecutive ascending numbers or two digit consecutive descending number, the difference between any two successive products is constant, that is 121.

$$11 \times 12 = 132$$
$$11 \times 23 = 253$$
$$11 \times 34 = 374$$
....

For example $374 - 253 = 121$.

When the number 111 is multiplied by two digit consecutive ascending or descending number like 12, 23, 34, 45, 56, 67, 78 and 89, the answer differs from previous one, only by 1,221.

$$111 \times 12 = 1,332$$
$$111 \times 23 = 2,553$$
$$111 \times 34 = 3,774$$
....

Similarly if we continue the multiplication of 1111 with two digit consecutive ascending number, the difference between the successive multiplications will be 12,221 and the multiplication of 11,111 with two digit consecutive ascending number, the difference between the successive multiplications will be 122,221 and so on.

If the repunit number 11 is multiplied by three digit consecutive ascending number 123, 234, 345 and so on the answer differs from previous one, only by 1,221. If the repunit number 111 is multiplied by 123, 234, 345 and so on the answer differs from previous one, only by 12,321. Similarly when the repunit number 1,111 is multiplied by 123, 234, 345 and so on the answer differs from previous one, only by 123321 and so on.

If the repunit number of 'n' digit is multiplied by 'n' digit of consecutive ascending or descending number, the maximum digit coming in the difference of products will be equal to digital sum of the repunit number. For example if 1111 is multiplied by 4321, 5432, 6543 and so on, the maximum digit coming in the difference of products will be 1234321. Similarly if the repunit number of 'n' digit is multiplied by 'n-1' digit of consecutive ascending or descending number, the maximum digit coming in the difference of products will have 1 less than digital sum of the repunit number. For example if 1111 is multiplied by 321, 432, 543 and so on, the maximum digit coming in the difference of products will be 123321. Likewise if the repunit number of 'n' digit is multiplied by 'n-2' digit of consecutive ascending or descending number, the maximum digit coming in the difference of products will have 2 less than digital sum of the repunit number. For example if 1111 is multiplied by 321, 432, 543…, the maximum digit coming in the difference of products will be 12221 and so on.

If we multiply 11 by repeated digits of 2 such as 2, 22, 222… then the answer will contain the first and last digits as 2 and 4s between the two 2s. The number of 4s is given by 1 less than the number of 2s present in the multiplier. For example $11 \times 222 = 2442$.

All the Repunit numbers are palindrome numbers, but the reverse is not true.

All repdigit numbers are multiples of repunits.

All repunit numbers are odd repunit numbers as well as they are all super odd repunit numbers.

The repunit numbers which are also prime numbers are called as prime repunit numbers and the repunit numbers which are composite are called as composite repunit numbers. For example 11 is a prime repunit number and 111 is a composite repunit number.

The numbers in the series 1, 11, 111, 1111, 11111, etc... are all triangular numbers in base-9.

Fun triangles with ones: -
Fun triangle (a)

$$1 \times 1 = 1$$
$$11 \times 11 = 121$$
$$111 \times 111 = 12321$$
$$1111 \times 1111 = 1234321$$
$$11111 \times 11111 = 123454321$$

……………………....................

Fun triangle (b)

$$11 \times 1 = 11$$
$$111 \times 11 = 1221$$
$$1111 \times 111 = 123321$$
$$11111 \times 1111 = 12344321$$
$$111111 \times 11111 = 1234554321$$

…………………………………

Fun triangle (c)

$$(7+4)^0 \ = 1$$
$$(7+4) \ \ = 11$$
$$57+54 \ \ = 111$$
$$557+554 \ = 1,111$$
$$5557+5554 = 11,111$$

………………………

Fun triangle (d)

$$6 - 5 = 1$$
$$56^2 - 45^2 = 1,111$$
$$556^2 - 445^2 = 111,111$$
$$5556^2 - 4445^2 = 11,111,111$$
$$55556^2 - 44445^2 = 1,111,111,111$$

…………………………………..

Fun triangle (e)

$$(10^2 - 1)/9 = 11$$
$$(60^2 - 51^2)/9 = 111$$
$$(560^2 - 551^2)/9 = 1,111$$
$$(5560^2 - 5551^2)/9 = 11,111$$
$$(55560^2 - 55551^2)/9 = 111,111$$

……………………………………….

Fun triangle (f)

$$6^2 + 75 = 111$$
$$66^2 + 6755 = 11,111$$
$$666^2 + 667555 = 1,111,111$$
$$6666^2 + 66675555 = 111,111,111$$
$$66666^2 + 6666755555 = 11,111,111,111$$

……………………………………………….

Fun triangle (g)

$$7^2 + 62 = 111$$
$$67^2 + 6622 = 11,111$$
$$667^2 + 666222 = 1,111,111$$
$$6667^2 + 66662222 = 111,111,111$$
$$66667^2 + 6666622222 = 11,111,111,111$$

………………………………………………….

Fun triangle (h)

$$3 \times 37 = 111$$
$$33 \times 3367 = 111,111$$
$$333 \times 333667 = 111,111,111$$
$$3333 \times 33336667 = 111,111,111,111$$
$$33333 \times 3333366667 = 111,111,111,111,111$$

……………………………………………..

Fun triangle (i)

$$(3 \times 4) - 1 = 11$$
$$(33 \times 34) - 11 = 1,111$$
$$(333 \times 334) - 111 = 111,111$$
$$(3333 \times 3334) - 1111 = 11,111,111$$
$$(33333 \times 33334) - 11111 = 1,111,111,111$$

……………………………………………………………

Fun triangle (j)

$$2 + 3^2 = 11$$
$$22 + 33^2 = 1,111$$
$$222 + 333^2 = 111,111$$
$$2222 + 3333^2 = 11,111,111$$
$$22222 + 33333^2 = 1,111,111,111$$

………………… …………………………

Fun triangle (k)

$$(10^1 - 1)/9 = 1$$
$$(10^2 - 1)/9 = 11$$
$$(10^3 - 1)/9 = 111$$
$$(10^4 - 1)/9 = 1,111$$
$$(10^5 - 1)/9 = 11,111$$

………………………………

Fun triangle (l)

$$1_{10} = 1_2$$
$$3_{10} = 11_2$$
$$7_{10} = 111_2$$
$$15_{10} = 1111_2$$
$$31_{10} = 11111_2$$

…………………..

Fun triangles with ones in various numbering systems are given as follows.

$$1_{10} = 1_3$$
$$4_{10} = 11_3$$
$$13_{10} = 111_3$$
$$40_{10} = 1111_3$$
$$121_{10} = 11111_3$$

…………………..

$$1_{10} = 1_4$$
$$5_{10} = 11_4$$
$$21_{10} = 111_4$$
$$85_{10} = 1111_4$$
$$341_{10} = 11111_4$$

…………………..

$$1_{10} = 1_5$$
$$6_{10} = 11_5$$
$$31_{10} = 111_5$$
$$156_{10} = 1111_5$$
$$781_{10} = 11111_5$$

…………………..

$$1_{10} = 1_6$$
$$7_{10} = 11_6$$
$$43_{10} = 111_6$$
$$259_{10} = 1111_6$$
$$1555_{10} = 11111_6$$
....

$$1_{10} = 1_7$$
$$8_{10} = 11_7$$
$$57_{10} = 111_7$$
$$400_{10} = 1111_7$$
$$2801_{10} = 11111_7$$
....

$$1_{10} = 1_8$$
$$9_{10} = 11_8$$
$$73_{10} = 111_8$$
$$585_{10} = 1111_8$$
$$4681_{10} = 11111_8$$
....

$$1_{10} = 1_9$$
$$10_{10} = 11_9$$
$$91_{10} = 111_9$$
$$820_{10} = 1111_9$$
$$7381_{10} = 11111_9$$
....

and so on.

Features of number 2: -

The number 2 is an important number based upon which we can classify the natural numbers as even number and odd number.

With the base value 2, the numbering system developed is called as binary numbering system. It is the widely used numbering system in computer applications.

The numbers formed with the repeated digit of two are repdigit numbers of two. For example 22, 222, 2222…

The sum of $1+2+3+4+5+6+7+8+9 = 45$. If we multiply 123456789 by 2 we get 246913578, the sum of which is again 45.

Using the number 2 and the four arithmetic operations only we can get all the whole numbers as follows.

$$2+2-2-(2/2) =1$$
$$2+2+2-2-2 =2$$
$$2+2-2+(2/2) =3$$

$$2\times2\times2-2-2 =4$$
$$2+2+2-(2/2) =5$$
$$2+2+2+2-2 =6$$

$$(22/2)-2-2 =7$$
$$2\times2\times2+2-2 =8$$
$$2\times2\times2 + (2/2) =9$$
$$2 - 2 =0$$

Features of number 3: -

The number 3 is called as triangular number. Triangle is a closed geometrical figure with the least number of sides and the number of sides is 3.

With the base value 3, the numbering system developed is called as ternary numbering system.

The numbers formed with the repeated digit of three are repdigit numbers of three. For example 33, 333, 3333…

Features of number 4: -

The number 4 is called as square number.

It is the first perfect square in the real number sequence.

The numbers formed with the repeated digit of four are repdigit numbers of four. For example 44, 444, 4444…

Features of number 5: -

The number 5 is called as pentagonal number.

With the base value 5, the numbering system developed is called as pental numbering system.

The numbers formed with the repeated digit of five are repdigit numbers of five. For example 55, 555, 5555…

The sequence of fives 5, 55, 555… are upside down numbers.

An easy method to multiply any number made of digit 5, 'n' times by any given number like 55555×7 is multiply 7 by 5 and adds digits of the resultant. Then insert addition sum 'n-1' times in between the resultant to get the answer. That is in the above case 7×5 gives 35, and then $3 + 5 = 8$. Now insert 8 in between 3 & 5 which is the resultant of 5×7 for 4 times to get the answer 388885.

Features of number 6: -

The number 6 is called as hexagonal number.

The numbers formed with the repeated digit of six are repdigit numbers of six. For example 66, 666, 6666…

Features of number 7: -

The number 7 is called as heptagonal number.

The numbers formed with the repeated digit of seven are repdigit numbers of seven. For example 77, 777, 7777…

Features of number 8: -

The number eight is a first even cube number.

It is the only single digit cube number.

It is a Fibonacci number.

If eight is multiplied with the ascending number excluding 8 itself gives a descending number as follows. That is $12345679 \times 8 = 98765432$.

With the base value 8, the numbering system developed is called as octal numbering system.

The numbers formed with the repeated digit of eight are repdigit numbers of eight. For example 88, 888, 888…

The multiplication of eight in the following manners gives special triangles.

(i)
$$1 \times 8 + 1 = 9$$
$$12 \times 8 + 2 = 98$$
$$123 \times 8 + 3 = 987$$
$$1234 \times 8 + 4 = 9876$$
$$12345 \times 8 + 5 = 98765$$
$$123456 \times 8 + 6 = 987654$$
$$1234567 \times 8 + 7 = 9876543$$
$$12345678 \times 8 + 8 = 98765432$$
$$123456789 \times 8 + 9 = 987654321$$

(ii)
$$9 \times 0 + 8 = 8$$
$$9 \times 9 + 7 = 88$$
$$98 \times 9 + 6 = 888$$
$$987 \times 9 + 5 = 8888$$
$$9876 \times 9 + 4 = 88888$$
$$98765 \times 9 + 3 = 888888$$
$$987654 \times 9 + 2 = 8888888$$
$$9876543 \times 9 + 1 = 88888888$$
$$98765432 \times 9 + 0 = 888888888$$

(iii) The following operation of '8's gives palindrome number.
$$8x8 + 13 = 77$$
$$88x8 + 13 = 717$$
$$888x8 + 13 = 7117$$
$$8888x8 + 13 = 71117$$
$$88888x8 + 13 = 711117$$

……………………………………

(iv) Enigma of 8

$$12345679 \times 9 = 111111111$$
$$12345679 \times 18 = 222222222$$
$$12345679 \times 27 = 333333333$$

$$12345679 \times 36 = 444444444$$
$$12345679 \times 45 = 555555555$$
$$12345679 \times 54 = 666666666$$

$$12345679 \times 63 = 777777777$$
$$12345679 \times 72 = 888888888$$
$$12345679 \times 81 = 999999999$$

Further the enigma of 8 can be extended for $0.12345679 \times 9 = 1.11111111$
$$0.12345679 \times 18 = 2.22222222$$
$$0.12345679 \times 27 = 3.33333333$$
....
up to $0.12345679 \times 81 = 9.99999999$

(v) The multiplication of numbers with equal 1's and 8's gives the following

$$1 \times 8 = 8$$
$$11 \times 88 = 968$$
$$111 \times 888 = 98568$$
$$1111 \times 8888 = 9874568$$
$$11111 \times 88888 = 987634568$$
$$111111 \times 888888 = 98765234568$$
$$11111111 \times 8888888 = 9876541234568$$

....

Fun triangles with multiplication: -

Fun triangle (a)

$$4 \times 4 = 16$$
$$34 \times 34 = 1156$$
$$334 \times 334 = 111556$$
$$3334 \times 3334 = 11115556$$
$$33334 \times 33334 = 1111155556$$

....

Fun triangle (b)

$$7 \times 7 = 49$$
$$67 \times 67 = 4489$$
$$667 \times 667 = 444889$$
$$6667 \times 6667 = 44448889$$
$$66667 \times 66667 = 4444488889$$

....

Fun triangle (c)

$$6 \times 7 = 42$$
$$66 \times 67 = 4422$$
$$666 \times 667 = 444222$$
$$6666 \times 6667 = 44442222$$
$$66666 \times 66667 = 4444422222$$

....

Fun triangle (d)

$$7 \times 9 = 63$$
$$77 \times 99 = 7623$$
$$777 \times 999 = 776223$$
$$7777 \times 9999 = 77762223$$
$$77777 \times 99999 = 7777622223$$

…………………..…………………….

Fun triangle with multiplication and addition

$$1 \times 7 + 3 = 10$$
$$14 \times 7 + 2 = 100$$
$$142 \times 7 + 6 = 1000$$
$$1428 \times 7 + 4 = 10,000$$
$$14285 \times 7 + 5 = 100,000$$
$$142857 \times 7 + 1 = 1,000,000$$
$$1428571 \times 7 + 3 = 10,000,000$$
$$14285714 \times 7 + 2 = 100,000,000$$
$$142857142 \times 7 + 6 = 1,000,000,000$$
$$1428571428 \times 7 + 4 = 10,000,000,000$$

……………………………………..

In the above case 142857 repeats in the multiplication step that is in the each multiplication series and 326451 repeats in the addition step that is in the each addition series.

NUMBER 9:– A MARVEL

Number 9 is the largest single digit number in the decenary system. Likewise 99, 999, 9999… is the largest two digit, three digit, four digit number respectively.

It is the first odd square number.

With the base value 9, the numbering system developed is called as nonary numbering system.

The numbers formed with the repeated digit of nine are repdigit numbers of nine. For example 99, 999, 9999…

Number 9 is sum of the first 3 factorials: $1!+2!+3! = 9$.

$1^3+2^3 = 9 = (1+2)^2$.

The nines odd powers end with 9 and its even powers end with 1. For example $9^2=81$ and $9^3=729$.

$9 = 1^3+2^2+3^1+4^0$.

The multiplication of any number with 9 will give a product whose summation of the digits (digit sum) is always 9 or multiples of 9. For example $1234\times9=11,106$ (sum of the digits of the product $1+1+1+0+6=9$)

The converse of the above fact is also true. That is when we subtract any number from its sum total of its digits or digit sum, the result is always divisible by 9. For example 8247 and the sum of its digits is $8+2+4+7=21$ and now $8247-21=8226$. This is divisible by 9.

Note:-Digit sum is the result we get when we add all the digits in a number.

Sum of any nine consecutive numbers will give the resultant with number digit 9. For example $21+22+23+24+25+26+27+28+29=225$ and the number digit of the resultant is 9.

When we subtract any number from any of its transposed number (if the number transposed is greater than the original number, subtract the original number from the transposed number) then the resultant will be a divisor of 9. For example 213 and the transposed numbers are 132, 321, 231, 123 and 312. Now $213-132=81$ this is divisible by 9. Again $312-213=99$ this is divisible by 9 and like this we can extend for other numbers.

Any number when added to 9 will have the resultant having the sum of the digits equal to the sum of the digits of the addendum added to 9. For example $1132+9=1141$. Here the sum of the digits of addendum 1132 is 7 and the sum of the digits of resultant 1141 is 7.

Note:-For single digit numbers from 1 to 9 when added to 9 the resultant will have the sum of the digits equal to the addendum that is the number itself. For example $8+9=17$.

The resultant formed from the addition of two 9's is palindrome of the resultant formed from the multiplication. That is $9 + 9 = 18$, the palindrome number of addition resultant is $81 = 9 \times 9$. This property is called as Weird property.

Magic with nines addition

$$9 + 9 = 18$$
$$99 + 99 = 198$$
$$999 + 999 = 1998$$
$$9999 + 9999 = 19998$$
$$99999 + 99999 = 199998$$

……………………………………

Magic with nines multiplication

$$9 \times 9 = 81$$
$$99 \times 99 = 9801$$
$$999 \times 999 = 998001$$
$$9999 \times 9999 = 99980001$$
$$99999 \times 99999 = 9999800001$$

……………………………..

Note: - To get the square of a number which contains only 9s, write from right to left first 1, then zeros, one less than the number of 9s and then write 8. Finally write as many 9s as the zeros.

Another fun with nines multiplication

$$9 \times 9 = 81$$
$$99 \times 9 = 891$$
$$999 \times 9 = 8991$$
$$9999 \times 9 = 89991$$
$$99999 \times 9 = 899991$$

………………..

Further the funs with nine multiplications are

$$0 \times 9 + 1 = 1$$
$$1 \times 9 + 2 = 11$$
$$12 \times 9 + 3 = 111$$
$$123 \times 9 + 4 = 1111$$
$$1234 \times 9 + 5 = 11111$$
$$12345 \times 9 + 6 = 111111$$
$$123456 \times 9 + 7 = 1111111$$
$$1234567 \times 9 + 8 = 11111111$$
$$12345678 \times 9 + 9 = 111111111$$
$$123456789 \times 9 + 10 = 1111111111$$

In the following number triangle, addendum is 8 and the sum of the digits of the resultant is 8.

$$1 \times 9 - 1 = 08$$
$$21 \times 9 - 1 = 188$$
$$321 \times 9 - 1 = 2888$$
$$4321 \times 9 - 1 = 38888$$
$$54321 \times 9 - 1 = 488888$$
$$654321 \times 9 - 1 = 5888888$$
$$7654321 \times 9 - 1 = 68888888$$
$$87654321 \times 9 - 1 = 788888888$$
$$987654321 \times 9 - 1 = 8888888888$$

From the resultant we can see that the first digit is obtained by subtracting 1 from the first digit of the multiplicand followed by writing first digit's time of multiplicand by 8.

Another fun triangle is

$$0 \times 9 + 8 = 8$$
$$9 \times 9 + 7 = 88$$
$$98 \times 9 + 6 = 888$$
$$987 \times 9 + 5 = 888\ 8$$
$$9876 \times 9 + 4 = 888\ 88$$
$$98765 \times 9 + 3 = 888\ 888$$
$$987654 \times 9 + 2 = 888\ 888\ 8$$
$$9876543 \times 9 + 1 = 888\ 888\ 88$$
$$98765432 \times 9 + 0 = 888\ 888\ 888$$
$$987654321 \times 9 - 1 = 888\ 888\ 888\ 8$$
$$9876543210 \times 9 - 2 = 888\ 888\ 888\ 88$$

The multiplication of 1089 with 9 is the reverse of the number and similarly the multiplication of 10989 with 9 is the reverse of the number and this can be extended to 109989, 1099989...

That is $1089 \times 9 = 9801$
$$10989 \times 9 = 98901$$
$$109989 \times 9 = 989901$$
$$1099989 \times 9 = 9899901$$
$$10999989 \times 9 = 98999901$$

....

Likewise the division of 1089 by 9 gives a palindrome number as follows.

$$1089 / 9 = 121$$
$$10989 / 9 = 1,221$$
$$109989 / 9 = 12,221$$
$$1099989 / 9 = 122,221$$
$$10999989 / 9 = 1,222,221$$

....

The multiplication of 9 with the following repunit numbers give

$$111111111 \times 9 = 0999999999$$
$$222222222 \times 9 = 1999999998$$
$$333333333 \times 9 = 2999999997$$

$$444444444 \times 9 = 3999999996$$
$$555555555 \times 9 = 4999999995$$
$$666666666 \times 9 = 5999999994$$

$$777777777 \times 9 = 6999999993$$
$$888888888 \times 9 = 7999999992$$
$$999999999 \times 9 = 8999999991$$

From the resultant obtained we can see that the sum of the first and last digit is same and is 9. For example consider the product 5999999994 and the sum of the first and last digit is 5+4=9. Also the first digit is increasing and the last digit is decreasing while the middle digits remain the same.

$$123456789 \times 9 = 1111111101$$
$$123456789 \times 18 = 2222222202$$
$$123456789 \times 27 = 3333333303$$

$$123456789 \times 36 = 4444444404$$
$$123456789 \times 45 = 5555555505$$

$$123456789×54=6666666606$$

$$123456789×63=7777777707$$
$$123456789×72=8888888808$$
$$123456789×81=9999999909$$

Also another curious relation is,

$$987654321× 9=08888888889$$
$$987654321×18=17777777778$$
$$987654321×27=26666666667$$

$$987654321×36=35555555556$$
$$987654321×45=44444444445$$
$$987654321×54=53333333334$$

$$987654321×63=62222222223$$
$$987654321×72=71111111112$$
$$987654321×81=80000000001$$

Multiplication of a number with series of 9: -

Case I: - When a number is multiplied by 9 or 99 or 999 and so on and if the number of digits in the multiplicand is equal to the number of digits of the multiplier then to get the product write 1 less than the multiplicand in the left hand side and subtract each digit of the LHS from 9 then write respectively in the right hand side.

For example $345×999=344654$

Case II: - When a number is multiplied by 9 or 99 or 999 and so on and if the number of digits in the multiplicand is less than the number of digits of the multiplier then add zero or zeros to get the number of digits same. Then to get the product of the multiplication proceed as per case I.

For example $345×9999=0345×9999=03449655$

Case III: - When a number is multiplied by 9 or 99 or 999 and so on and if the number of digits in the multiplicand is more than the number of digits of the multiplier then put as many zero or zeros as there are in the multiplier and from the result if we subtract the multiplicand we will get the resultant.

For example $345×99=34500-345$
$$=34155$$
$$111×9=1110-111$$
$$=999$$

Consider the numbers 58132764, 76125483, 72645831, 16583742 and 81274365.

Multiply all these numbers by 9, $58132764×9=523194876$
$$72645831×9=653812479$$
$$76125483×9=685129347$$

$$16583742×9=149253678$$
$$81274365×9=731469285$$

The specialty of the above multiplication is, each of the four numbers mentioned above has eight digits and they are from 1 to 8. No digit is repeated. When these numbers are multiplied by 9, the products have digits from 1 to 9 and here also no digit is repeated.

The set of fractions 1/9, 1/99, 1/999… will give infinite recurring fractions 0.1, 0.01, 0.001… respectively. Based on this method a simple way of converting a decimal into a recurring infinite decimal is as follows. 7/9, 8/99, 24/999… will give infinite fractions 0.7, 0.08, 0.024… respectively.

The division of 1÷9=0.111…

2÷9=0.222…
3÷9=0.333…

4÷9=0.444…
5÷9=0.555…
6÷9=0.666…

7÷9=0.777…
8÷9=0.888…
9÷9=0.999…

Of course 9÷9=1 but is actually equal to 0.999… and this is proved as follows.

Consider N=0.999…
Then 10N=9.999…
Now 10N-N=9
That is 9N=9 or N=1=0.999…

Multiplication table for 9: -

$1 \times 9 = \ /9 = (1-1)/9-(1-1) = 9$
$2 \times 9 = 1/8 = (2-1)/9-(2-1) = 18$
$3 \times 9 = 2/7 = (3-1)/9-(3-1) = 27$

$4 \times 9 = 3/6 = (4-1)/9-(4-1) = 36$
$5 \times 9 = 4/5 = (5-1)/9-(5-1) = 45$
$6 \times 9 = 5/4 = (6-1)/9-(6-1) = 54$

………………………………………

In the products we see the unit digits go like 9,8,7,…,0 while the tens digits go like 0,1,2,3,…9 This pattern repeats further and is characteristic for multiplication table of 9. The same argument can be extended for multiplication of a number with 99, 999, 9999, 99999…

For example $11 \times 99 = 10/89 = (11-1)/99-(11-1) = 1089$
$12 \times 99 = 11/88 = (12-1)/99-(12-1) = 1188$
$13 \times 99 = 12/87 = (13-1)/99-(13-1) = 1287$

$14 \times 99 = 13/86 = (14-1)/99-(14-1) = 1386$
$15 \times 99 = 14/85 = (15-1)/99-(15-1) = 1485$

………………………………………………………

Note: - The above method is applicable only if the multiplicand and multiplier have equal number of digits. That is if the multiplicand is single digit, double digit, three digit… then the multiplier should also be a single digit, double digit, three digit… respectively.

Using the basic arithmetic operations, any number can be expressed with the help of nine only as follows.

$1 = (9/9)^9$
$2 = (9 + 9) / 9$
$3 = \sqrt9 + 9 - 9$

$4 = \sqrt9 + 9 / 9$
$5 = \sqrt9! - (9 / 9)$

$$6 = 9 - 9 / \sqrt{9}$$

$$7 = 9 - \sqrt{9} + (9 / 9)$$
$$8 = 9 - 9 / 9$$
$$9 = {}^9\sqrt{(9^9)}$$

$$10 = 9 + (9 / 9)$$
$$11 = 99 / 9$$
$$12 = 9 + 9 / \sqrt{9} \text{ and so on.}$$

If we multiply the cyclic number 142857 by 7 gives 142857×7=999999. Also the addition of digits of the multiplicand is (1+4+2+8+5+7=27 and again 2+7=9).

It is to be noted that the numbers 9, 99, 999, 9999… are all Kaprekar numbers.

Also the numbers 9, 99, 999, 9999… are all trimorphic numbers.

The squares of the numbers 9, 99, 999, 9999… have special property described as follows.

$$9^2 = 81 \text{ and } 8 + 1 = 9$$
$$99^2 = 9801 \text{ and } 98 + 01 = 99$$
$$999^2 = 998001 \text{ and } 998 + 001 = 999$$
$$9999^2 = 99980001 \text{ and } 9998 + 0001 = 9999$$

……………………………………………………………

Reciprocal number and nine:

A curious division of the numbers $(1/37) = 0.027027027…$
$(1/27) = 0.037037037…$
Also $(1/37) \times (1/27) = (1/999)$

An easy way of dividing a number by 9 is, keep the left hand digit of the number as such and go on adding the digits from left side onwards. Now in the resultant number the right hand digit is the remainder of the division and the left out digits will give the quotient. For example 1302/9. Now keep left digit 1 followed by 4 (that is 1+3) followed by 4 (that is 1+3+0) and then finally by 6 (that is 1+3+0+2) Now 144 is the quotient and 6 is the remainder.

If you divide a number by the amount of 9s corresponding to its number of digits, the number is turned into a repeating decimal. For example 274/999 = 0.274274274274...

The recurring decimal (1/17) gives a special phenomenon where if we split the recurrence into two parts and add, we get eight 9's.

           *   *

$(1/17) = 0.0588235294117647$

The asterisk is put to indicate the recurrence. Now split the decimal into two parts and add

            05882352
           <u>94117647</u>
           <u>99999999</u>

Similarly the recurring decimal (1/19) gives a special phenomenon where if we split the recurrence into two parts and add we get nine 9's.

          *   *

$(1/19) = 0.052631578947368421$

Now split the decimal into two parts and add

           052631578
           <u>947368421</u>
           <u>999999999</u>

Relation between nine's power and reciprocal number.

Nine's power	Reciprocal number
$1/9^2 = 1/81$	0.0123456790123456790…
$1/99^2 = 1/9801$	0.00010203040506070809…
$1/999^2 = 1/998001$	0.00000100200300400500 6…
$1/9999^2 = 1/99980001$	0.000000010002000300040 0050006…
……….	…………………………..

It is to be noted that 9 is the binary complement of 6 because $9_{10} = 1001_2$ and $6_{10} = 0110_2$. Now $15_{10} = 1111_2$.

A special multiplication with consecutive digits of number and nine is

$01 \times 9 = 9$ (Digital sum is 9)

$12 \times 9 = 108$ (Digital sum is 9)

$23 \times 9 = 207$ (Digital sum is 9)

………………………………..

$121 \times 9 = 1089$ $(1 + 0 + 8 + 9 = 18; 1 + 8 = 9)$

$234 \times 9 = 2106$ $(2 + 1 + 0 + 6 = 9)$

$345 \times 9 = 3105$ $(3 + 1 + 0 + 5 = 9)$

……………………………………

$1234 \times 9 = 11106$ (Digital sum is 9)

$2345 \times 9 = 21105$ (Digital sum is 9)

$3456 \times 9 = 31104$ (Digital sum is 9)

…………………

The following table shows multiplication of nines with sequence of numbers. From the resultant obtained we can see that the sum of the first and last digit is same and is 9. Also the first digit is increasing and the last digit is decreasing while the middle digits remain the same.

Multiplication with 9	Multiplication with 99	Multiplication with 999	Multiplication with 9999
9×1=09	99×1=099	999×1=0999	9999×1=09999
9×2=18	99×1=198	999×2=1998	9999×2=19998
9×3=27	99×1=297	999×3=2997	9999×3=29997
9×4=36	99×1=396	999×4=3996	9999×4=39996
9×5=45	99×1=495	999×5=4995	9999×5=49995
9×6=54	99×1=594	999×6=5994	9999×6=59994
9×7=63	99×1=693	999×7=6993	9999×7=69993
……	………	………	…….and so on.

Relation between number 3 and 9:-

The number 3 is also closely related with 9. Since the minimum number of sides required to form a closed figure is three, the square of three is a special number. So the number nine can be humbly called as the progeny of three.

All the numbers which are divisible by 9 are also divisible by 3, but the reverse is not true.

The relationship between 3rd multiples and nine is tabulated as follows.

3rd Multiples	Reflux number of the product	
$3^2=09$	9	9+0=9
$33^2=1089$	9801	98+01=99
$333^2=110889$	988011	988+011=999
$3333^2=11108889$	98880111	9888+0111=9999
$33333^2=1111088889$	9888801111	98888+01111=9999
…………….	…………….	…………….

Relation between number 18 and 9:-

The number 18 can be called as brother of 9. The product obtained by multiplying with 18 by any number will have the sum of the digits as 9 or multiples of 9. For example $18 \times 11 = 198$(sum of the digits of the resultant 198 is 18).

Another example is $18 \times 18 = 324$(sum of the digits of the resultant 324 is 9).

Likewise further powers of 18 also give the resultant whose sum of the digits equal to 9 or multiples of 9. That is $18^4 = 104976$ (sum of the digits of the resultant is $1+0+4+9+7+6=27$). Now multiply the numbers 58132764, 76125483, 72645831, 16583742 and 81274365 with 18. The product will have ten digits from 0 to 9 once and only once. Again here no digits are repeated.

$$58132764 \times 18 = 1046389752$$
$$72645831 \times 18 = 1307624958$$

$$76125483 \times 18 = 1370258694$$
$$81274365 \times 18 = 1462938570$$

The number 18 is the only number whose twice the sum of its digits $[2(1+8)]$ is equal to the original number.

Specialty of 365: -

$$365 = 10^2 + 11^2 + 12^2$$
$$365 = 13^2 + 14^2$$

Specialty of 666: -

666 is a repdigit number.

It is a triangular number.

It is the 36th triangular number and again number 36 is a triangular number.

It is a Harshad number. That is 666 is divisible by its sum of digits 18.

$$666 = 6^3 + 6^3 + 6^3 + 6 + 6 + 6$$
$$666 = 2^2 + 3^2 + 5^2 + 7^2 + 11^2 + 13^2 + 17^2$$

$$666 = 1^3 + 2^3 + 3^3 + 4^3 + 5^3 + 6^3 + 5^3 + 4^3 + 3^3 + 2^3 + 1^3$$
$$666 = 1 + 2 + 3 + 4 + 567 + 89$$
$$666 = 123 + 456 + 78 + 9$$

$$666 = 1 + 2 + 3 + ... + 36$$
$$666 = 9 + 87 + 6 + 543 + 21$$
$$666 = 223 + 443$$

$$666 = 3^6 - 2^6 + 1^6$$
$$666 = 1+2+3+4+5+6+7....+35+36$$

Specialty of 142857: -

The number 142857 is the recurring decimal obtained from the resultant of the infinite fraction 1/7. The recurring decimal of this infinite fraction has the number digit 9. That is 1/7=0.142857… and the number digit of recurring decimals is 9. The number 142857 is divisible by 9.

The recurring decimal 142857 from the resultant of the infinite fraction has 6 digits, the same as the difference between the denominator and the numerator of the fraction.

The multiples of 1/7 are simply repeated copies of the corresponding multiples of 142857:

$$1 \div 7 = 0.142857$$
$$2 \div 7 = 0.285714$$
$$3 \div 7 = 0.428571$$

$$4 \div 7 = 0.571428$$
$$5 \div 7 = 0.714285$$
$$6 \div 7 = 0.857142$$

$$7 \div 7 = 0.999999$$
$$8 \div 7 = 1.142857$$
$$9 \div 7 = 1.285714$$

....

Multiply 142857 by any number from 1 to 6. In each answer you find the same digits. If the products are written in the form of a circle, the order of the digits remains the same and so the number 142857 is called as a perfect cyclic number.

$$142857 \times 1 = 142857$$

$$142857 \times 2 = 285714$$
$$142857 \times 3 = 428571$$

$$142857 \times 4 = 571428$$
$$142857 \times 5 = 714285$$
$$142857 \times 6 = 857142$$

Another oddity is, if we add the products of the number when multiplied by 1, 2, 3, 4, 5 and 6 horizontally and vertically, the digits all add up to 27.

1	4	2	8	5	7
2	8	5	7	1	4
4	2	8	5	7	1
5	7	1	4	2	8
7	1	4	2	8	5
8	5	7	1	4	2

For example if we add vertically the fifth column 5 1 7 2 8 4 we get 27. If we add horizontally the second row 2 8 5 7 1 4 we get 27.

Also $142857 \div 2 = 71428.5$
$$142857 \div 5 = 28571.4$$

The summation of the first three digits 142 and the second three digits 857 gives 999. Also the summation of $14+28+57=99$.

The multiplication of $142857 \times 7 = 999,999$. Multiplying by a multiple of 7 will result in 999999 through this process $142857 \times 14 = 1,999,998$
$$142857 \times 21 = 2,999,997$$
$$142857 \times 28 = 3,999,996$$

$$142857 \times 35 = 4,999,995$$
$$142857 \times 42 = 5,999,994$$
$$142857 \times 49 = 6,999,993$$

$$142857 \times 56 = 7,999,992$$
$$142857 \times 63 = 8,999,991$$
$$142857 \times 70 = 9,999,990$$
$$\dots\dots\dots\dots\dots\dots\dots$$

$142857 \times 7^2 = 6999993$ and if we add the first and last digit of the product will give 999999.

$$142857 \times 7^4 = 342999657$$
$$342 + 999,657 = 999,999.$$

If we multiply by an integer greater than 7, there is a simple process to get to a cyclic permutation of 142857. By adding the first six digits (ones through hundred thousands digits from right hand side of the number) to the remaining digits and repeating this process until you have only the six digits left, it will result in a cyclic permutation of 142857.

$$142857 \times 8 = 1142856$$
$$1 + 142856 = 142857$$
$$142857 \times 815 = 116428455$$

$$116 + 428455 = 428571$$
$$857^2 - 142^2 = 734449 - 20164 = 714285 \text{ again the same set of digits}$$
are repeating.
$142857^2 = 20408122449$ and now $20408 + 122449 = 142857$.

The summation of the digits in the number 142857 is 27 and is the divisor of the original number. So 142857 is a harshad number.

The addition 142857 143
285714 286
571428 572
1000000001

Special multiplications: -

(a) The specialty of multiplication of 15,873 with multiples of 7 is

$15,873 \times 7 = 111,111 = 1(7) \times 15,873 = 1(111,111)$
$15,873 \times 14 = 222,222 = 2(7) \times 15,873 = 2(111,111)$
$15,873 \times 21 = 333,333 = 3(7) \times 15,873 = 3(111,111)$

$15,873 \times 28 = 444,444 = 4(7) \times 15,873 = 4(111,111)$
$15,873 \times 35 = 555,555 = 5(7) \times 15,873 = 5(111,111)$
$15,873 \times 42 = 666,666 = 6(7) \times 15,873 = 6(111,111)$

$15,873 \times 49 = 777,777 = 7(7) \times 15,873 = 7(111,111)$
$15,873 \times 56 = 888,888 = 8(7) \times 15,873 = 8(111,111)$
$15,873 \times 63 = 999,999 = 9(7) \times 15,873 = 9(111,111)$

(b) The specialty of 65359477124183 when multiplied by 17 and multiples of 17 is

$65359477124183 \times 17 = 1,111,111,111,111,111$
$65359477124183 \times 34 = 2,222,222,222,222,222$
$65359477124183 \times 51 = 3,333,333,333,333,333$

$65359477124183 \times 68 = 4,444,444,444,444,444$
$65359477124183 \times 85 = 5,555,555,555,555,555$
$65359477124183 \times 102 = 6,666,666,666,666,666$

$65359477124183 \times 119 = 7,777,777,777,777,777$
$65359477124183 \times 136 = 8,888,888,888,888,888$
$65359477124183 \times 153 = 9,999,999,999,999,999$

(c) $5363222357 \times 2071723 = 11,111,111,111,111,111$
(d) $10726444714 \times 2071723 = 22,222,222,222,222,222$
(e) $5363222357 \times 6215169 = 33,333,333,333,333,333$

(f) $1763664903 \times 7 = 12345654321$
(g) $135 = (1+3+5)(1 \times 3 \times 5)$
(h) $144 = (1+4+4)(1 \times 4 \times 4)$

(i) $1120 = (1 \times 2 \times 3 \times 4 \times 5 \times 6 \times 7 \times 8) / (1 + 2 + 3 + 4 + 5 + 6 + 7 + 8)$.
(j) $8064 = (1 \times 2 \times 3 \times 4 \times 5 \times 6 \times 7 \times 8 \times 9) / (1+2+3+4+5+6+7+8+9)$.
(k) $2808 = (9 \times 10 \times 11 \times 12 \times 13) / (9 + 10 + 11 + 12 + 13)$.

(l) $1,221 = 1 \times 11 \times 111$.
$9,768 = 2 \times 22 \times 222$.
$32,967 = 3 \times 33 \times 333$.
(m) $1500 = (5+1) \times (5+5) \times (5+0) \times (5+0)$.

(n) $1706 = (5 \times 6 \times 7 \times 8) + (5 + 6 + 7 + 8)$.

(o) $3054 = (6 \times 7 \times 8 \times 9) + (6 + 7 + 8 + 9)$.

(p) $7488 = (12 \times 13 \times 14 \times 15 \times 16) / (12 + 13 + 14 + 15 + 16)$.

(q) $2919 = (2 + 9 + 1 + 9) \times (29 + 91 + 19)$.

(r) $7958 = (8 \times 9 \times 10 \times 11) + (8 + 9 + 10 + 11)$.

(s) $4004 = (10 \times 11 \times 12 \times 13 \times 14) / (10 + 11 + 12 + 13 + 14)$.

(t) $3920 = (5+3) \times (5+9) \times (5+2) \times (5+0)$.

(u) $4320 = (6+4) \times (6+3) \times (6+2) \times (6+0)$.

(v) $4752 = (4+4) \times (4+7) \times (4+5) \times (4+2)$.

(w) $1232 = (7 \times 8 \times 9 \times 10 \times 11) / (7 + 8 + 9 + 10 + 11)$.

(x) $7326 = 1 \times 22 \times 333$.

(y) $7986 = 11 \times 22 \times 33$.

(z) $1992 = 8 \times 3 \times 83$.

(aa) $1209 = 1 \times 3 \times 13 \times 31$.

(ab) $5856 = 1 \times 6 \times 16 \times 61$.

(ac) 1214 is a number whose product of digits is equal to its sum of digits. All of its cyclic numbers and transpose numbers obey this property. That is 2114, 2141, 1124, 1142, 1412, 1421… obey this property.

Parasitic number: -

The eighteen decimal number 421,052,631,578,947,368 when multiplied by 2 gives the resultant where the last digit of the multiplicand comes first followed by the same set of digits as in the multiplicand. Numbers with such property are called as parasitic numbers. That is $421{,}052{,}631{,}578{,}947{,}368 \times 2 = 842{,}105{,}263{,}157{,}894{,}736$.

An n-parasitic number (in base 10) is a positive natural number which can be multiplied by n by moving the rightmost digit of its decimal representation to the front. Here n is itself a single-digit positive natural number. In other words, the decimal representation undergoes a right circular shift by one place. For example, $4 \bullet 128205 = 512820$, so 128205 is 4-parasitic. Most authors do not allow leading zeros to be used, and this article follows that convention. So even though $4 \bullet 025641 = 102564$, the number 025641 is not 4-parasitic.

So 179487 is a 4-parasitic number with unit digit 7. Other n-parasitic integers can be built by concatenation. That is 179487179487, 179487179487179487 etc are 4-parasitic number with unit digit 7.

Some other examples of this type are

$263{,}157{,}894{,}736{,}842{,}105 \times 2 = 526{,}315{,}789{,}473{,}684{,}210$.

$105{,}263{,}157{,}894{,}736{,}842 \times 2 = 210{,}526{,}315{,}789{,}473{,}684$

$1{,}034{,}482{,}758{,}620{,}689{,}655{,}172{,}413{,}793 \times 3 = 3{,}103{,}448{,}275{,}862{,}068{,}965{,}517{,}241{,}379$.

$1{,}014{,}492{,}753{,}623{,}188{,}405{,}797 \times 7 = 7{,}101{,}449{,}275{,}362{,}318{,}840{,}579$.

Funny multiplications: -

The thirteen decimal number 4,109,589,041,096 when multiplied by 83 gives the resultant where the unit digit of the multiplier comes first followed by the same set of digits as in the multiplicand and finally the tenth digit of the multiplier comes last. That is $4{,}109{,}589{,}041{,}096 \times 83 = 341{,}095{,}890{,}410{,}968$.

The number digit of the number $123{,}456{,}789$ is 9 and when the number $123{,}456{,}789$ is multiplied by 2, we get $246{,}913{,}578$ again having all the digits from 1 to 9 in some order and the digits are not repeated. The number digit of this number is 9, same as that of the multiplicand.

The common factor in the following multiplications is that each of the digits from 1 to 9 appears once and only once in the resultant, multiplicand and multiplier.

(1) $483 \times 12 = 5796$

(2) $138 \times 42 = 5796$

(3) $297 \times 18 = 5346$

(4) $198 \times 27 = 5346$

(5) $159 \times 48 = 7632$

(6) $186 \times 39 = 7254$

(7) $157 \times 28 = 4396$

(8) $1738 \times 4 = 6952$

(9) $1963 \times 4 = 7852$

(10) $51249876 \times 3 = 153{,}749{,}628$

(11) $32547891 \times 6 = 195{,}287{,}346$

(12) $16583742 \times 9 = 149{,}253{,}678$

The common factor in the following multiplications is that each of the digits from 0 to 9 appears once and only once in the resultant, multiplicand and multiplier.

$58401 = 63 \times 927$ $32890 = 46 \times 715$

$26910 = 78 \times 345$ $19084 = 52 \times 367$

$17820 = 36 \times 495$ $17820 = 45 \times 396$

$16038 = 27 \times 594$ $16038 = 54 \times 297$

$15678 = 39 \times 402$ $65821 = 7 \times 9403$

$65128 = 7 \times 9304$ $34902 = 6 \times 5817$

$36508 = 4 \times 9127$ $28651 = 7 \times 4093$

$28156 = 4 \times 7039$ $27504 = 3 \times 9168$

$24507 = 3 \times 8169$ $21658 = 7 \times 3094$

$20754 = 3 \times 6918$ $20457 = 3 \times 6819$

$17082 = 3 \times 5694$ $15628 = 4 \times 3907$

In the following curious multiplications the multiplicand and the multiplier can be rearranged in such a way that the product remains same.

$2 \times 819 = 9 \times 182$ $2198 \times 9 = 9891 \times 2$

$3 \times 728 = 8 \times 273$ $4 \times 217 = 7 \times 124$

$4 \times 427 = 7 \times 244$ $4 \times 637 = 7 \times 364$

$4 \times 847 = 7 \times 484$ $5 \times 546 = 6 \times 455$

6 x 455 = 5 x 546	7 x 124 = 4 x 217
7 x 244 = 4 x 427	7 x 364 = 4 x 637
8 x 273 = 3 x 728	9 x 182 = 2 x 819
59 x 25 = 5 x 295	2 x 7138 = 83 x 172
3297 x 8 = 8792 x 3	4132 x 7 = 7231 x 4
4264 x 7 = 7462 x 4	4396 x 7 = 7693 x 4
5495 x 6 = 6594 x 5	6594 x 5 = 5495 x 6
7231 x 4 = 4132 x 7	7462 x 4 = 4264 x 7
7693 x 4 = 4396 x 7	8792 x 3 = 3297 x 8
9891 x 2 = 2198 x 9	1 x 6264 = 4 x 6 x 261
1 x 9168 = 8 x 6 x 191	4 x 3149 = 94 x 134
2 x 3168 = 8 x 6 x 132	3 x 3464 = 4 x 6 x 433
4 x 7866 = 6 x 6 x 874	3 x 21525 = 525 x 123
3 x 42525 = 525 x 243	3 x 63525 = 525 x 363
3 x 84525 = 525 x 483	8 x 22287 = 782 x 228
8 x 23575 = 575 x 328	8 x 46575 = 575 x 648
8 x 69575 = 575x968	49 x 2994 = 499 x 294
59 x 2995 = 599 x 295	97 x 6769 = 967 x 679

Factorials: -

The successive multiplications of real numbers from 1 onwards up to a particular number are known as the factorial of that particular number. It is denoted by the symbol "!". That is for getting factorial for n^{th} number is n! = $1 \times 2 \times 3 \times 4 \times \ldots \times n$ where 'n' is a real number. For example 5! = $1 \times 2 \times 3 \times 4 \times 5$ = 120 and 120 is the factorial of 5. The factorial sequence is 1, 2, 6, 24, 120, 720, 5040, 40320, 362880, 3628800…

There are 7 factorials below 10,000.

Factorial numbers are always even numbers.

Factorial numbers will not become consecutive or following numbers.

Factorial numbers are used to find out the numbers of permutations or combinations for 'n' number of objects.

The factorial of a number $n! = \dfrac{(n+1)!}{(n+1)}$ where 'n' is a real number.

for example $4! = \dfrac{(4+1)!}{(4+1)} = \dfrac{5!}{5} = \dfrac{120}{5} = 24$

Catalan numbers: -

The numbers generated by the formula $C_n = (2n)!/(n! \times (n+1)!)$ where 'n' is a real number, are called as Catalan numbers. For example the fifth Catalan number C_5 = 10! / (5! × 6!) = 3628800 / (120 × 720) = 42. The Catalan number sequence is 1, 2, 5, 14, 42, 132, 429, 1430, 4862, 16796, 58786...

The Catalan numbers which are odd are called as odd Catalan number and the Catalan numbers which are even are called as even Catalan numbers. For example 5 is an odd Catalan number and 42 is an even Catalan number.

Further if the even Catalan number contains all of its digits as even, then it is a super even Catalan number. For example 4862 is a super even Catalan number and 1430 is not a super even Catalan number.

Some curious facts with addition of factorials:

33 = 1! +2! +3! +4!

23 = 1! + (2! + 2!) + (3! + 3! + 3!).

145 = 1! + 4! + 5!

153 = 1!+2!+3!+4!+5!

801 = (7! + 8! + 9! + 10!) / (7 × 8 × 9 × 10).

873 = 1!+2!+3!+4!+5!+6!

5161 = 5! + (1+6)! + 1.

5162 = 5! + (1+6)! + 2.

5163 = 5! + (1+6)! + 3.

5164 = 5! + (1+6)! + 4.

5165 = 5! + (1+6)! + 5.

5166 = 5! + (1+6)! + 6.

5167 = 5! + (1+6)! + 7.

5169 = 5! + (1+6)! + 9.

5913 = 1! + 2! + 3! + 4! + 5! + 6! + 7!

5914 = 0! + 1! + 2! + 3! + 4! + 5! + 6! + 7!

Some curious facts with multiplication of factorials:

6! × 7! = 10!

5! × 3! = 6!

3! × 5! × 7! = 10!

2! × 47! × 4! = 48!

2! × 287! × 4! × 3! = 288!

Specialty of 3rd multiples: -

3×37= 111= 1(3)×37= 1(111) and 1+1+1=3

6×37= 222= 2(3)×37= 2(111) and 2+2+2=6

9×37= 333= 3(3)×37= 3(111) and 3+3+3=9

12×37= 444= 4(3)×37= 4(111) and 4+4+4=12

15×37= 555= 5(3)×37= 5(111) and 5+5+5=15

18×37= 666= 6(3)×37= 6(111) and 6+6+6=18

21×37= 777= 7(3)×37= 7(111) and 7+7+7=21

24×37= 888= 8(3)×37= 8(111) and 8+8+8=24

27×37= 999= 9(3)×37= 9(111) and 9+9+9=27

This is called as triplication.

Thus the prime number 37 can divide 111, 222, 333, 444, 555, 666, 777, 888, 999 and multiples of 10 of these numbers like 1110, 2220, etc. without any remainder.

Note that if a three digit number of the form ABC is a multiple of 37, then the cyclic numbers of ABC that is BCA and CAB are also divisible by 37. For example 37 × 4 = 148 and the cyclic numbers of 148 which is 481 and 814 are also divisible by 37. That is 37 × 13 = 481 and 37 × 22 = 814.

$$3 \times 37037 = 111,111$$
$$6 \times 37037 = 222,222$$
$$9 \times 37037 = 333,333$$

$$12 \times 37037 = 444,444$$
$$15 \times 37037 = 555,555$$
$$18 \times 37037 = 666,666$$

$$21 \times 37037 = 777,777$$
$$24 \times 37037 = 888,888$$
$$27 \times 37037 = 999,999$$

This is called as hexation. The same can be extended to nonation, dodecation and so on by multiplying the third multiple numbers correspondingly with 37037037, 37037037037...

When the number 3367 is multiplied by 33, 66, 99 and so on up to 297 also gives interesting results.

$$3367 \times 33 = 111,111$$
$$3367 \times 66 = 222,222$$
$$3367 \times 99 = 333,333$$

$$3367 \times 132 = 444,444$$
$$3367 \times 165 = 555,555$$
$$3367 \times 198 = 666,666$$

$$3367 \times 231 = 777,777$$
$$3367 \times 264 = 888,888$$
$$3367 \times 297 = 999,999$$

$$3367 \times 330 = 1,111,110$$

....

The number 12345679 when multiplied by multiples of three numbers it gives special type of characteristic numbers as below.

$$12345679 \times 3 = 37037037$$
$$12345679 \times 6 = 74074074$$
$$12345679 \times 9 = 111111111$$

$$12345679 \times 12 = 148148148$$
$$12345679 \times 15 = 185185185$$
$$12345679 \times 18 = 222222222$$

$$12345679 \times 21 = 259259259$$
$$12345679 \times 24 = 296296296$$
$$12345679 \times 27 = 333333333$$

$$12345679 \times 30 = 370370370$$
$$12345679 \times 33 = 407407407$$

$$12345679 \times 36 = 444444444$$

$$12345679 \times 39 = 481481481$$
$$12345679 \times 42 = 518518518$$
$$12345679 \times 45 = 555555555$$

$$12345679 \times 48 = 592592592$$
$$12345679 \times 51 = 629629629$$
$$12345679 \times 54 = 666666666$$

$$12345679 \times 57 = 703703703$$
$$12345679 \times 60 = 740740740$$
$$12345679 \times 63 = 777777777$$

$$12345679 \times 66 = 814814814$$
$$12345679 \times 69 = 851851851$$
$$12345679 \times 72 = 888888888$$

$$12345679 \times 75 = 925925925$$
$$12345679 \times 78 = 962962962$$
$$12345679 \times 81 = 999999999$$

$$12345679 \times 84 = 1037037036$$
$$12345679 \times 87 = 1074074073$$
$$12345679 \times 90 = 1111111110$$

$$12345679 \times 93 = 1148148147$$
$$12345679 \times 96 = 1185185184$$
$$12345679 \times 99 = 1222222221$$

....

If the number $12,345,679 \times 999,999,999 = 12345678987654321$

It is to be noted that 3 is the only natural number whose sum of its preceding numbers is equal to 3 itself, that is 1+2=3.

The 3rd multiple of 3 that is the square of 3 is in fact a marvel and its characteristics are described in Number 9:– A marvel.

The product of consecutive numbers: -

The product of two consecutive numbers is always an even number.

$$6 \times 7 = 42$$
$$5 \times 6 = 30$$
$$8 \times 9 = 72$$

From the above products subtract the multiplicand that is (42-6), (30-5) and (72-8) respectively. We get 36, 25 and 64 and these are the squares of the multiplicands. To the above products add the multiplier that is (42+7), (30+6) and (72+9) respectively.

We get 49, 36 and 81 and these are the squares of multipliers. The reason can be explained as follows.

$n \times (n+1) = n^2 + n$

$n^2 + n = n^2 + n$ (subtract n from both sides)

$n^2 = n^2$

Similarly $(n-1) \times n = n^2 - n$ (add n to both sides)

$$n^2 - n + n = n^2 - n + n$$
$$n^2 = n^2$$

The product of two or more positive consecutive integers will not be a power number. For example 99x100=9900; 105×106×107 will not be a power number.

Now take three consecutive numbers and see the product.

For example 13×14×15=2730 and now add the middle consecutive number to this product, we get 2730+14=2744 and this is the cube of the middle consecutive number.

That is $14^3 = 2744$

This can be explained as follows.

$$(n-1)n(n+1) = (n^3 - n)(n+1)$$
$$= n^3 - n^2 + n^2 - n$$
$$= n^3 - n$$

Now add n to the product we get n^3

Now consider four consecutive numbers and see the product.

For example 4×5×6×7=840 and add 1 to this product. We get 841.

Now 841 is a square number and the square root of this number is 29.

Now add 1 to 29. We get 30 and this is the product of the two middle consecutive numbers. That is 5×6=30.

5. POWERS

The multiplication of one number with the same number is called as 'to the power of two' or the square of the number.

For example if 3 is the number then $3 \times 3 = 9$ is its square. This equation can be written as $3^2 = 9$ and is read as 3 to the power of 2 is equal to 9 or 3 squared is equal to 9. Here 3 is called as the square root of 9 and it is denoted as $_2\sqrt{9} = 3$. In this equation $\sqrt{}$ is called as radical and 9 is called as radicand. The above equation can also be written as $9^{1/2} = 3$.

The square number series is 4, 9, 16, 25, 36, 49, 64, 81, 100, 121…

Note: - Although 1 is a square number it cannot be considered as a perfect square since $1^n = 1$ and so 1 can be called as a power number rather than a square number. Also $0^n = 0$ and hence 0 can also be considered as a special power number like 1.

Similarly the product of one number with the same number three times is called as 'to the power of three' or the cube of the number.

For example if 3 is the number then $3 \times 3 \times 3 = 27$ is its cube. This equation can be written as $3^3 = 27$ and is read as 3 to the power of 3 is equal to 27 or 3 cubed is equal to 27. Here 3 is called as cube root of 27 and is denoted as $_3\sqrt{27} = 3$ and also can be written as $27^{1/3} = 3$.

Likewise the multiplication of a number with the same number four times is called as the fourth power of the number and so on.

Here the expression $_2\sqrt{9}$ is called as surd of the order 2 and the expression $_3\sqrt{27}$ is called as surd of the order 3 and so on.

The Power behind the powers: -

Here lies the story of powers.

Once a wise man visited a King's court with a chess board. He asked the King to give two grains for the first square of the chess board, four grains for the second square, eight for the third, and sixteen for the fourth and so on up to the 64th square of the chess board. The King without knowing the consequences, ordered his ministers to make arrangements to provide the same. After sometime the ministers feebly explained to the King that it was not possible to give grains because the amount to be paid was increasing enormously for 2^{30} it was 1,073,741,824 and for 2^{40} it was 1,099,511,627,776 and increases further gigantically as they proceed. Also the ministers explained that the country will be bankrupt if the King wanted to complete the task.

Such is the power of the powers.

Surd: -

The expression '$_n\sqrt{a}$' is called a surd of the order "n".

If two surds are of the same order, then the one whose radicand is larger is the larger of the two. For example $_3\sqrt{27} > _3\sqrt{19}$.

Two or more surds having the same radicand are called as like surds and if they have different radicands they are called as unlike surds. For example $_3\sqrt{27}$ and $5\sqrt{27}$ are like surds. $_3\sqrt{27}$ and $_3\sqrt{8}$ are unlike surds.

$_n\sqrt{a} \times _n\sqrt{b} = _n\sqrt{ab}$ and $(_n\sqrt{a})/(_n\sqrt{b}) = _n\sqrt{(a/b)}$.

Digital root rule of powers: -

(a)The digit sum of the square root when multiplied by itself (that is the square of the digit sum's square root) should equal to the digit sum of the square.

For example $23^2 = 529$. The digit sum of 23 is 5 and when 5 is multiplied by itself the answer is 25. The digit sum of 25 is 7 and the digit sum of 529 is also 7.

The digit sum of the cube root when multiplied by itself and once again by itself (that is the cube of the digit sum's cube root) should be equal to the digit sum of the cube. For example $13^3 = 2197$. The digit sum of 13 is 4 and when 4 is multiplied by itself and again by itself the answer is 64. The digit sum of 64 is 1 and the digit sum of 2197 is also 1. Likewise we can extend the concept for higher power numbers.

(b) For a number to be a square number the digital root of the number should be 1, 4 or 9. For example 102110022102101100 is not a square since its digital root is 6.

(c) Even numbered powers will have the sum of the digits that is number digits as 1, 4, 7 or 9 only. For example $4^4 = 256$ and the sum of the digits of 256 is 4.

(d) Sum of the digits of a cubic number that is number digit will always end with 1, 8 or 9 and also it is seen that the order of 1, 8 or 9 is coming serially in a cyclic manner. For example $1^3=1$; $2^3=8$; $3^3=9$; $4^3=64$ (digital root is 1); $5^3=125$ (digital root is 8) and so on.

(e) No power number's sum of the digits that is number digit will end with 3.

Power table

N	n^1	n^2	n^3	n^4	n^5	n^6	n^7	n^8	n^9	n^{10}
1	1	1	1	1	1	1	1	1	1	1
2	2	4	8	16	32	64	128	256	512	1024
3	3	9	27	81	243	729	2187	6561	19683	59049
4	4	16	64	256	1024	4096	18384	65536	262144	1048576
5	5	25	125	625	3125	15625	78125	390625	1953125	9765625
6	6	36	216	1296	7776	46656	279936	1679616	10077696	60466176
7	7	49	343	2401	16807	117649	823543	5764801	40353607	282475249
8	8	64	512	4096	32768	262144	2697152	16777216	134217728	1073741824
9	9	81	729	6561	59049	531441	4782969	43046721	387420489	3486784401
10	10	10^2	10^3	10^4	10^5	10^6	10^7	10^8	10^9	10^{10}

Note: - For 10^x put 1 followed by $1 \times x$ number of zeros in the power number. For 100^x put 1 followed by $2 \times x$ number of zeros and for 1000^x put 1 followed by $3 \times x$ number of zeros and so on.

Power root table

N	$1\sqrt{n}$	$2\sqrt{n}$	$3\sqrt{n}$	$4\sqrt{n}$	$5\sqrt{n}$	$6\sqrt{n}$	$7\sqrt{n}$	$8\sqrt{n}$	$9\sqrt{n}$	$10\sqrt{n}$
1	1	1.00	1.00	1.00	1.00	1.00	1.00	1.00	1.00	1.00
2	2	1.41	1.26	1.19	1.15	1.12	1.10	1.09	1.08	1.07
3	3	1.73	1.44	1.32	1.24	1.20	1.17	1.15	1.13	1.12
4	4	2.00	1.54	1.41	1.32	1.26	1.22	1.19.	1.17	1.15
5	5	2.23	1.71	1.19	1.38	1.31	1.26	1.22	1.19	1.17
6	6	2.45	1.82	1.56	1.43	1.32	1.29	1.25	1.22	1.19
7	7	2.64	1.91	1.63	1.47	1.38	1.32	1.27	1.24	1.21
8	8	2.83	2.00	1.68	1.51	1.14	1.34	1.30	1.26	1.23
9	9	3.00	2.08	1.73	1.55	1.44	1.37	1.32	1.28	1.24
10	10	3.16	2.15	1.17	1.58	1.47	1.39	1.33	1.29	1.26

Some easy methods of finding out powers: -

To find out the square of a number which ends with 5 such as 25, 325,etc., an easy method is put the square of 5 as 25 at the right end and the product of the left out number with its consecutive number on the left side as such. For example the square of 105 is 25 put at the right end and the product of left out number 10 with its consecutive number 11 that is 110 at the left side and we get 11025.

Note: - We can apply the same technique for multiplication of any two numbers provided the following two conditions are satisfied. One is the resultant of last digit of the two numbers should be 10 and the second is the remaining digits should be same. For example $91 \times 99 = 9009$

$$32 \times 38 = 1216$$

To find out the square of a number which ends with 3, an easy method is, put the square of 3 as 9 at the right end and the product of the left out number by 6 penultimate to 9. If the product by 6 is a two digit number and above, put the unit decimal number only and carry over the remaining number and add to the square of the left out number and put this resultant in the extreme left side as such. For example the square of 123 is 9 put at the right end and the product of left out number 12 by 6 is 72. Here put the unit decimal number 2 penultimate to 9 and carry over the remaining number 7. Then add the carry over number with the square of left out number that is (12^2) 144 and put the resultant 151 at the extreme left to get the result 15129.

To find out the square of a number which ends with 4 the same method described to find out the square of a number ending with 3 can be adopted. Only difference is put the square of 4 as 6 at the right end and carry over 1 and add the product of left out number by 8. If the addition of 1 and the product is a two digit number and above, put the unit decimal number only penultimate to 6 and carry over the remaining number and add to the square of the left out number and put this resultant in the extreme left side as such. For example the square of 74 is 6 put at the right end and the product of left out number by 8 is 56 added with 1 is 57. Here put 7 penultimate to 6 and carry over the remaining 5 and add to the square of the left out number to get 54. Now put this number at the extreme left to get the square of 74. That is 5476.

If from the sequence 1,2,3,4,5,6,…we delete every n^{th} term, then every $(n-1)^{st}$ term, then every $(n-1)^{nd}$ term, … and lastly every 2nd term, computing the partial sums after each round of deletions, then the final sequence of partial sums is the sequence of nth powers. For example fifth power sequence is calculated as follows.

Step 1:- write the real number sequence 1,2,3,4,5,6,7,8,9,10,11,12,13,14,…

Step 2:- delete every fifth number we get 1,2,3,4,6,7,8,9,11,12,13,14,…

Step 3:- partial sums of the above sequence 1,3,6,10,16,23,31,40,51,63,76,90,…

Step 4:- delete every fourth term of the above sequence 1,3,6, 16,23,31,51,63,76…

Step 5:- partial sums of the above sequence 1, 4, 10, 26, 49, 80, 131,194, 270…

Step 6:- delete every third term of the above sequence 1, 4, 26, 49, 131, 194…

Step 7:- partial sums of the above sequence 1, 5, 31, 80, 211, 405…

Step 8:- delete every second term of the above sequence 1, 31, 211…

Step 9:- partial sums of the above sequence 1, 32, 243… and now this sequence is the fifth power sequence. Likewise we can calculate for the other power sequences.

A simple method of calculating the fourth powers of numbers is described as follows. If the natural numbers are arranged into groups like

(1), (2,3), (4,5,6), (7,8,9,10), (11,12,13,14,15), (16,17,18,19,20,21)…

Now strike out the second groups alternately we get

(1), (4,5,6), (11,12,13,14,15)…

From the above sequence we can find out the fourth power of any number as follows.

For getting 1^4 add the first group number only that is 1

For getting 2^4 add the first and second group numbers only that is 1+4+5+6 = 16

For getting 3^4 add the first, second and third group numbers only and so on.

General properties of power numbers: -

The product of a square number with the same or another square number is always a square number. That is the product of nth power number with the same or another n^{th} power number is always an nth powered number. For example $16 \times 16 = 256$ and is $4^2 \times 4^2 = 16^2$. Another example is $25 \times 4 = 100$ and is $5^2 \times 2^2 = 10^2$.

The multiple of a square number with a non square number two times is always an another square number. Similarly the multiples of a cubic number with a non cubic number three times are always a cubic number. Likewise the multiples of a fourth powered number with a non fourth powered number four times are always a fourth powered number and so on. For example $27 \times 2 = 54$; $54 \times 2 = 108$; $108 \times 2 = 216$. That is $3^3 \times 2^3 = 6^3$.

The multiples of the square number with the non square number are called as off square numbers. For example $16 \times 3 = 48$; $16 \times 10 = 160$. Here 48 and 160 are off square numbers. Similarly the multiples of the cube numbers with the non cubic number are called as off cube numbers and so on.

If the square of a number has 2n (where n indicates the number of digits) digits, that is even number of digits then its square root will have 'n' digits. If the square number has odd number of 'n' digits, then its square root will have [(n+1)/2] digits. Similarly if the cubic number has 3n digits then its cubic root will have n digits and so on. For example $4096 = 8^4$. Here the fourth powered number has 4 digits and its fourth power root has single digit.

The number of digits in a cube root of a number is the same as the number of 3 digit groups in the given number (from left to right for decimals and from right to left for non decimals) including a single or two digit group left out in the 3 digit grouping. For example 437,240 will have two digits in its cubic root.

25,072,306 will have three digits in its cubic root

45 will have single digit in its cubic root

Note: - The above rule is applicable for other powers also. For example the square root of a number can be found by grouping the number in 2's and the fourth power root of a number is found by grouping the digits in 4's and so on.

Any number to the power of a square number is always a square number. Similarly for cubes and n^{th} powers. For example $21^4 = 194,481 = 441^2$ and $5^9 = 1,953,125 = 125^3$.

The power of an odd number will be an odd number and the power of an even number will be an even number. For example $3^2 = 9$ and $4^2 = 16$.

If $x = y \times z$ then $x^n = y^n \times z^n$, where 'n' is a positive number. For example $10 = 2 \times 5$ then $10^9 = 2^9 \times 5^9 = 512 \times 1,953,125$. Similarly $6 = 3 \times 2$ then $6^3 = 3^3 \times 2^3 = 27 \times 8$.

The even numbers which are power numbers are called as even power numbers. For example 64 is an even power number.

The power numbers which are odd are called as odd power numbers. For example 121 is an odd power number.

In the power number series if we categorize the numbers as odd (O) number and even (E) number the sequence is O, E, O, E, O, E… and it repeats further as we move. For example in the case of 7th power number series 1, 128, 2187, 18384… and follows the order O, E, O, E, O, E…

The power numbers which are alternating are called as power alternating number. For example 121 is a square alternating number and 27 is a cube alternating number and so on.

Obviously pronic numbers will not become power numbers.

Repunit numbers and repdigit numbers will not become power numbers.

Properties of square number: -

The square number series is 1, 4, 9, 16, 25, 36, 49, 64, 81, 100, 121…

The square number which is even is called as even square number and the square number which is odd is called as odd square number. For example 121 is an odd square number and 100 is an even square number.

In the square number series, if we categorize the numbers as odd (O) number and even (E) number the sequence is O, E, O, E, O, E… and it repeats further as we move.

Further if the square number contains all the digits as even, then it is a super even square number. For example 64 is a super even square number. Also it is to be noted that super even square numbers will have their square roots as super even numbers, but the reverse is not true. That is all the super even square numbers are produced from super even numbers but all the super even numbers will not produce super even square numbers. For example 22 the super even number will produce 484, the super even square number. But 24 the super even number will produce 576 is not a super even square number.

The digital sum of a square number will always be 1, 4, 7 or 9.

The numbers which are both squares and cubes will have the peculiar property of the cubed square number series or the squared cube number series.

That is 1, 64, 729, 4096, 15625…

$= 1^3, 4^3, 9^3, 16^3, 25^3….$

$= 1^2, 8^2, 27^2, 64^2, 125^2…$

If the difference of cubes of two consecutive integers is equal to square of an integer, then the resultant integer is the sum of the squares of two consecutive integers. For example $8^3 - 7^3 = 169$ and this is the square of 13 which can be represented as $2^2 + 3^2$.

The difference between the squares of consecutive numbers is $(n+1)^2 - n^2 = 2n+1$ where 'n' is a positive integer. For example the squares of the numbers 50 and 51 differ by 101.

The real numbers which are not square numbers between n^2 and $(n+1)^2$ is 2n. For example between 45^2 and 46^2 there are 90 non square numbers present.

If the square of a number is equal to the sum of two squares, then the number itself is a sum of two square numbers. For example $5^2 = 3^2 + 4^2$; $5 = 1^2 + 2^2$.

All odd squares leave a remainder of 1 when divided by 8. For example (121/8) leaves a remainder of 1.

The sum of two odd squares will not be a square number. This rule is used to find out the Pythagorean triples. For example 121 + 49 = 170 and the resultant 170 is not a square number.

The square number when represented as dots will form a geometrical square. Similarly for cubic numbers it will form a cube and for a fourth powered number it will form a tetrahedron and so on.

For example in the case of square numbers 1   4   9   16

Diagrammatically the square numbers can be represented geometrically as follows.

1

1	2
3	4

1	2	3
4	5	6
7	8	9

1	2	3	4
5	6	7	8
9	10	11	12
13	14	15	16

Note:-Each row or column or diagonal of the square or cube or tetrahedron etc represents its square root or cubic root or fourth powered root respectively.

In the multiplication table the square numbers are diagonally arranged.

Multiplication	1	2	3	4	5	6	7	8	9	10
1	**1**	2	3	4	5	6	7	8	9	10
2	2	**4**	6	8	10	12	14	16	18	20
3	3	6	**9**	12	15	18	21	24	27	30
4	4	8	12	**16**	20	24	28	32	36	40
5	5	10	15	20	**25**	30	35	40	45	50
6	6	12	18	24	30	**36**	42	48	54	60
7	7	14	21	28	35	42	**49**	56	63	70
8	8	16	24	32	40	48	56	**64**	72	80
9	9	18	27	36	45	54	63	72	**81**	90
10	10	20	30	40	50	60	70	80	90	**100**

In the above multiplication table 1, 4, 9, 16, 25, 36... are square numbers. This can be written as 1^2, 2^2, 3^2, 4^2, 5^2, 6^2... Since we are getting the square numbers by multiplying a number with the same number two times, another option is there, that is adding a number with the same number two times as 2, 4, 6, 8, 10, 12, 14... and this can be written as 1_1, 2_2, 3_2, 4_2, 5_2, 6_2...

The representation of all the multiples as well as power numbers can be written as above. For example 5_3 means (5+5+5=) 15 and 5^3 means (5×5×5=) 125.

The general formula to find out the total number of geometrical squares is $1^2+2^2+3^2+...+n^2$ where 'n' is the number of units of small squares forming the side in the square. That is in figure 1, (the number of units of small squares forming the side in the square is 3) total number of squares is ($1^2+2^2+3^2$) 14 and in figure 2, total number of squares is 30.

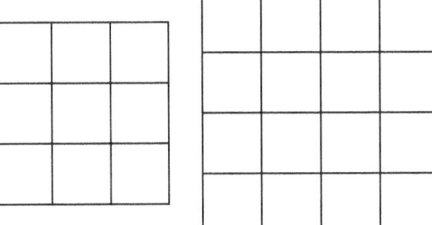

Figure 1   Figure 2

The power numbers are orderly arranged combination of natural numbers.

In the case of square numbers, the following type of orderly arranged combination of natural numbers is present.

$$1^2= 1 =1$$
$$2^2= 4 =1+2+1$$
$$3^2= 9 =1+2+3+2+1$$
$$4^2=16=1+2+3+4+3+2+1$$
$$5^2=25=1+2+3+5+4+3+2+1$$

....

Diagrammatically the square numbers can be represented geometrically as follows.

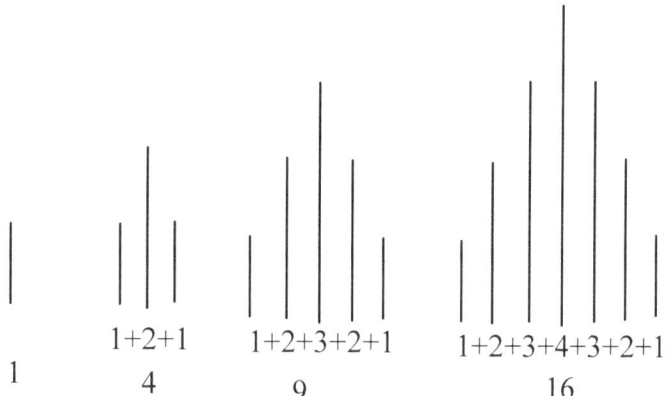

Another relationship between the squares and the sequence of odd numbers is as follows.

$$1^2= 1 =1$$
$$2^2= 4 =1+3$$
$$3^2= 9 =1+3+5$$
$$4^2=16=1+3+5+7$$
$$5^2=25=1+3+5+7+9$$

....

A square number cannot end with odd number of zeros. For example $20^2 = 400$.

A square cannot end in 6 with its previous digit being even. For example 196. The converse of the above concept is also true. That is a number could not be a square number with 6 in the unit decimal place followed by an even number in the tenth place.

Every square number has an odd number of factors whereas every non square number has an even number of factors. For example 16 has the factors 1, 2, 4, 8 and 16 (Total number of factors is 5) while 24 has the factors 1, 2, 3, 4, 6, 8, 12 and 24 (Total number of factors is 8). Also it is to be stated that the sum of the factors of the perfect square number is an odd number and the sum of the factors of the non square number is an even number. That is in the above case the sum of the factors of 16 is 31 and the sum of the factors of 24 is 60.

Another way of expressing the formation of square numbers is as follows. That is $1^2, 2^2, 3^2, 4^2$... is obtained by $0^2 + (0 + 0 + 1) = 1^2$, $1^2 + (1 + 1 + 1) = 2^2$, $2^2 + (2 + 2 + 1) = 3^2$, $3^2 + (3 + 3 + 1) = 4^2$...

Diagrammatically the formation of square numbers is represented as follows.

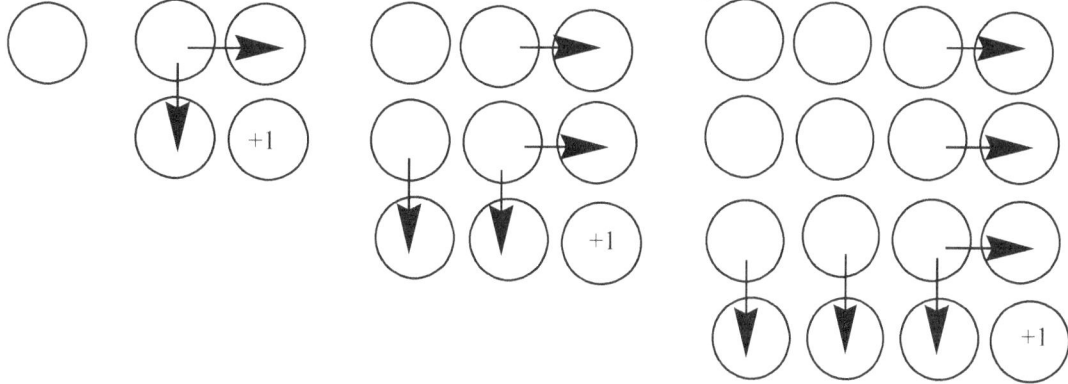

Every positive integer can be written as the sum of four perfect squares. For example $75 = 64 + 9 + 1 + 1$.

2 to the power of 2 again to the power of 'n' followed by the addition with 1, that is of the form $2^{2^n}+1$ ends in the digit 7 for all the positive integers where n>1 (Fermat number). For example if n = 3, then 257 is obtained.

Multiply any four consecutive numbers and add number 1 to the product. The resultant would be a complete square. For example $(3\times4\times5\times6) +1=360+1=361=19^2$. Another example is $(11\times12\times13\times14)+1=24024+1=24025=155^2$.

The multiplication of powers of 2 with 7 gives a relationship as follows.

$$7 \times 2^1 = 14 \times 2^0 = 7 \times 2 = 14$$
$$7 \times 2^2 = 14 \times 2^1 = 14 \times 2 = 28$$
$$7 \times 2^3 = 14 \times 2^2 = 28 \times 2 = 56$$

$$7 \times 2^4 = 14 \times 2^3 = 56 \times 2 = 112$$
$$7 \times 2^5 = 14 \times 2^4 = 112 \times 2 = 224$$
$$7 \times 2^6 = 14 \times 2^5 = 224 \times 2 = 448$$

$$7 \times 2^7 = 14 \times 2^6 = 448 \times 2 = 896$$
$$7 \times 2^8 = 14 \times 2^7 = 896 \times 2 = 1792$$
$$7 \times 2^9 = 14 \times 2^8 = 1792 \times 2 = 3584$$

....

Likewise we can extend for the multiplication of power numbers with a constant number.

If the square number is palindromic in nature means, it is called as palindromic square number. For example 121, 484, 676, 10201, 12321…

Repunit numbers and repdigit numbers will not become square numbers.

If the square number is alternating in nature means, it is called as alternating square number. For example 256, 1296… are some of the alternating square numbers.

Relationship between square numbers and cube numbers: -

The summation of the cubic numbers is correspondingly equal to the sum of the arithmetic progressive squared.

$$1= \quad 1=1^3 \qquad = (1)^2$$
$$1+8= \quad 9=1^3+2^3 \qquad = (1+2)^2$$
$$1+8+27= 36=1^3+2^3+3^3 \qquad = (1+2+3)^2$$
$$1+8+27+64=100=1^3+2^3+3^3+4^3 = (1+2+3+4)^2$$
$$1+8+27+64+125=225=1^3+2^3+3^3+4^3+5^3= (1+2+3+4+5)^2$$

....

Relationship between square numbers and whole numbers: -

If the square numbers are arranged in a spiral form starting from whole number onwards, the following interesting pattern is observed. In that all the square numbers will fall in a diagonal line. In the square diagonal line all the odd square numbers like 1, 9, 25, 49… will be in one radius and all the even square numbers like 4, 16, 36, 64… will be in the opposite of that.

Whole number spiral

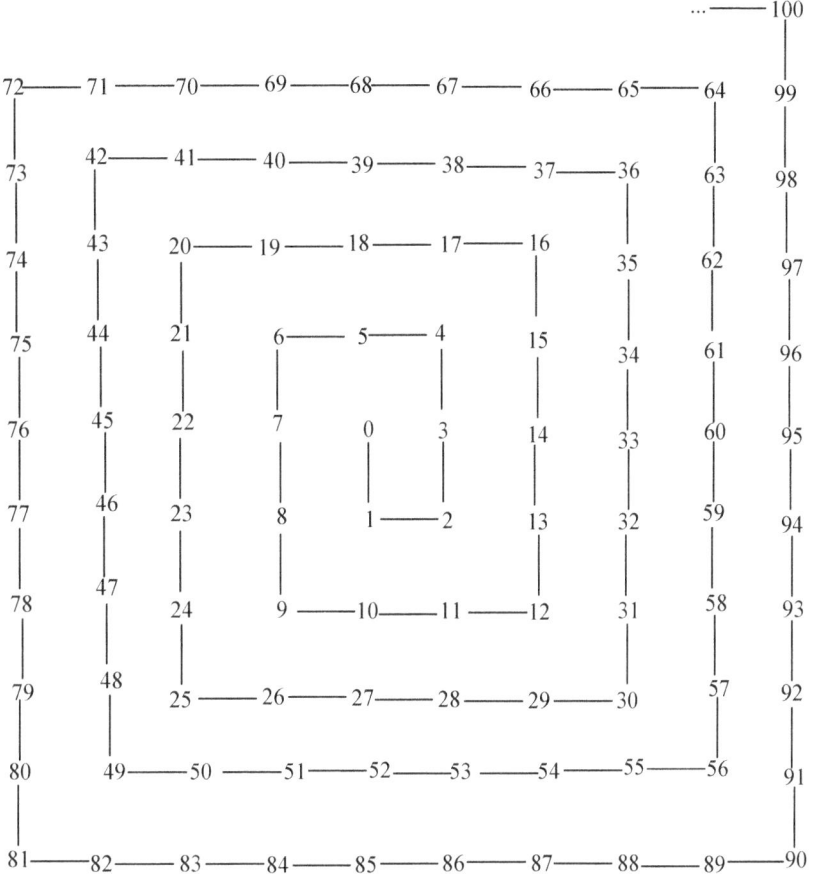

Square number sequence in the whole number spiral is represented as follows.

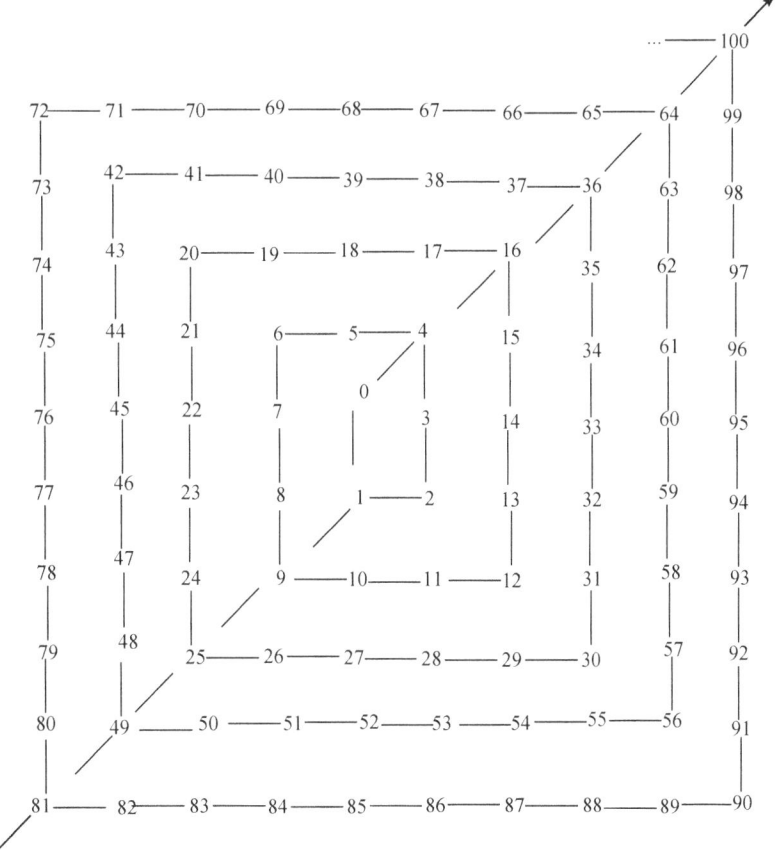

Properties of cube number: -

The cube number series is 1, 8, 27, 64, 125, 216, 343, 512, 729, 1000, 1331…

The cube number which is even is called as even cube number and the cube number which is odd is called as odd cube number. For example 343 is an odd cube number and 216 is an even cube number.

In the cube number series if we categorize the numbers as odd (O) number and even (E) number the sequence is O, E, O, E, O, E… and it repeats further as we move.

Further if the cube number contains all the digits as even, then it is a super even cube number and if the cube number contains all the digits as odd then it is a super odd cube number. For example 1000 is a super even cube number and 1331 is a super odd cube number.

In the case of cube numbers, the following type of orderly arranged combination of natural numbers is present.

$$1^3 = 1 = 1$$
$$2^3 = 8 = 2(1) + 2(3)$$
$$3^3 = 27 = 3(1) + 3(3) + 3(5)$$
$$4^3 = 64 = 4(1) + 4(3) + 4(5) + 4(7)$$
$$5^3 = 125 = 5(1) + 5(3) + 5(5) + 5(7) + 5(9)$$

……………………………………………..

Diagrammatically the cube numbers can be represented geometrically as follows.

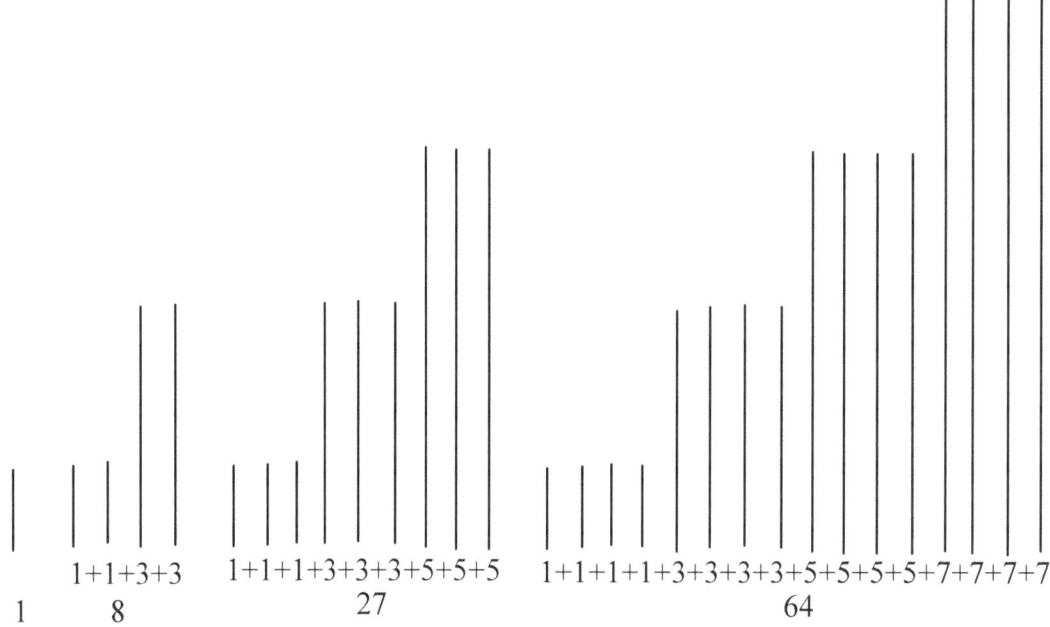

1+1+3+3	1+1+1+3+3+3+5+5+5	1+1+1+1+3+3+3+3+5+5+5+5+7+7+7+7
1 8	27	64

Every odd cube number n^3 is a sum of three consecutive odd numbers whose middle term is n^2 with (n^2-2) as first term and (n^2+2) as last term.
For example $3^3 = 27 = 7 + 9 + 11$

The digit in the tenth place of any cubic number (power of 3) is always an even number. For example $4^3 = 64$; $9^3 = 729$
The vice versa of the above concept that is the tenth place of any cubic number 3^n (where 'n' is a positive integer) is always an even number. For example $3^8 = 6561$.

Every positive integer can be written as the sum of nine (or fewer) positive cubes. This upper limit of nine cubes cannot be reduced because, for example, 23 cannot be written as the sum of fewer than nine positive cubes, that is
$23 = 2^3 + 2^3 + 1^3 + 1^3 + 1^3 + 1^3 + 1^3 + 1^3 + 1^3$.

The cube of any integer can be represented as the difference of squares of two integers. For example the cube number 8 can be represented as the difference of square numbers 9 and 16.

If the cube number is palindromic in nature means it is called as palindromic cube number. For example 9009 is a palindromic cubic number.

Trimorphic number: -

A number 'n' is called as a trimorphic number if n^3 ends with n. For example, 99 is trimorphic number since $99^3 = 970299$. The trimorphic number sequence is 1, 4, 5, 6, 9, 24, 25, 49, 51, 75, 76, 99, 125, 249, 251, 375, 376, 499, 501, 624, 625, 749, 751, 875, 999…

The trimorphic number which is odd is called as odd trimorphic number and the trimorphic number which is even is called as even trimorphic number. For example 625 is an odd trimorphic number and 376 is an even trimorphic number.

Further if the odd trimorphic number contains all of its digits as odd, then it is a super odd trimorphic number and if the even trimorphic number contains all of its digits as even, then it is a super even trimorphic number. For example 375 is a super odd trimorphic number and 624 is a super even trimorphic number.

The numbers 9, 99, 999, 9999… are trimorphic numbers.

The trimorphic numbers which are square numbers are called as trimorphic square numbers. For example 4, 9, 25, 49, 625… are trimorphic square numbers.

Table of the end digit of powers and its properties: -

Construction of a table with the real number series against the ending digit of the number rose to the power gives some salient features.

Number	square	cube	4th	5th	6th	7th	8th	9th	10th
1	1	1	1	1	1	1	1	1	1
2	4	8	6	2	4	8	6	2	4
3	9	7	1	3	9	7	1	3	9
4	6	4	6	4	6	4	6	4	6
5	5	5	5	5	5	5	5	5	5
6	6	6	6	6	6	6	6	6	6
7	9	3	1	7	9	3	1	7	9
8	4	2	6	8	4	2	6	8	4
9	1	9	1	9	1	9	1	9	1
10	0_2	0_3	0_4	0_5	0_6	0_7	0_8	0_9	0_{10}

Note: - The ending digits for other decades 11 to 20, 21 to 30… are the same as for the above decade.

From the table we can observe the followings

(i)Factors about the last digit of a square number:

The last digit or unit digit of the square of a number will always end with 1, 4, 9, 6, 5 or 0 only. Also the sequence repeats for the squares of higher numbers.

Note: - The digits 2, 3, 7 and 8 do not appear as the last digit of a square number.

The power of 2 series is 2, 4, 8, 16, 32, 64, 128, 256… The last digit of the powers of 2 repeats as 2, 4, 8, 6, 2, 4, 8, 6… and it repeats in the same order as we move to the higher powers of 2.

Note: - For a number to be a sum of consecutive integers, it is necessary and sufficient that it shall not be a power of 2. For example 16 is a power of 2, and it cannot be represented as a sum of any two consecutive integers.

The number 128 is the only multidigit power of 2 in which each digit is also a power of 2. That is $128 = 2^7$ and $1 = 2^0$; $2 = 2^1$; and $8 = 2^3$.

If we arrange the numbers from 0 to 10 and their squares correspondingly, the end digits will be 0, 1, 4, 9, 16, 25, 36, 49, 64, 81 and 00 respectively. Similarly if we arrange the numbers from 50 to 40 and their squares correspondingly, the end digits will be 0, 1, 4, 9, 16, 25, 36, 49, 64, 81 and 00 respectively. This type of similarity is observed if we go from 50 to 60 and 100 to 90 with their corresponding squares respectively. Further his type of similarity is observed if we go from 100 to 110 and 150 to 140 with their corresponding squares respectively and so on.

(ii) Factors about the last digit of a cube number:

In the cube column all the digits 1 to 9 appear in the order 1, 8, 7, 4, 5, 6, 3, 2, 9 corresponding to the numbers 1 to 9. The extreme pairs (1, 9), (8, 2), (7, 3) and (4, 6) are such that one digit in each pair is the complement of 10 of the other digit in the same pair.

The power of 3 series is 3, 9, 27, 81, 243, 729... The last digit of the powers of 3 repeats as 3, 9, 7, 1, 3, 9, 7, 1... and it repeats in the same order as we move to the higher powers of 3.

A cube number does not end with two zeros.

(iv) Factors about the last digit of a fourth power number:

For the fourth power the endings are 1, 6, 1, 6, 5, 6, 1, 6, 1 so that there are only two digits involved that is 1 and 6 which bear symmetry on either side of 5.

(v) Factors about the last digit of a fifth power number:

Fifth powers of numbers have the end digit from 1 to 9 and then 0 in an order. This type of sequence is recurring as we proceed further.

That is $1^5 = 1$; $2^5 = 32$; $3^5 = 243$; $4^5 = 1024$; $5^5 = 3125$; $6^5 = 7776$; $7^5 = 16807$; $8^5 = 32768$; $9^5 = 59049$; $10^5 = 100000$; $11^5 = 161051$; $12^5 = 248832$; $13^5 = 371293$; $14^5 = 537824$; $15^5 = 759375$; ....

The endings in respect of 1st, 5th, ninth, thirteenth, seventeenth, twenty first and twenty fifth roots are like fifth roots, in that, the power and the root number have the same terminal digit number. This can be extended to 6th, 7th and 8th powers which are exactly the same as those under 2nd, 3rd and 4th powers respectively. From this we can find the generalized statement that the ending digits of the powers of the real number 'n' will be same as that of the powers of its 'n+4' number.

The powers of the numbers ending with 4 will have their ending digits repeating in pairs (6,4) and for the powers of the numbers ending with 9 will have their ending digits repeating in pairs of (1,9), one digit in the pair being the complement of the other.

The number ending with zero will have the power number ending with zero only. Also the number of zero in the power number is equal to the power raised on the number. The subscript in the above table shows the number of zeros in the power number.

Consider the power numbers 1^1, 2^2, 3^3, 4^4, 5^5, 6^6, 7^7, 8^8, 9^9...n^n. On expansion these numbers can be written as 1, 4, 27, 256, 3125, 46656, 823543, 1677216, 387420489... The digits in the unit place of these numbers are 1, 4, 7, 6, 5, 6, 3, 6, 9, 0... The sequence repeats after 20th term. The twenty terms are 1, 4, 7, 6, 5, 6, 3, 6, 9, 0, 1, 6, 3, 6, 5, 6, 7, 4, 9 and 0. This is verified by expanding 11^{11}, 12^{22}, 13^{33}, 14^{44}...

If $x^p = q$ and q has 'r' number of digits then the number of digits in 'x' is [(p+r-1)/p]. For example $x^4 = 332\ 150\ 625$. Here p = 4 and q = 9, applying the above formula we get x = 3. In fact $135^4 = 332\ 150\ 625$.

No cube number can be written as a sum of two other cube numbers and vice versa. For example the sum of 12^3+11^3 cannot be written as another cube number.

9^n ends with the unit digit as 1 or 9 depending on whether 'n' is odd or even. If 'n' is odd 9^n ends with the unit digit 9 and vice versa.

Specialty of 3rd powers: -

(a)A number formed with 3^n identical digits is divisible by 3^n. For example, if n=2 then $3^2 = 9$. So the numbers with nine identical digits is divisible by 9. That is $222,222,222 \div 9 = 24,691,358$ and $777,777,777 \div 9 = 86419753$ and so on.

(b)A pyramid of 3rd power numbers can be arranged as follows

$$3^2 = 09$$
$$33^2 = 1089$$
$$333^2 = 110889$$
$$3333^2 = 11108889$$
$$33333^2 = 1111088889$$

....

It is to be noted that the number of ones and the number of eights in the product is same.

(c)The square of numbers having even number of digit 3 has the property

$$33^2 = 65^2 - 56^2$$
$$3333^2 = 6565^2 - 5656^2$$
$$333333^2 = 656565^2 - 565656^2$$

....

(d)Sum of the digits of the powers of the 3rd multiples is always 9.

3rd multiples	3rd multiples power	Sum of the digits of 3rd multiples power
3	$(3)^2 = 9$	9
6	$(3+3)^2 = 36$	3+6=9
9	$(3+3+3)^2 = 81$	8+1=9
12	$(3+3+3+3)^2 = 144$	1+4+4=9
15	$(3+3+3+3+3)^2 = 225$	2+2+5=9
...	....	....

Specialty of square roots: -

The continuous square root of any number will lead to one. For example square root of 21 = 4.58; square root of 4.58 =2.14; square root of 2.14=1.46; square root of 1.46= 1.21; square root of 1.21=1.09; square root of 1.09=1.04; square root of 1.04=1.02; square root of 1.02=1.01; square root of 1.01=1.006; square root of 1.006=1.002; square root of 1.002=1.001; square root of 1.007, square root of 1.0003, square root of 1.0002, square root of 1.00009, square root of 1.000046, square root of 1.0002, square root of 1.00001, square root of 1.000005, square root of 1.000003, square root of 1.000001, 1.0000007, square root of 1.0000004, square root of 1.0000002, square root of 1.0000001=1.

The square root of 0.11111… is 0.33333…

A comparative study of ending digits of square and its square root: -

If the square ends in 1 then the last digits of its square root will either be 1 or 9.

If the square ends in 4 then the last digits of its square root will either be 2 or 8.

If the square ends in 5 then the last digits of its square root will either be 5 or 00.

If the square ends in 6 then the last digits of its square root will either be 4 or 6.

If the square ends in 9 then the last digits of its square root will either be 3 or 7.

The square of any integer is either a multiple of 4 or one more than multiple of 4.

A comparative study of ending digits of cube and its cube root: -

If a cubic number ends in 1, 4, 5, 6, 9 or 0 its cubic root will also end in the same number respectively. For example the cubic root of 64 is 4.

If a cubic number ends in 2, the unit digits of the cubic root will end in 8 and vice versa.

If a cubic number ends in 7, the unit digits of the cubic root will end in 3 and vice versa.

Automorphic number or curious number:-

Let 'x' be a number with the property that x^n ends in the same digit as 'x', that is, if the number of digits in 'x' is k, then the last k digits of x^n are the same as those of 'x', then 'x' is called as an automorphic number. Or simply we can say that if the ending digit of a number is 0, 1, 5 or 6, the power of that number too ends in 0, 1, 5 or 6 respectively. For example any number ends with twenty five to the power 'n' also ends with twenty five. That is $225^3 = 11390625$.

Single digit automorphic numbers are 0, 1, 5 and 6. For example $10^2 = 100$; $11^3 = 1331$; $125^3 = 1953125$ and $126^3 = 2000376$.

Two digit automorphic numbers are 00, 01, 25 and 76.

Like the two digit number 25 which is giving this type of power phenomenon, three digit number 625 gives the same type of phenomenon. For example $3625^2 = 13140625$

From the above type of phenomenon we can say that any number ends with seventy six to the power 'n' also ends with seventy six. That is $276^2 = 76176$. Also 76^n end with 76. That is $76^2 = 5776$

$$76^3 = 438976$$
$$76^4 = 33362176$$
$$76^5 = 2535525376$$

………………………

Like the two digit number 76 which is giving the power phenomenon, three digit number 376 gives the same type of phenomenon. For example $3376^2 = 11397376$

Similarly the strings of following numbers express this type of power phenomenon.

…………7109376

Properties of fifth power numbers: -

For every positive integral value of 'n' the number $3(1^5+2^5+3^5+…+n^5)$ is divisible by $1^3+2^3+3^3+…+n^3$. For example if n = 3 then $3(1+32+243) = 828$ and $1+8+27 = 36$. Now $828 \div 36 = 23$.

The sum of the 5th powers of real numbers up to 'n' with 7th powers is equal to twice the sum of the 4th powers of real numbers up to 'n'.

That is $(1^5+2^5+3^5+…+n^5) + (1^7+2^7+3^7+…+n^7)$ is always equal to $2(1+2+3+…+n)^4$.

For example if n = 4 then $(1^5+2^5+3^5+4^5) + (1^7+2^7+3^7+4^7) = 2(1+2+3+4)^4$

$$1300 + 18,700 = 20,000$$
$$20,000 = 20,000$$

Special power triangle: -

Number of decimals 'n'	Number of decimals (2n-1)
1^2	1
11^2	121
111^2	12321
$1,111^2$	1234321
$11,111^2$	123454321
$111,111^2$	12345654321
$1,111,111^2$	1234567654321
$11,111,111^2$	123456787654321
$111,111,111^2$	12345678987654321

The resultants in this pyramid are palindrome numbers. But further proceeding of the number $1,111,111,111^2$ will not give a palindrome number.

The highest digit coming in the resultant is equal to the sum of the digits of its square root. For example $111^2 = 12321$, the highest digit of the resultant is 3, is equal to the sum of the digits of its square root $(1+1+1=3)$.

Other interesting facts are

$$121 = (22×22)/(1+2+1)$$
$$12321 = (333×333)/(1+2+3+2+1)$$
$$1234321 = (4444×4444)/(1+2+3+4+3+2+1)$$
$$23454321 = (55555×55555)/(1+2+3+4+5+4+3+2+1)$$

....

$1234567898765432 1=(999999999×999999999)/(1+2+3+4+5+6+7+8+9+8+7+6+5+4+3+2+1).$

Further if we multiply the repdigit of 1 with higher order repdigit of 1, we get palindromes of the shape 123….321 as follows.

$$11x111 = 1221$$
$$111x1,111 = 123321$$
$$1,111x11,111 = 12344321$$
$$11,111x111,111 = 1234554321$$

....

$$111\ 111\ 111\ x\ 1\ 111\ 111\ 111=123456789987654321$$

The powers of 11 up to 4th power, is a palindrome. For example $11^4=14641$ and $111^3=1367631$. The number 11^5 is not a palindrome number.

The resultant of the power numbers contains odd number of digits only.

Similar type of power triangle is observed in the following case.

Number of decimals 'n'	Number of decimals (2n-1)
11^2	121
101^2	10201
10101^2	102030201
1010101^2	1020304030201
101010101^2	10203040504030201
10101010101^2	102030405060504030201
1010101010101^2	1020304050607060504030201
101010101010101^2	10203040506070807060504030201
10101010101010101^2	102030405060708090706050403020 1

Here also further squaring of 1010101010101010101 will not give palindrome number.

Other special type of power triangles are given below.

$$13^2 = 169$$
$$103^2 = 10609$$
$$1003^2 = 1006009$$
$$10003^2 = 100060009$$
$$100003^2 = 10000600009$$

$$..$$

$$11^2 = 121$$
$$101^2 = 10201$$
$$1001^2 = 1002001$$
$$10001^2 = 100020001$$
$$100001^2 = 10000200001$$

$$.................................$$

Palindromic cubes

$$11^3 = 1331$$
$$101^3 = 1030301$$
$$1001^3 = 1003003001$$
$$10001^3 = 1000300030001$$

$$..$$

The fourth power Palindromic triangle is

$$11^4 = 14641$$
$$101^4 = 104060401$$
$$1001^4 = 1004006004001$$
$$10001^4 = 1000400060004000 1$$

$$..$$

It is to be noted that further process of fifth powers will not give fifth power Palindromic numbers. Also the group of zeros inserted is equal to the group of zeros in the base. The number of groups (of zeros) is equal to the power.

Magic square of square numbers
   A third order magic square comprising square numbers only is as follows.

156^2	936^2	1288^2
1416^2	548^2	504^2
728^2	1176^2	804^2

In the above magic square, the magic square sum is 2,559,376 whether added in any row or any column of the square. That is,

24336	876096	1658944
2005056	300304	254016
529984	1382976	646416

Armstrong number: -
   The addition of the cubes of each individual digit for a three digit number is equal to the number itself means, then it is called as an Armstrong number. Likewise for the four digit number, the addition of the fourth powers of each individual digit is equal to the number itself and so on.
   For example (a) $1^1=1$
   (b) $153=1^3+5^3+3^3=1+125+27$
   (c) $370=3^3+7^3+0^3=27+343+0$

   (d) $371=3^3+7^3+1^3=27+343+1$

(e) $407=4^3+0^3+7^3=64+0+343$

(f) $1634=1^4+6^4+3^4+4^4=1+1296+81+256$

(g) $8208=8^4+2^4+0+8^4=4096+16+0+4096$

(h) $9474=9^4+4^4+7^4+4^4=6561+256+2401+256$

(i) $54748=5^5+4^5+7^5+4^5+8^5=3125+1024+16807+1024+32768$

(j) $92727=9^5+2^5+7^5+2^5+7^5=59049+32+16807+32+16807$

(k) $93084=9^5+3^5+0+8^5+4^5=59049+243+0+32768+1024$.

(l) $548834=5^6+4^6+8^6+8^6+3^6+4^6$

Note: Except 1, all the single digit numbers are not Armstrong numbers. There is no two digit number having this property.

The order of the Armstrong number is the power raised in the number. For example 1634 is a fourth order Armstrong number.

If the Armstrong number is odd, it is an odd Armstrong number and if it is even, it is an even Armstrong number. For example 153 is an odd Armstrong number and 1634 is an even Armstrong number.

Dudeney number: -

If the digital sum of a cube number is equal to the number itself or the cube root means it is called as Dudeney number. For example $27^3 = 19683$ and the digital sum of 19683 is $1+9+6+8+3 = 27$.

Other five such examples are 1^3, 8^3, 17^3, 18^3, and 26^3. It is easy to be noted that there is an upper limit for Dudeney numbers: for example, if it had 7 digits, its cube root would have to be at least 100 (because $100^3=1\,000\,000$) but 7 digits can't add up to any more than $7 \times 9=63$.

Happy number: -

A number is said to be a happy number if we square its digits, and add them together, and then take the result and square its digits and add them together, and keep doing that over and over again and come down to the number 1. For example 32 is a happy number.

$32: 3^2+2^2=9+4=13$

$13: 1^2+3^2=1+9=10$

$10: 1^2+0^2=1+0=1$

Obviously 13 and 10 are also happy numbers.

The happy number series is 1, 7, 10, 13, 19, 23, 28, 31, 32, 44, 49, 68, 70, 79, 82, 86, 91, 94, 97, 100, 103, 109, 129, 130, 133, 139, 167, 176, 188, 190, 192, 193, 203, 208, 219, 226, 230, 236, 239, 262, 263, 280, 291, 293, 301, 302, 310, 313, 319, 320, 326, 329, 331, 338, 356, 362, 365, 367, 368, 376, 379, 383, 386, 391, 392, 397, 404, 409, 440, 446, 464, 469, 478, 487, 490, 496, 536, 556, 563, 565, 566, 608, 617, 622, 623, 632, 635, 637, 638, 644, 649, 653, 655, 656, 665, 671, 673, 680, 683, 694, 700, 709, 716, 736, 739, 748, 761, 763, 784, 790, 793, 802, 806, 818, 820, 833, 836, 847, 860, 863, 874, 881, 888, 899, 901, 904, 907, 910, 912, 913, 921, 923, 931, 932, 937, 940, 946, 964, 970, 973, 989, 998, 1000…

The happy number which is also an odd number is called as odd happy number and the happy number which is also an even number is called as even happy number. For example 31 is an odd happy number and 32 is an even happy number.

Further if the happy number contains all the digits as even, then it is a super even happy number and if the happy number contains all the digits as odd then it is a super odd happy number. For example 226 is a super even happy number and 193 is a super odd happy number.

The happy number which is also a square number is called as happy square number. For example 49 is a happy square number.

Happy primes: Happy numbers that are prime are called as happy prime numbers. The happy prime number sequence is 7, 13, 19, 23, 31, 79, 97, 103, 109, 139, 167, 193…

The happy number which is also a composite number is called as composite happy number. For example 82 is a composite happy number.

The happy number which is palindromic in nature is called as palindromic happy number. For example 262 is a palindromic happy number.

The happy number which is repdigit in nature is called as repdigit happy number. For example 888 is a repdigit happy number.

The happy number which is ascending in nature is called as happy ascending number and the happy number which is descending in nature is called as descending happy number. For example 139 is an ascending happy number and 310 is a descending happy number.

Happy Pythagorean triplets: -The Pythagorean triplets which are happy numbers are called as happy Pythagorean triplets. For example (700, 3465, 3535) (748, 8211, 8245) (910, 8256, 8306) (940, 2109, 2309)…

Unhappy number: -

A number is said to be an unhappy number if we square its digits, and add them together, and then take the result and square its digits and add them together, and keep doing that over and over again, not coming down to the number 1. For example 25 is an unhappy number.

$25: 2^2 + 5^2 = 4 + 25 = 29$

$29: 2^2 + 9^2 = 4 + 81 = 85$

$85: 8^2 + 5^2 = 64 + 25 = 89$

$89: 8^2 + 9^2 = 64 + 81 = 145$

$145: 1^2 + 4^2 + 5^2 = 1 + 16 + 25 = 42$

$42: 4^2 + 2^2 = 16 + 4 = 20$

$20: 2^2 + 0^2 = 4 + 0 = 4$

The unhappy number sequence is 2, 3, 4, 5, 6, 8, 9, 11, 12, 14, 15, 16, 17, 18, 20, 21, 22, 24, 25, 26, 27, 29, 30, 33, 34, 35, 36, 37, 38, 39, 40, 41, 42, 43, 45, 46, 47, 48, 50…

If the unhappy number is ought to be a prime number, then it is called as prime unhappy number and if the unhappy number is composite number, then it is called as composite number. For example 29 is prime unhappy number and 85 is composite unhappy number.

The unhappy number which is also an odd number is called as odd unhappy number and the unhappy number which is also an even number is called as even unhappy number. For example 25 is an odd unhappy number and 20 is an even unhappy number.

Further if the unhappy number contains all the digits as even, then it is a super even unhappy number and if the unhappy number contains all the digits as odd then it is a super odd unhappy number. For example 46 is a super even unhappy number and 39 is a super odd unhappy number.

Lucky number: -

A lucky number is a natural number which is generated by a sieve similar to that of Eratosthenes sieve that generates the primes.

Write the natural number series 1, 2, 3, 4, 5, 6, 7, 8, 9, 10, 11, 12, 13, 14, 15, 16, 17…

Keeping the first integer eliminate all the second term in the above series

1, 3, 5, 7, 9, 11, 13, 15, 17, 19, 21, 23, 25…

Now keeping the first and second integers, eliminate all the third term in the series

1, 3, 7, 9, 13, 15, 19, 21, 25…

Now keep the first, second and third integers and eliminate all the fourth term

1, 3, 7, 9, 13, 15, 21, 25…

As this procedure is repeated indefinitely, the survivors are the lucky numbers:

1, 3, 7, 9, 13, 15, 21, 25, 31, 33, 37, 43, 49, 51, 63, 67, 69, 73, 75, 79, 87, 93, 99...

It is obvious that all the lucky numbers are odd numbers as we removed all the second terms in the natural number series.

Further if the lucky number contains all the digits as odd then it is a super odd lucky number. For example 79 is a super odd lucky number and 87 is not a super odd lucky number.

A lucky prime is a lucky number that is prime. The lucky prime sequence is 3, 7, 13, 31, 37, 43, 67, 73, 79, 127, 151, 163, 193… These numbers are also called as prime lucky numbers. The lucky numbers which are also composite numbers are called as composite lucky numbers. For example 9, 15, 21, 25, 33… are composite lucky numbers.

Palindromic lucky numbers are the lucky numbers which are palindromic in nature. For example 151 is a palindromic lucky number.

Repdigit lucky numbers are the lucky numbers which are repdigit in nature. For example 99 is a repdigit lucky number.

Friedman number: -

A Friedman number is an integer which, in a given base, is the result of an expression using all its own digits in combination with any of the four basic arithmetic operators ($+, -, \times, \div$) and sometimes exponentiation. For example, 347 is a Friedman number since $347 = 7^3 + 4$. The first few base 10 Friedman numbers are 25, 121, 125, 126, 127, 128, 153, 216, 289, 343…

Parentheses can be used in the expressions, but only to override the default operator precedence, for example, in $1024 = (4 - 2)^{10}$. Leading zeros cannot be used, since that would also result in trivial Friedman numbers, such as $001729 = 1700 + 29$.

A nice Friedman number is a Friedman number where the digits in the expression can be arranged to be in the same order as in the number itself. For example, we can arrange $127 = 2^7 - 1$ as $127 = -1 + 2^7$. The first nice Friedman numbers are 127, 343, 736, 1285, 2187, 2502, 2592, 2737, 3125, 3685…

Zeroless pandigital Friedman numbers are also known. Two of them are: $123456789 = ((86 + 2 \times 7)^5 - 91) / 3^4$, and $987654321 = (8 \times (97 + 6/2)^5 + 1) / 3^4$. Only one of the 81 known zeroless pandigital Friedman numbers is nice: $268435179 = -268 + 4^{(3 \times 5 - 17)} - 9$.

All powers of 5 are Friedman numbers.

Roman numerals which can be expressed in the Friedman type are Roman Friedman number. For example 8, since VIII = (V - I) × II.

The repunit numbers which can be expressed in the nice Friedman type are repunit nice Friedman numbers. For example $99999999 = (9 + 9/9)^{9-9/9} - 9/9$.

Vampire numbers: -

Vampire numbers are a type of Friedman numbers where the only operation is a multiplication of two numbers with the same number of digits.

For example $1260 = 21 \times 60$. Some other examples are $1395 = 15 \times 93$

$$1435 = 41 \times 35$$
$$1530 = 51 \times 30$$

Special numbers:-

$512 = 8^3$ and $5+1+2 = 8$
$4913 = 17^3$ and $4+9+1+3 = 17$
$5832 = 18^3$ and $5+8+3+2 = 18$

$17576 = 26^3$ and $1+7+5+7+6 = 26$
$19,683 = 27^3$ and $1+9+6+8+3 = 27$

$1 = 1^2$
$5 = 1^2 + 2^2$
$14 = 1^2 + 2^2 + 3^2$
$30 = 1^2 + 2^2 + 3^2 + 4^2$
$55 = 1^2 + 2^2 + 3^2 + 4^2 + 5^2$

....

$1 = 1^4$
$17 = 1^4 + 2^4$
$98 = 1^4 + 2^4 + 3^4$
$354 = 1^4 + 2^4 + 3^4 + 4^4$
$979 = 1^4 + 2^4 + 3^4 + 4^4 + 5^4.$

....

$1 = 1^5$
$33 = 1^5 + 2^5.$
$276 = 1^5 + 2^5 + 3^5.$
$1300 = 1^5 + 2^5 + 3^5 + 4^5.$
$4425 = 1^5 + 2^5 + 3^5 + 4^5 + 5^5.$

....

$1 = 1^6$
$65 = 1^6 + 2^6$
$794 = 1^6 + 2^6 + 3^6.$
$4890 = 1^6 + 2^6 + 3^6 + 4^6$
$20515 = 1^6 + 2^6 + 3^6 + 4^6 + 5^6$

....

$1 = 1^7$
$129 = 1^7 + 2^7$
$2316 = 1^7 + 2^7 + 3^7.$
$18700 = 1^7 + 2^7 + 3^7 + 4^7$
$96825 = 1^7 + 2^7 + 3^7 + 4^7 + 5^7$

....

$$1 = 1^8$$
$$257 = 1^8 + 2^8$$
$$6818 = 1^8 + 2^8 + 3^8$$
$$72354 = 1^8 + 2^8 + 3^8 + 4^8$$
$$462979 = 1^8 + 2^8 + 3^8 + 4^8 + 5^8$$

....

$$1 = 2^0$$
$$3 = 2^0 + 2^1$$
$$7 = 2^0 + 2^1 + 2^2$$
$$15 = 2^0 + 2^1 + 2^2 + 2^3$$
$$31 = 2^0 + 2^1 + 2^2 + 2^3 + 2^4$$

....

$$1 = 3^0$$
$$4 = 3^0 + 3^1$$
$$13 = 3^0 + 3^1 + 3^2$$
$$40 = 3^0 + 3^1 + 3^2 + 3^3$$
$$121 = 3^0 + 3^1 + 3^2 + 3^3 + 3^4$$

....

$$3 = 3^1$$
$$12 = 3^1 + 3^2$$
$$39 = 3^1 + 3^2 + 3^3$$
$$120 = 3^1 + 3^2 + 3^3 + 3^4$$
$$363 = 3^1 + 3^2 + 3^3 + 3^4 + 3^5$$

....

$$1 = 4^0$$
$$5 = 4^0 + 4^1$$
$$21 = 4^0 + 4^1 + 4^2$$
$$85 = 4^0 + 4^1 + 4^2 + 4^3$$
$$341 = 4^0 + 4^1 + 4^2 + 4^3 + 4^4$$

....

$$4 = 4^1$$
$$20 = 4^1 + 4^2$$
$$84 = 4^1 + 4^2 + 4^3$$
$$340 = 4^1 + 4^2 + 4^3 + 4^4$$
$$1364 = 4^1 + 4^2 + 4^3 + 4^4 + 4^5$$

....

$$1 = 6^0$$
$$7 = 6^0 + 6^1$$
$$43 = 6^0 + 6^1 + 6^2$$
$$259 = 6^0 + 6^1 + 6^2 + 6^3$$
$$1555 = 6^0 + 6^1 + 6^2 + 6^3 + 6^4$$

....

$$1 = 1^0$$
$$3 = 1^0 + 2^1$$
$$12 = 1^0 + 2^1 + 3^2$$
$$76 = 1^0 + 2^1 + 3^2 + 4^3$$
$$701 = 1^0 + 2^1 + 3^2 + 4^3 + 5^4$$
$$8477 = 1^0 + 2^1 + 3^2 + 4^3 + 5^4 + 6^5$$

....

$$1 = 1^1$$
$$5 = 1^1 + 2^2$$
$$32 = 1^1 + 2^2 + 3^3$$
$$288 = 1^1 + 2^2 + 3^3 + 4^4.$$
$$3413 = 1^1 + 2^2 + 3^3 + 4^4 + 5^5.$$

....

$$5 = 1^1 \times 2^2$$
$$108 = 1^1 \times 2^2 \times 3^3$$
$$27,648 = 1^1 \times 2^2 \times 3^3 \times 4^4$$

....

Further fun with powers

The following equality has all the digits from 1 to 10

$$2^7 4^3 9^5 = 6^{10} 8^1$$

$$1^1 + 2^1 + 6^1 = 4^1 + 5^1$$
$$1^2 + 2^2 + 6^2 = 4^2 + 5^2$$

$$8970 = 8 + 9^4 + 7^4 + 0.$$
$$8971 = 8 + 9^4 + 7^4 + 1.$$
$$8972 = 8 + 9^4 + 7^4 + 2.$$

$$8973 = 8 + 9^4 + 7^4 + 3$$
$$8974 = 8 + 9^4 + 7^4 + 4.$$
$$8975 = 8 + 9^4 + 7^4 + 5.$$

$$8976 = 8 + 9^4 + 7^4 + 6.$$
$$8977 = 8 + 9^4 + 7^4 + 7$$
$$8978 = 8 + 9^4 + 7^4 + 8.$$
$$8979 = 8 + 9^4 + 7^4 + 9.$$

$$4150 = 4^5 + 1^5 + 5^5 + 0.$$
$$4151 = 4^5 + 1^5 + 5^5 + 1$$
$$4152 = 4^5 + 1^5 + 5^5 + 2$$

$$4153 = 4^5 + 1^5 + 5^5 + 3.$$
$$4154 = 4^5 + 1^5 + 5^5 + 4.$$
$$4155 = 4^5 + 1^5 + 5^5 + 5.$$

$$4156 = 4^5 + 1^5 + 5^5 + 6.$$
$$4157 = 4^5 + 1^5 + 5^5 + 7.$$
$$4158 = 4^5 + 1^5 + 5^5 + 8.$$
$$4159 = 4^5 + 1^5 + 5^5 + 9.$$

$$8200 = 8 + 2^{13} + 0 + 0.$$
$$8201 = 8 + 2^{13} + 0 + 1.$$
$$8202 = 8 + 2^{13} + 0 + 2.$$

$$8203 = 8 + 2^{13} + 0 + 3.$$
$$8204 = 8 + 2^{13} + 0 + 4.$$
$$8205 = 8 + 2^{13} + 0 + 5.$$

$$8206 = 8 + 2^{13} + 0 + 6.$$
$$8207 = 8 + 2^{13} + 0 + 7.$$
$$8208 = 8 + 2^{13} + 0 + 8.$$
$$8209 = 8 + 2^{13} + 0 + 9.$$

(a) $4475 = 6^2 + 7^3 + 8^4$.

(b) $1033 = 8^1 + 8^0 + 8^3 + 8^3$.

(c) $7122 = 7^2 + 8^3 + 9^4$.

(d) $89 = 8^1 + 9^2 = 8 + 81$

(e) $598 = 5^1 + 9^2 + 8^3$.

(f) $135 \ = 1^1 + 3^2 + 5^3$

(g) $518 = 5^1 + 1^2 + 8^3$.

(h) $175 \ = 1^1 + 7^2 + 5^3$

(i) $1306 = 1^1 + 3^2 + 0^3 + 6^4$.

(j) $2427 = 2^1 + 4^2 + 2^3 + 7^4$.

(k) $1676 = 1^1 + 6^2 + 7^3 + 6^4$.

(l) $3412 = 2^2 + 3^3 + 4^4 + 5^5$.

(m) $698 = 3^2 + 4^3 + 5^4$.

(n) $3153 = 1^1 + 3^3 + 5^5$.

(o) $1371 = 1^2 + 37^2 + 1^2$.

(p) $2642 = 5^2 + 6^3 + 7^4$.

(q) $2^2 - 1^2 = 3 = 2 + 1$

(r) $3312 = 33^2 + 12^2$.

(s) $8833 = 88^2 + 33^2$.

(t) $3468 = 68^2 - 34^2$.

(u) $1370 = 1^2 + 37^2 + 0^2$.

(v) $1233 = 12^2 + 33^2$.

(w) $1386 = 1^2 + 3^4 + 8^1 + 6^4$.

(x) $6^3 = 3^3 + 4^3 + 5^3$

(y) $9^3 = 1^3 + 6^3 + 8^3$

(z) $90^3 = 25^3 + 38^3 + 87^3$

(aa) $1634 = 1^4 + 6^4 + 3^4 + 4^4$

(ab) $4675 = 2^4 + 3^4 + 4^4 + 5^4 + 6^4 + 7^4$.

(ac) $4624 = 4^4 + 4^6 + 4^2 + 4^4$.

(ad) $4355 = 2^4 + 3^5 + 4^6$.

(ae) $3786 = 3^4 + 7^4 + 8 + 6^4$.

(af) $9474 = 9^4 + 4^4 + 7^4 + 4^4$

(ag) $4160 = 4^3 + 16^3 + 0^3$.

(ah) $4161 = 4^3 + 16^3 + 1^3$.

(ai) $666 = 1^3 + 2^3 + 3^3 + 4^3 + 5^3 + 6^3 + 5^3 + 4^3 + 3^3 + 2^3 + 1^3$

(aj) $1850 = (10^3 + 10^4 + 10^5) / (3 \times 4 \times 5)$.

(ak) $2213 = 2^3 + 2^3 + 13^3$.

(al) $3408 = 3^3 + 4^4 + 5^5$.

(am) $1715 = 1^3 \times 7^3 \times 1 \times 5$.

(an) $8465 = 4^3 + 5^4 + 6^5$.

(ao) $9250 = (10^3 + 10^4 + 10^5 + 10^6) / (3 \times 4 \times 5 \times 6)$.

(ap) $8208 = 8^4 + 2^4 + 0^4 + 8^4$

(aq) $978 = 2^4 + 3^4 + 4^4 + 5^4$.

(ar) $8771 = 2^4 + 3^4 + 4^4 + 5^4 + 6^4 + 7^4 + 8^4$.

(as) $23 = 1^4 + 2^3 + 3^2 + 4^1 + 5^0$

(at) $66 = 1^5 + 2^4 + 3^3 + 4^2 + 5^1 + 6^0$.

(au) $594 = 1^5 + 2^9 + 3^4$.

(av) $4424 = 2^5 + 3^5 + 4^5 + 5^5$.

(aw) $144^5 = 27^5 + 84^5 + 110^5 + 133^5$.

(ax) $548834 = 5^6 + 4^6 + 8^6 + 8^6 + 3^6 + 4^6$.

(ay) $3996 = (6^6 + 6^7 + 6^8 + 6^9) / (6 \times 7 \times 8 \times 9)$.

(az) $4889 = 2^6 + 3^6 + 4^6$.

(ba) $165033 = 16^3 + 50^3 + 33^3$

(bb) $732 = 1^7 + 2^6 + 3^5 + 4^4 + 5^3 + 6^2 + 7^1$.

(bc) $1000165033 = 1000^3 + 16^3 + 50^3 + 33^3$

(bd) $746 = 1^7 + 2^4 + 3^6$.

(be) $38 = 2^2 + 3^2 + 5^2$

(bf) $1741725 = 1^7 + 7^7 + 4^7 + 1^7 + 7^7 + 2^7 + 5^7$.

(bg) $3212 = 3^7 + 2^9 + 1^7 + 2^9$.
(bh) $41 = 4^2 + 5^2$
(bi) $6912 = 6 \times 9 \times 1 \times 2^7$.

(bj) $2^9 \times 5^9 = 512 \times 1953125 = 1,000,000,000$.
(bk) $2592 = 2^5 \times 9^2 = 32 \times 81$.
(bl) $99 = 2^3 + 3^3 + 4^3$

(bm) $3435 = 3^3 + 4^4 + 3^3 + 5^5$
(bn) $43 = 4^2 + 3^3$
(bo) $144^5 = 27^5 + 84^5 + 110^5 + 133^5$.

(bp) $2780 = 1^8 + 2^7 + 3^6 + 4^5 + 5^4 + 6^3 + 7^2 + 8^1$.
(bq) $77 = 4^2 + 5^2 + 6^2$.
(br) $3366 = (1^9 + 2^9 + 3^9) / (1 \times 2 \times 3)$.

(bs) $6481 = (3^{12} + 1) / (3^4 + 1)$.
(bt) $55^2 = 44^2 + 33^2$
(bu) $20615673^4 = 2682440^4 + 15365639^4 + 18796760^4$

(bv) $422481^4 = 95800^4 + 217519^4 + 41560^4$
(bw) $43^7 = 271818611107 = (2+7+1+8+1+8+6+1+1+1+0+7)^7$
(bx) $53^7 = 1,174,711,139,837 = (1+1+7+4+7+1+1+1+3+9+8+3+7)^7$

Special power numbers: -

The number 16 is the only number of the form $x^y = y^x$ (that is $2^4 = 4^2$) with x and y being different integers.

The two whole numbers with the difference of their squares is a cube and the difference of their cubes is a square is 10 and 6. That is $10^2 - 6^2 = 64 = 4^3$

$$10^3 - 6^3 = 784 = 28^2.$$

Super squares: - The square numbers 1, 16, 81, 256, 625, 1296, 2401… can be expressed as square to square of their roots respectively. These numbers are called as super square numbers. That is $(1^2)^2$, $(2^2)^2$, $(3^2)^2$, $(4^2)^2$, $(5^2)^2$…

The number 26 is the only positive number which lies between a square and a cube.

The number 567 has the property that, it and its square together use the digits from 1 to 9 only once. That is $567^2 = 321489$.

Another example is $854^2 = 729316$.

The numbers 33, 39, 57, 73 and 85 have the special property described as follows. That is $33 = 17^2 - 16^2$ and $17 + 16 = 33$

$39 = 20^2 - 19^2$ and $20 + 19 = 39$

$57 = 29^2 - 28^2$ and $29 + 28 = 57$

$73 = 37^2 - 36^2$ and $37 + 36 = 73$

$85 = 43^2 - 42^2$ and $43 + 42 = 85$

The numbers 1, 133, 315, 803, 1148, 1547 and 2196 have special property. That is the number can be expressed as follows

$(1)(1^2) = 1$

$(1 + 3 + 3)(1^2 + 3^2 + 3^2) = 133$
$(3 + 1 + 5)(3^2 + 1^2 + 5^2) = 315$
$(8 + 0 + 3)(8^2 + 0^2 + 3^2) = 803$
$(1 + 1 + 4 + 8)(1^2 + 1^2 + 4^2 + 8^2) = 1148$
$(1 + 5 + 4 + 7)(1^2 + 5^2 + 4^2 + 7^2) = 1547$
$(2 + 1 + 9 + 6)(2^2 + 1^2 + 9^2 + 6^2) = 2196$

The number 1127 has the property that if each digit is replaced by its square, the resulting number is also a square. That is 11449 is a square number ($107^2 = 11449$). Some more numbers with the same property are 1281, 1522, 1641, 1805, 2405, 2722, 2966, 3203, 9305, 2722, 4205, 4300, 4402, 5204, 5309, 6100, 8003, 8401, 8663 and 9602.

The number 3115 has the property that if each digit is replaced by its square, the resulting number is a cube. That is 91125 is the cube of 45.

The number 8821 has the property that if each of its digits is replaced by its cube, the resulting number is a square. That is 51251281 is the square of 7159.

The number 1681 is a square ($=41^2$) and each of its two 2-digit parts that is 16 and 81 are also squares.

The number 2178 is the only number known which when multiplied by its reverse yields a 4th power. That is $2178 \times 8712 = 18974736 = 66^4$.

The number 2326 is the smallest number whose cube contains every digit at least once. That is the cube 12584301976 is a pandigital number.

The number 2353 has the property that $588^2 + 2353^2 = 5882353$ and $9412^2 + 2353^2 = 94122353$.

The number 10036224($=3168^2$) is a square number and its reflux number that is 42263001($=6501^2$) is also a square number.

The numbers 11 and 4631 have cube with only odd digits, that is 1331 and 99,317,171,591.

The square number which is having its sum of the digits equal to the square root of that number is 81. That is the sum of the digits of 81 is 9 which is nothing but the square root of 81.

The square numbers 25 and 36 have special characteristic behavior that they are having the last digit of the number 5 and 6 as their square roots respectively.

The following seven consecutive squares which have the property that whose digits sum is a square are 81, 100, 121, 144, 169, 196 and 225.

The only number whose power values like square, cube and n^{th} power (where n is a real number) same is 1. So the number 1 is called as a unit number. That is $1^2=1^n=1$.

The only number which has the same value of square and its sum value is 2. That is $2\times2=4$ and $2+2=4$. Similarly $4\div2=2$ and $4-2=2$.

The lowest square number containing all the nine digits once and only once is 139854276, the square of 11826 and the highest square number under the same conditions is 923187456, the square of 30384.

The four numbers with the property that the sum of every two and the sum of all four are perfect squares are a=10430; b=3970; c=2114 and d=386.

$a + b = 14,400 = 120^2$	$a + c = 12,544 = 112^2$
$a + d = 10,816 = 104^2$	$b + c = 6,084 = 78^2$
$b + d = 4,356 = 66^2$	$c + d = 2,500 = 50^2$

$a + b + c + d = 16900 = 130^2$.

The following squares have the special property that they are having consecutive numbers.

$$82,81 = 91^2$$
$$183,184 = 428^2$$
$$328,329 = 573^2$$

$$528,529 = 727^2$$
$$715,716 = 846^2$$
$$6099,6100 = 7810^2$$

$$13224,13225 = 36365^2$$
$$40495,40496 = 63636^2$$

Kaprekar number: -

When an 'n' digit number is squared and in the square number, if the right hand digit/s is added to the left hand digit/s and is equal to the original number that is the square root itself means, then it is called as Kaprekar number. Since here we are splitting the number into two parts, the numbers can be called as Kaprekar diads.

The Kaprekar number series is 1, 9, 45, 55, 99, 703, 999, 2223, 2728, 5292, 7272, 7777, 9999, 857143…

For example 9 and the square of it is 81. Now the addition of the right hand digit and the left hand digit in the squared number is 8+1=9.

Other examples are $45^2 = 2025$ and $20+25=45$
$$55^2 = 3025 \text{ and } 30+25=55$$
$$99^2 = 9801 \text{ and } 98 + 01 = 99$$
$$703^2 = 494209 \text{ and } 494+209=703$$

Likewise if we split the resultant into three parts and the resultant is equal to the original number itself means, it is called as Kaprekar triads or Kaprekar triples. For example $8^3=512$; Here 5+1+2=8.

Other examples are $45^3=91125$; Here 9+11+25=45
$$297^3=26198073; 26+198+073=297$$

Similarly we can get Kaprekar tetrads and so on.

The palindromic Kaprekar number is a Kaprekar number which is palindromic in nature. For example 55, 999, 7777…

The Kaprekar number which is also odd number is called as odd Kaprekar number and the Kaprekar number which is an even number is called as even Kaprekar number. For example 2223 is an odd Kaprekar number and 2728 is an even Kaprekar number.

Powerful number: -

When the sum of individual digit of a number is raised to power is equal to the original number, then it is called as powerful number.

For example $3435 = 3^3 + 4^4 + 3^3 + 5^5$.

Relationship between power numbers and the arithmetic sequences: -

The relationship between the arithmetic sum of odd numbers and the sequence of arithmetic cubes is as follows.

Odd number series 1, 3, 5, 7, 9, 11, 13, 15, 17, 19, 21, 23, 25, 27, 29...

The above series can be arranged as 1; 3+5; 7+9+11; 13+15+17+19; 21+23+25+27+29...

Now $1 = 1^3 = 1$
$$8 = 2^3 = 3+5$$

$27=3^3=7+9+11$
$64=4^3=13+15+17+19$
$125=5^3=21+23+25+27+29$

…….…………………

Another relationship between the arithmetic sum of discrete numbers and the sequence of cubes is as follows.

$$1 = 1$$
$$2 + 3 + 4 = 1 + 8$$
$$5 + 6 + 7 + 8 + 9 = 8 + 27$$
$$10 + 11 + 12 + 13 + 14 + 15 + 16 = 27 + 64$$

…………………………………………..

The sequence of cubes is equal to the sequence multiplication of arithmetic progression.

$$(1\times1)=1^3=1$$
$$(2\times1) + (2\times3) =2^3=8$$
$$(3\times1) + (3\times3) + (3\times5) =3^3=27$$
$$(4\times1) + (4\times3) + (4\times5) + (4\times7) =4^3=64$$
$$(5\times1) + (5\times3) + (5\times5) + (5\times7) + (5\times9) =5^3=125$$

………………………………………………

Another relationship between the fourth power of numbers and the arithmetic sum of odd numbers is as follows.

$1^4=1$
$2^4=1+3+5+7$
$3^4=1+3+5+7+9+11+13+15+17$
$4^4=1+3+5+7+9+11+13+15+17+19+21+23+25+27+29+31$

……………………………………………………..

For real number series the first common difference is 1,
$\qquad$ that is 0, 1, 2, 3, 4, 5, 6, 7…
$\qquad$ the first common difference is   1, 1, 1, 1…
For squares the second common difference is 2, that is 0, 1, 4, 9, 16, 25, 36…
$\qquad$ the first common difference is   1, 3, 5, 7, 9, 11…
$\qquad$ the second common difference is   2, 2, 2, 2…
Likewise we can extend for the other power number sequences.
For cubes the common third difference is 6
For fourth powers the common fourth difference is 24
For fifth powers the common fifth difference is 120

………………………………………………..

From the above data 1, 2, 6, 24, 120, 720…..
That is (1×1), (1×2), $(1\times2\times3)$, $(1\times2\times3\times4)$, $(1\times2\times3\times4\times5)$ …
That is 1!, 2!, 3!, 4!, 5!...
The square number series are 1, 4, 9, 16, 25, 36, 49, 64, 81, 100, 121, 144...
The multiplication of $1\times1= 1 =1^2$
$\qquad 1\times4= 4 =2^2$
$\qquad 1\times4\times9=36=6^2$
$\qquad 1\times4\times9\times16=576=24^2$
$\qquad 1\times4\times9\times16\times25=14400=120^2$

…………………………………..

From the above data 1^2, 2^2, 6^2, 24^2, 120^2, 720^2…..
That is $(1\times1)^2$, $(1\times2)^2$, $(1\times2\times3)^2$, $(1\times2\times3\times4)^2$, $(1\times2\times3\times4\times5)^2$…

Pythagorean triples: -

If $x^2+y^2=z^2$ where x, y and z are real numbers then x, y and z are called as Pythagorean triples. For example 9, 12 and 15 are called as Pythagorean triples. Likewise (5, 12, 13) and (9, 12, 15) are also Pythagorean triples.

$$3^2+4^2=5^2 \quad 5^2+12^2=13^2 \quad 9^2+12^2=15^2$$
$$9+16=25 \quad 25+144=169 \quad 81+144=225$$

The general equation for finding out Pythagorean triples is $(2n+1)^2 + [2n(n+1)]^2 = [2n(n+1)+1]^2$ where n›0. For example, the third equation is n=3 and substituting in the formula we get $7^2+24^2=25^2$.

Also it is to be noted that $z^2 = m^2 + n^2$ That is if the square of a number is equal to the sum of two square numbers, then the number itself is a sum of two square numbers. For example in the above case $15^2 = 12^2 + 9^2$

$$225 = 144 + 81$$
$$225 = 225$$

Further it is to be noted that either of x or y in $x^2+y^2=z^2$ and either of m or n in $z^2 = m^2 + n^2$ is odd and another should be even. It is not possible to have both the numbers in (m, n) pair or (x, y) pair as odd number or even number.

Pythagorean Theorem gives the relationship between the sides and hypotenuse of the right angled triangle. Thus $3^2+4^2=5^2$; $5^2+12^2=13^2$ and so on. Like Pythagorean triples, Pythagorean quadruples are available. For example replace $5^2 = 3^2 + 4^2$ in the second equation we get $3^2 + 4^2 + 12^2 = 13^2$.

Some other quadruples are $2^2+3^2+6^2=7^2$

$$1^2+70^2+70^2=99^2$$
$$8^2+49^2+64^2=81^2$$
$$4^2+5^2+20^2=21^2$$
$$1^2+4^2+8^2=9^2$$

Likewise Pythagorean quintuples ($5^2+6^2+8^2+10^2=15^2$), Pythagorean hexaples ($1^2+5^2+7^2+10^2+15^2=20^2$) and so on are also available.

Rachnisky sequence: -

If $(x-1)^2 + x^2 + (x+1)^2 = (x+2)^2 + (x+3)^2$ then the sequence is known as Rachinsky sequence. For example $10^2 + 11^2 + 12^2 = 13^2 + 14^2$

$$(-2)^2 + (-1)^2 + 0^2 = 1^2 + 2^2$$

Power numbers against digits of the number

Number	square	cube	4th power	5th power	6th power	7th power	8th power	9th power	10th power
Single digit	2 (4,9)	1 (8)	--	--	--	--	--	--	--
Double digit	6 (16,25,36,49 64,81)	2 (27, 64)	2 (16, 81)	1 (32)	1 (64)	--	--	--	--
Triple digit	22(100,121, 144,169,196, 225,256,289, 324,361,400, 441,484,529, 576,625,676, 729,784,841, 900,961)	5 (125, 216, 243, 512, 729)	2 (256, 625)	1 (243)	1 (729)	1 (128)	1 (256)	1 (512)	--

Chenshuwen's equal sums of like powers

(i) $1^a+19^a+20^a+51^a+57^a+80^a+82^a = 2^a+12^a+31^a+40^a+69^a+71^a+85^a$

where a=1, 2,3,4,5 and 6

(ii) $975^b+224368^b+300495^b+366448^b = 37648^b+202575^b+337168^b+344655^b$

where b=2, 3 and 4

(iii) $1^c+12^c+25^c+66^c+91^c+130^c+174^c+213^c+238^c+279^c+292^c+303^c$

$= 4^c+6^c+31^c+58^c+105^c+117^c+187^c+199^c+246^c+273^c+298^c+300^c$

where c=1,2,3,4,5,6,7,8,9,10 and 11

(iv) In $1^x + 2^x + 3^x + \ldots + 10^x$ for all values of x from 1 to 9 the resultant is divisible by 11. For example if x = 1 then we get the resultant 55 which is divisible by 11.

Leyland number: -

A number which can be expressed as a sum of $x^y + y^x$ where x and y are integers greater than 1, is called as Leyland number. For example 57 is a Leyland number because $2^5+5^2=57$. Leyland numbers sequence is: 17 $(= 2^3 + 3^2)$, 32, 54, 57, 100, 145, 177, 320, 368, 512, 593, 945, 1124, 1649, 2169, 2530, 4240, 5392, 6250, 7073, 8361...

The Leyland number which is odd is called as odd Leyland number and the Leyland number which is even is called as even Leyland number. For example 54 is an even Leyland number and 57 is an odd Leyland number.

Further if the odd Leyland number contains all of its digits as odd, then it is a super odd Leyland number and if the even Leyland number contains all of its digits as even, then it is a super even Leyland number. For example 177 is a super odd Leyland number and 4240 is a super even Leyland number.

The Leyland number which is prime is called as prime Leyland number and the Leyland number which is composite is called as composite Leyland number. For example the following numbers 17, 593, 32993, 2097593, 8589935681, 59604644783353249,523347633027360537213687137, 4314398832739895727934241975037460019... corresponding to 3^2+2^3, 9^2+2^9, 15^2+2^{15}, 21^2+2^{21}, 33^2+2^{33}, 24^5+5^{24}, 56^3+3^{56}, $32^{15}+15^{32}$ are prime Leyland number.

The square numbers in the Leyland number series is called as square Leyland numbers. For example 100 is a square Leyland number.

The special Leyland numbers $71= 2^3 \times 3^2$

$79= 2^7 - 7^2$

RAMANUJAN NUMBER

Ramanujan number is defined as the number which can be represented as the summation of two positive numbers or with their powers in two different ways.

It can be classified into first order Ramanujan number, second order Ramanujan number, third order Ramanujan number and so on based on the power raised on that number.

First order Ramanujan number is defined as the number which can be represented as the summation of real numbers in two different ways. For example $1+4 = 2+3 = 5$
In general terms the above equation can be represented as $L_1+L_2 = R_1+R_2 = M$. The numbers 5, 6, 7, 8 and 9 are single digit First order Ramanujan numbers.

Now place the digits of R_1 and R_2 adjacent to L_1 and L_2 and similarly place the digits of L_1 and L_2 adjacent to R_1 and R_2 to get again the Ramanujan number.
That is $L_1R_1+L_2R_2 = R_1L_1+R_2L_2$
In the above case $12+43 = 21+34 = 55$

The Ramanujan number thus obtained is a palindrome number. Also it is seen that the components of the number in the left hand side of the equation is palindrome to the components of the number in the right side of the equation. That is 12 is palindrome to 21. The numbers 33, 44, 55, 66, 77, 88 and 99 are double digit First order Ramanujan numbers.

For second order Ramanujan number summation of positive squares is considered. For example $338 = 7^2+17^2 = 13^2+13^2$

The second order Ramanujan number can be found by the formula $13[(n+1)^2+1]$ where $n = 1, 2, 3….$
For example if $n = 1$ means $65 = 1^2+8^2 = 4^2+7^2$

Another test for second order Ramanujan number is the summation of the numbers in the series can be divided by 10. For example in the above case $1+8+4+7=20$ and the resultant 20 can be divided by 10.

For third order Ramanujan number summation of positive cubes is considered. That is Ramanujan number of third order is defined as the summation of cubes of the two different numbers in two different ways. The first such number is 1729.

$$That is\ 1729 = 9^3+10^3 = 1^3+12^3$$
$$= 789+1000 = 1+1728$$

The next such number is 4104 and it can be represented as $2^3 + 16^3 = 9^3 + 15^3$.
Then comes $13832 = 24^3 + 2^3 = 20^3 + 18^3$ and $20683 = 27^3 + 10^3 = 24^3 + 19^3$.

Features of the number 1729:-

The number 1729 is an odd number and peculiarly the divisors of the numbers are odd numbers. That is 1, 7, 13, 19, 91, 133, 247 and 1729 is the divisors of 1729 and is all odd numbers. Also in these divisors, except 91 all are prime numbers.

The divisors of 1729 except 247 can be expressed as the summation of cubes of two different numbers or the difference of cubes two different numbers.

$1=1^3+0^3$
$7=2^3-1^3$
$19=3^3-2^3$

$91=3^3+4^3$
$133=2^3+5^3$

The difference between the first four divisors of the number 1729 is 6. That is 1,7,13 and 19.

The multiplication of the first four divisors of the 1729 is equal to 1729. That is $1 \times 7 \times 13 \times 19 = 1729$.

The multiplication of all the divisors of 1729 is $1 \times 7 \times 13 \times 19 \times 91 \times 133 \times 247 \times 1729 = 1729^4$. And deleting the first four divisors and keeping the multiplication of the remaining divisors of 1729 is $91 \times 133 \times 247 \times 1729 = 1729^3$. And still further deleting the last divisor and keeping the multiplication of the remaining divisors of 1729 is $91 \times 133 \times 247 = 1729^2$

The resultant obtained by the summation of the divisors of 1729 can also be expressed as the sum of the cubes of the two different numbers or the difference of the cubes of two different numbers.

That is $1+7+13+19+91+133+247 = 511 = 8^3 - 1^3$

$1+7+13+19+91+133+247+1729 = 2240 = 12^3 + 8^3 = 14^3 + 2^3 - 8^3$

We can state that $9^3 = 25^2 + 10^2 + 2^2$ So the basic Ramanujan equation can be written as $1729 = 9^3 + 10^3 = 10^3 + 25^2 + 10^2 + 2^2$

Other cube relatives for the number 1729 are

$1729 = 24^3 + 11^3 + 9^3 - 27^3$

$\quad\quad = 33^3 + 16^3 + 10^3 - 34^3$

$\quad\quad = 16^3 + 10^3 + 2^3 - 15^3$

The divisors of 1729 that is 1, 7,13,19,91,133,247 and 1729 when arranged in increasing order and multiplied as follows will give the resultant 1729.

$1 \times 1729 = 1729$

$7 \times 247 = 1729$

$13 \times 133 = 1729$

$19 \times 91 = 1729$

The divisors of 1729 have one extra to 6 or one extra to multiples of 6.

That is $7 = (6 \times 1) + 1$

$\quad\quad 13 = (6 \times 2) + 1$

$\quad\quad 19 = (6 \times 3) + 1$

$\quad\quad 91 = (6 \times 15) + 1$

$\quad\quad 133 = (6 \times 22) + 1$

$\quad\quad 1729 = (6 \times 288) + 1$

1729 is a Harshad number. That is the resultant of the sum of the digits of 1729 is a divisor of 1729 itself. That is $1+7+2+9 = 19$ and 19 is one of the divisors of 1729.

When the number 1729 is divided by the numbers from 2 to 9 except 5 and 7, the remainder is one only. That is

$\quad\quad 1729 \div 2 = 864$ and the remainder is 1

$\quad\quad 1729 \div 3 = 576$ and the remainder is 1

$\quad\quad 1729 \div 4 = 432$ and the remainder is 1

$\quad\quad 1729 \div 6 = 288$ and the remainder is 1

$\quad\quad 1729 \div 8 = 216$ and the remainder is 1

$\quad\quad 1729 \div 9 = 192$ and the remainder is 1

1729 can be expressed as the difference of squares of two numbers.

That is $1729 = 55^2 - 36^2$

$\quad\quad = 73^2 - 60^2$

$\quad\quad = 127^2 - 120^2$

$\quad\quad = 865^2 - 864^2$

Also we can see that the difference of numbers used in the above equations gives the divisor of 1729. That is $55 - 36 = 19$; $73 - 60 = 13$; $127 - 120 = 7$ and $865 - 864 = 1$.

1729 cannot be related by the difference of cubes of two numbers.

1729 can be related by the squares of three different numbers as follows.

That is $1729 = 37^2+19^2-1^2$

$$= 41^2 + 7^2 - 1^2$$
$$= 38^2 + 17^2 - 2^2$$
$$= 42^2 + 1^2 - 6^2$$

1729 is a centered cube number, as well as dodecagonal number, a 24-gonal and 84-gonal number.

1729 is one of the positive integers (with the others being 81, 1458 and the trivial case 1) which when its digits are added together, produces a sum which, when multiplied by its reversal, yields the original number. That is $1+7+2+9 = 19$ and the reversal of 19 is 91. Now $19 \times 91 = 1729$.

Ramanujan equation:-

Consider the following equations

$1^2+1=2 \times 1^2$...........Equation (1)

$7^2+1=2 \times 5^2$...........Equation (2)

$41^2+1=2 \times 29^2$.........Equation (3)

Now call the numbers in square form as x and y and generalize the above equations as follows $x^2+1=2 \times y^2$

Take any equation, multiply x of it by 6 and subtract the x of the previous equation. For example x of the second equation is 7 and multiplication with 6 gives 42. Now subtract 1, the x value of the previous equation from 42 will give 41 which is x of the next equation.

It applies to y also. That is $6 \times 5 - 1 = 29$

We can get any number of equations by using the above method.

$[(41 \times 6)-7]^2+1 = 2 \times [(29 \times 6)-5]^2$

$[246-7]^2+1 = 2 \times [174-5]^2$

$239^2+1 = 2 \times 169^2$... Equation (4)

Another method of getting the similar type of equations is as follows.

$3x+4y = x$ of the next equation

$2x+3y = y$ of the next equation

For example the equation (4) can be obtained as follows.

$[(3 \times 41) + (4 \times 29)]^2+1 = 2 \times [(2 \times 41) + (3 \times 29)]^2$

$[123+116]^2+1 = 2 \times [82+87]^2$

$239^2+1 = 2 \times 169^2$... Equation (4)

We can keep an equation where $x = y = 1$ as the first equation and further we can develop the equations.

Woodall number: -

It is the natural number of the form $(n \times 2^n)-1$. The Woodall number sequence is 1, 7, 23, 63, 159, 383, 895, 2047, 4607, 10239, 22527, 49151, 106495, 229375...

It is to be noted that all Woodall numbers are odd numbers. Further if the odd Woodall number contains all digits as odd, then it is a super odd Woodall number. For example 159 is a super odd Woodall number and 49151 is not a super odd Woodall number.

All Woodall numbers are prime numbers.

The Woodall number which is palindromic in nature is called as palindromic Woodall number. For example 383 is a palindromic Woodall number.

Cullen number: -

A Cullen number is a number of the form $2^n n + 1$. For example 3, 9, 25, 65, 161, 385, 897, 2049, 4609, 10241, 22529, 49153, 106497, 229377...

It is to be noted that the corresponding Woodall number can be obtained by subtracting 2 from the Cullen number.

All Cullen numbers are odd numbers.

The Cullen numbers which are also square numbers are called as square Cullen numbers. For example 9 and 25 are some of the square Cullen numbers.

The Cullen number which is palindromic in nature is called as palindromic Cullen number. For example 161 is a palindromic Cullen number.

Curzon number: -

A number 'n' is said to be Curzon number if $2n + 1$ divides $2^n + 1$. The Curzon number sequence is 1, 2, 5, 6, 9, 14, 18, 21, 26, 29, 30, 33, 41, 50, 53, 54, 65, 69, 74, 78, 81, 86, 89, 90, 98, 105...

The Curzon number which is odd is called as odd Curzon number and the Curzon number which is even is called as even Curzon number. For example 41 is an odd Curzon number and 54 is an even Curzon number.

Further if the odd Curzon number contains only odd digits then it is a super odd Curzon number and if the even Curzon number contains only even digits then it is a super even Curzon number. For example 86 is a super even Curzon number and 54 is not a super even Curzon number. Similarly 53 is a super odd Curzon number and 69 is not a super odd Curzon number.

The Curzon numbers which are also square numbers are called as square Curzon numbers. For example 9 and 81 are some of the square Curzon numbers.

The Curzon number which is also a repdigit number is called as repdigit Curzon number. For example 33 is a repdigit Curzon number.

The Curzon number which is a prime number is called as Curzon prime number and the Curzon number which is a composite number is called as Curzon composite number. For example 29 is a Curzon prime number 30 is a Curzon composite number.

Hogben number: -

It is the number equal to n^2-n+1. The Hogben numbers sequence is 1, 3, 7, 13, 21, 31, 43, 57, 73, 91, 111, 133, 157, 183, 211, 241, 273, 307, 343, 381, 421, 463, 507, 553, 601...

Hogben numbers are often called central polygonal numbers. That is all the Hogben numbers can be represented in one or other form of central polygonal number like central square number, central triangular number, central pentagonal number and so on.

All the Hogben numbers are odd numbers. Further if the Hogben number contains only odd digits then it is a super odd number. for example 157 is a super odd Hogben number.

The Hogben number which is ascending in nature is an ascending Hogben number and the Hogben number which is descending in nature is a descending Hogben number. For example 57 is an ascending Hogben number and 421 is a descending Hogben number.

The Hogben number which is prime is a Hogben prime number and the Hogben number which is composite is a Hogben composite number. For example 91 is a Hogben prime number and 91 is a Hogben composite number.

The Hogben number which is palindromic in nature is called as palindromic Hogben number. For example 343 is a palindromic Hogben number.

The Hogben number which is repunit in nature is called as repunit Hogben number. For example 111 is a repunit Hogben number.

In a spiral arrangement of the integers, Hogben numbers appear on the main diagonal.

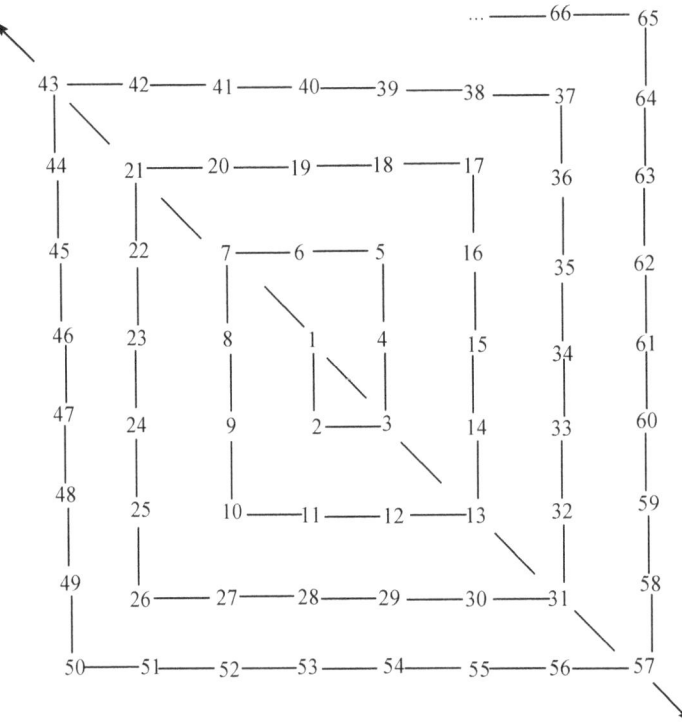

Complex number: -

A complex number consists of a real number and an imaginary number and is represented as 'a+ib' where 'a' is a real number and 'i' is an imaginary number. The imaginary part i=$\sqrt{-1}$ and i^2 = (-1).

Algebra: -

Arithmetic approach of powers involving addition, subtraction, multiplication and division using the alphabets is called as algebra.

For example $(a+b)^2 = a^2+2ab+b^2$

Let us consider a=2 and b=3

$(2+3)^2 = 2^2 + (2\times2\times3) + 3^2$

$5^2 = 4+12+9$

$25 = 25.$

6. DIVISION

Division is nothing but factorization. It is the reverse process of multiplication. Division is denoted by the symbol ÷ or simply by /. There are two types of division namely perfect division or proper fraction and imperfect division or improper fraction.

For example 30/3=10. Here 30 is called as dividend or numerator, 3 is called as divisor or denominator and 10 is called as quotient or resultant. Here there is no remainder left over. This type of division is called as perfect division.

But in 30/4=7(2/4) or 7.5

In the above case 2 is called as the remainder. This type of division is called as imperfect division or fractional division.

$$Dividend = Quotient + \frac{Re\,mainder}{Divisor}$$

$$\text{For example } \frac{30}{4} = 7 + \frac{2}{4}$$

Division rule: -

Unlike addition, subtraction and multiplication we have to take last decimal number of dividend and divide from left to right side of the number by the divisor. Suppose the left decimal number is smaller than the divisor then we have to take both the left decimal number and its next right side decimal number and so on. For example 12812÷2=6406

If two numbers when divided by another number say 'n' give the same remainder means, the difference between the two numbers will be perfectly divisible by 'n'. For example the remainder of 41÷2 is 1 and the remainder of 21÷2 is 1. Now the difference of 41 and 21 is divisible by 2.

Digital root rule of division: -

The digital root of the divisor multiplied by that of the quotient, added to the digital root of the remainder is equal to the digital root of the dividend. That is in the formula

$$Dividend = Quotient + \frac{Re\,mainder}{Divisor}$$

use digital root instead of the actual numbers.

For example 6432÷23 gives 279 as the quotient and 15 as the remainder. Now (5×9)+6=51 whose digital root 6 is the digital root of the dividend 6432.

The above formula can be used for checking the division performed, that is multiply the quotient and divisor followed by the addition of remainder. Now the resultant obtained should be the dividend. That is in the above case we have (279×23) + 15 = 6432.

The remainder theorem: -

The remainder obtained from the product of numbers when divided by another number say 'n' is same as the remainder of the product of the remainders of the numbers divided individually by the same number 'n'.

For example 20x41x19 when divided by 6 the remainder of the product is 4.

The remainder of 20÷6 is 2; 41÷6 is 5 and 19÷6 is 1. The product of the remainders is 2x5x1=10 and the remainder of the product 10÷6 is 4.

Types of fractional division: -

The fractional division can be further divided into two types namely definite fraction or finite fraction and indefinite fraction or infinite fraction. If the fractional division ends with a real number then it is called as a definite fraction. For example 30/4=7.5 is a definite fraction or finite fraction. But if the fractional division ends with a number and it continues and repeats itself means it is called as indefinite fraction or infinite fraction.

For example 30/11=2.72727272…or simply 2.72

Another example 10/3=3.333…… or simply 3.3

One more example 1/7=0.142857

Note:-The asterisk above the digit represents the recurrence of the remainder. Usually the first asterisk above the digit represents the number from where the recurring starts and another asterisk above the number indicates where the recurring ends and repeats further. The definite set of numbers which is recurring, for example 142857 in the above case is called as period of the fraction.

Based upon the recurrence, infinite fraction is classified into two types namely recurring infinite fraction and non recurring infinite fraction. An infinite fraction which recurs with a definite set of numbers is called as recurring infinite fraction. For example 1/7, 3/7, 3/10… are examples of recurring infinite fractions.

An infinite fraction which recurs without a definite set of numbers is called as non recurring infinite fraction. For example √2 = 1.41421356…
(22/7) or π value = 3.14159265358… are examples of non recurring infinite fractions.

In a fractional division except 2/5 when both the numerator and denominator contains prime number, it will give an infinite fraction.

For example 3/7=0.428571

Another example for infinite fractions are fractions whose denominators are 3's like 1/3, 1/33, 1/333… and the corresponding infinite fractions are 0.333, 0.030, 0.003… respectively. This infinite fraction can be represented by 0.3+0.03+0.003+…..Here 1/3 is called as rational number and 0.333 as a repeating decimal. Conversion of this type of repeating decimal number to rational number is as follows.

0.3=1(0.3+0.03+0.003+….)
  =1(0.3/0.9) or 1(3/9)
  =1/3

Another example to convert the repeating decimal to rational number

4.76=4(0.76+0.076+0.0076+….)
  =4(0.76/0.99) or 4(76/99)
  =472/99

Further infinite fractions are obtained from the fractions whose denominators are 1's like 1/11, 1/111, 1/1111… and the corresponding infinite fractions are 0.09, 0.009, 0.0009… respectively.

Note: - The above rule is applicable to all fractions whose denominators are 1's, 3's or 9's. For example in 8/99=0.08, 17/333=0.051 and 107/111=0.963 recurring fractions are obtained.

A decimal point of a number borders between the real number's units, tens, hundreds... from the left side to right side and the rational number's tens, hundreds, thousands... from the right side to left side. That is

...hundreds, tens, units. tens, hundreds, thousands...

For example 6537.3451

Diagrammatically the classification of a division can be mentioned as follows.

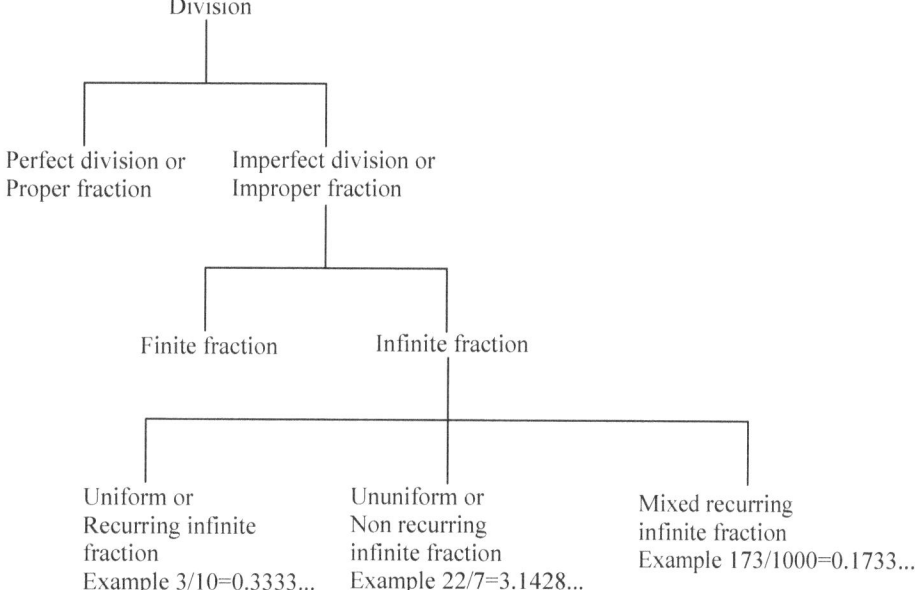

Note: - The recurring infinite fraction in which a particular digit only repeatedly coming is called as mixed infinite fractions. In the above case 173/1000=0.1733... the digit 3 only repeating and so it is a mixed infinite recurring fraction.

Simply in the fractions, if the numerator is greater than the denominator then it is a mixed fraction and if the numerator is lesser than the denominator then it will be either proper or improper fraction. That is the fraction which is having the numerator is greater than the denominator can also be expressed as mixed fraction along with a whole number as follows.

$$3½ = 7/2$$

Composite number:-

A number which is divisible by another number without any remainder apart from 1 and the number we started itself means it is called as a composite number or divisible number. For example 21 and this number can be divided by 1, 3, 7 and 21 without any remainder and these numbers are called as factors or divisors of 21.
The divisible number series is 4, 6, 8, 9, 12, 14, 15, 16, 18, 20...

For any 'n' positive integers there exists at least a subset of them whose sum is divisible by 'n'. For example let n = 6 and the integers be 7, 9, 10, 11, 13 and 14. Now (10, 14) a subset of them whose sum is 24 is divisible by 6. Another example let n = 8 and the numbers be 1, 7, 9, 13, 17, 27, 2 and 30. Now (7, 9) a subset of them whose sum is 16 is divisible by 8.

If a, b and c are the three consecutive numbers and x, y and z are their corresponding lowest possible divisors then the product of x, y and z will be a divisor for the product of a, b and c.

For example 110,111 and 112
2 is the divisor for both 110 and 112.
3 is the divisor for 111.

Now the product of 2×3 = 6 is the divisor of the product of 110×111×112=1367520. That is the resultant 1367520 is divisible by 6.

Smith number: -

A composite number is called as a Smith number if the sum of its digits equals the sum of all the digits appearing in its prime divisors. It is important to remember that, by definition, the factors are treated as digits. For example, 22 factors to 2 × 11 and yields three digits: 2, 1, 1. Therefore 22 is a Smith number because 2 + 2 = 2 + 1 + 1. For example 4, 22, 27, 58, 85, 94, 121, 166, 202, 265… are smith numbers.

The Smith number which is also an odd number is called as odd smith number and the smith number which is even is called as even smith number. For example 27 is an odd smith number and 22 is an even smith number.

Further if the even smith number contains all the digits as even, then it is a super even smith number. For example 202 is a super even smith number.

The smith number which is palindromic in nature is called as palindromic smith number. For example 121 is a palindromic smith number.

The square numbers in the smith number series are called as square smith number. For example 121 is a square smith number.

There are 376 Smith numbers below 10,000.

Reciprocal number: -

The number 1 divided by the rational number 'A' is called as the reciprocal number. For example ½ will give 0.5 and is called as reciprocal number.

The reciprocal number of 81 gives an interesting recurring decimal expansion 1/81 = 0.012345679012345679012345790... The result is actually a repeating decimal containing every digit from 0 to 9, except for 8.

HCF and LCM: -

The addition and subtraction of fractions can be done only if the two fractions have common denominator. If they are having different denominator we have to get common or equivalent denominator to facilitate the addition and subtraction of fractions.

$$\text{For example } \frac{3}{4}+\frac{5}{4}=\frac{3+5}{4}=\frac{8}{4}$$

If the denominators are different, we have to find out the least common denominator to make the fractions equivalent and then the arithmetic operations are carried over.

$$\text{For example } \frac{3}{4}+\frac{5}{2}=\frac{3+10}{4}=\frac{13}{4}$$

The least common denominator is also called as least common multiplier (LCM) or least common factor, because with that common denominator only we are getting the equivalent fractions by multiplication.

We can also make the equivalent fractions by finding out the highest common multiplier or highest common factor (HCF). By convention we do the arithmetic operations of fractions with LCM.

There are two types of finding out LCM. One is by common multiple method and another one by factorization method. For example the LCM of 63 and 84 can be found by the common multiple method as follows.

Multiples of 63 are 63, 126, 189, 252, 315…

Multiples of 84 are 84, 168, 252, 336, 420…

Now the least common multiple of the two numbers is seen as 252 is the LCM.

The LCM for 63 and 84 can be calculated by factorization method as follows.

$$3\,|\,63, 84$$
$$7\,|\,21, 28$$
$$\quad\;\, 3, 4$$

Now multiply 3 and 7, the common factors of the numbers will give 21, followed by the multiplication of 3 and 4, the remainders. That is $3\times7\times3\times4 = 252$ and the resultant 252 is called as the LCM of 63 and 84.

Whenever if you take LCM, leave or cancel the common factors of the denominator and then take the LCM.

For example $1/ (a+x)^2 + 1/ (a+x)(a-x) = 2a/a^2 – x^2$

Highest Common Factor or Greatest Common Divisor is shortly called as HCF or GCD. There are two types of finding out HCF. One is by common divisor method and another one by factorization method. For example HCF for 63 and 84 can be calculated common divisor method as follows.

Divisors of 63 are 1, 3, 9, 21 and 63
Divisors of 84 are 1, 2, 4, 6, 12, 21, 42 and 84

Now the highest common factor of the two numbers is seen as 21 is the HCF.
The HCF for 63 and 84 can be calculated by factorization method as follows.

$$3\,|\,63, 84$$
$$7\,|\,21, 28$$
$$\quad\;\, 3, 4$$

Now multiplication of 3 and 7 which are the common divisors of 63 and 84, that is 21 is the HCF of the numbers.

A check for LCM and HCF is product of HCF and LCM will give the product of the given numbers. For example in the above case $252 \times 21 = 63 \times 84 = 5292$.

Factorization: -

Factorization is nothing but splitting of a number into two parts, so that the multiplication of which will give the original number. Factorization is the reverse process of multiplication. The two parts are called as factors. For example, 24 can be factorized into 6 and 4. The two parts 6 and 4 are called as factors of 24.

The factors can be arranged in an increasing order and connected as a rainbow model of factorization as follows.

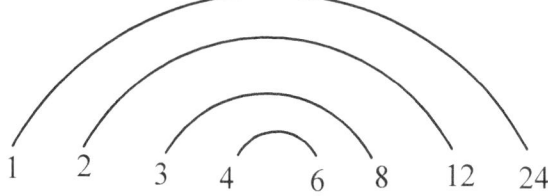

1   2   3   4   6   8   12   24

For all numbers, the number 1 and the original number itself will be factors and are called as trivial factors. In the above case, 1 and 24 are trivial factors of 24.

If the factor of a number is prime then it is called as a prime factor and if the factor of a number is composite then it is called as a composite factor. Similarly if the factor of a number is odd then it is called as odd factor and if the factor of a number is even then it is called as even factor.

A factor tree is formed by drawing all the factors of a number till we reach primes at the end. For example in the above case, the factor tree is drawn as follows.

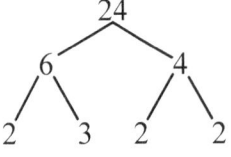

Prime numbers

Prime number is a number which does not have a factor or divisor apart from 1 and the number itself. This number can also be called as indivisible number. For example 31 is a prime number because it is not having any number to divide it without any remainder apart from 1 and 31 itself. Prime numbers are opposite to composite numbers. The prime number series is 2, 3, 5, 7, 11, 13, 17, 19, 23, 29, 31, 37, 41, 43, 47, 53, 59, 61, 67, 71, 73, 79, 83, 89, 97…

There are 1229 primes below 10,000. It is obvious that prime number cannot be a perfect number.

Note: -The number 1 is called as unit number and is neither a prime number nor a composite number.

Formation of prime number: -

A conventional method of finding out a prime number is by using the Eratosthenes sieve as follows. Starting from 2 cross out every second number (or multiples of 2) but not 2 itself. Move to 3 and cross out every third number but not 3 itself. Then with 4, cross out every fourth number and continue crossing out of numbers up to 10. This will eliminate all the numbers which are not prime numbers.

1	2	3	4	5	6	7	8	9	10
11	12	13	14	15	16	17	18	19	20
21	22	23	24	25	26	27	28	29	30
31	32	33	34	35	36	37	38	39	40
41	42	43	44	45	46	47	48	49	50
51	52	53	54	55	56	57	58	59	60
61	62	63	64	65	66	67	68	69	70
71	72	73	74	75	76	77	78	79	80
81	82	83	84	85	86	87	88	89	90
91	92	93	94	95	96	97	98	99	100

Now the prime numbers up to 100 are the numbers left out in the table. Likewise we can find the prime numbers by extending the table.

If we arrange the numbers in six columns, except 2 and 3 the prime numbers are seen only in the first and fifth columns.

1	2	3	4	5	6
7	8	9	10	11	12
13	14	15	16	17	18
19	20	21	22	23	24
25	26	27	28	29	30
31	32	33	34	35	36
37	38	39	40	41	42
43	44	45	46	47	48

However, just because a number is in the 1st or 5th columns doesn't mean that it is prime. All we're saying is that if a number is a prime, it can only be found in one of those two columns.

Prime numbers can also be found by using the following formulas

(i) $x^2 - x + 41$ where x = 0 to 40 gives prime number.

For example if x = 2 means $2^2 - 2 + 41 = 2 + 41 = 43$ and 43 is a prime number.

(ii) $2x^2 + 29$ where x = 0 to 28 gives prime number.

For example if x = 3 means $2(3)^2 + 29 = 18 + 29 = 47$.

(iii) If 'n' is a positive integer then $2^n - 1$ is a called as a Mersenne number. If the Mersenne number is a prime, then it is called as Mersenne prime. Obviously all the

Mersenne numbers are odd numbers. For example $2^7 - 1 = 128 - 1 = 127$ is a prime number and the prime numbers formed from Mersenne numbers are called as Mersenne prime numbers. 8191 is a Mersenne prime. The Mersenne prime numbers are 1, 3, 7, 17, 31…

Properties of prime numbers: -

Every even integer greater than 2 can be written as the sum of two prime numbers. For example 200=7+193. The combination of prime numbers will give either the prime number or the composite number. But the combination of composite numbers will not produce prime numbers. That is combination of prime numbers can produce all natural numbers. So prime numbers can also be called as Godly numbers.

The sum of two prime numbers is a composite number provided that one prime number should not be 2. If one of the prime number is 2 its addition with another prime number may give again a prime number or a composite number. For example 2 + 11 = 13 and 13 is a prime number but 2 + 13 = 15 gives a composite number.

The numbers 9, 98, 987, 9876, 98765, 987654, 9876543, 98765432 and 987654321 are all composite numbers. But the reverse of these numbers leaving 9, will give prime numbers.

There are 25 primes between 1 and 100. But there are only 9 primes between 9,999,900 and 10,000,000 and there are only 2 primes between 10,000,000 and 10,000,100. Those two primes are 10,000,019 and 10,000,079. The distribution of primes among natural numbers is very irregular.

There are two forms of prime numbers. All prime numbers except 2 are either of the form 4n-1 or 4n+1 where n>1. Of these, every prime number of the form 4n+1 can be expressed as the sum of two squares in one way only. These prime numbers can also be called as Pythagorean primes. For example $5=1^2+2^2$; $13=2^2+3^2$.

Note: All prime numbers of the form 4n+1 can be expressed as the sum of two square numbers but all numbers that can be expressed as the sum of two square numbers cannot be a prime number. For example $545=17^2+16^2$, but 545 is not a prime number.

Special prime numbers –

The prime numbers end with 1, 3, 7 or 9 only. 5 is the only prime number which ends in 5.

The prime number 73,939,133 has a special property that, if we remove a digit from the right end of this number each of the remaining numbers is also prime number. For example 7,393,913 is a prime number and so on.

Except 2 prime numbers are always odd numbers. That is, 2 is the only even prime number. Thus the sequence of primes excluding 2 is called as odd prime number sequence.

Further if the odd prime number contains all the digits as odd, then it is a super odd prime number. For example 97 is a super odd prime number and 41 is not a super odd prime number.

3, 5 and 7 are unique primes where these consecutive primes can be written as 3, 3 + 2 and 3 + 4.

Except 2 and 3, every prime number will be divisible by 6 if you either add or subtract 1 from the number. For example the number 17 plus 1, is divisible by 6 and the number 19 minus 1, is divisible by 6.

The prime number 223 is a special prime in which the sums of the nth powers of its digits are prime for all n between 1 and 6 inclusive: sum of digits (2+2+3) = 7, sum of squares of digits $(2^2+2^2+3^2) = 17$, sum of cubes of digits = 43, sum of fourth powers = 113, sum of fifth powers = 307 and sum of sixth powers = 857.

The following prime numbers have special property. The digital sum of these prime numbers have some special property.

(a) 15555…55551 (33-digits) The digital sum of this prime is 157, another prime (whose digit sum in turn is yet another prime is 13).

(b) 10220…02201 (55-digits) The digit sum of this prime is 110, which is the double of its number of digits.

(c) 14444…44441 (67-digits) The digit sum of this prime is 262, a peak palindrome.

(d) 18181…18181 (77-digits) The digit sum of this 77-digit prime is $343 = 7^3$.

(e) 31313…31313 (83-digits) The digit sum of this prime is the prime 167.

(f) 19999…99991 (87-digits) The digit sum of this prime is a palindrome, 767.

(g) 37777…77773 (87-digits) The digit sum of this prime is 601, another prime.

Sphenic number: -

A number is called as sphenic number if it is the product of 3 distinct primes. For example, 370 is a sphenic number because it is the product of the 3 primes 2, 5 and 37.

The sphenic number sequence is 30, 42, 66, 70, 78, 102, 105, 110, 114, 130, 138, 154, 165, 170, 174, 182, 186, 190, 195, 222, 230, 231, 238, 246, 255, 258, 266, 273, 282, 285, 286, 290, 310…

The sphenic number which is odd is an odd sphenic number and the sphenic number which is even is an even sphenic number. For example 105 is an odd sphenic number and 258 is an even sphenic number.

Further if the odd sphenic number contains all of its digits as odd, then it is a super odd sphenic number and if the even sphenic number contains all of its digits as even, then it is a super even sphenic number. For example 246 is a super even sphenic number and 195 is a super odd sphenic number.

If the sphenic number is palindromic in nature, then it is a palindromic sphenic number. For example 282 is a palindromic sphenic number.

If the sphenic number is repdigit in nature, then it is a repdigit sphenic number. For example 66 and 222 are some of the repdigit sphenic numbers.

Interprime number: -

An interprime number is a composite number that is the average of two consecutive primes, i.e., it is at equal distance from previous prime and next prime. For example, 21 is an interprime number since it is the average of the two consecutive primes 19 and 23.

The interprime number sequence is 4, 6, 9, 12, 15, 18, 21, 26, 30, 34, 39, 42, 45, 50, 56, 60, 64, 69, 72, 76, 81, 86, 93, 99, 102, 105, 108, 111, 120...

The interprime number which is odd is an odd interprime number and the interprime number which is even is an even interprime number. For example 45 is an odd interprime number and 108 is an even interprime number.

Further if the odd interprime number contains all of its digits as odd, then it is a super odd interprime number and if the even interprime number contains all of its digits as even, then it is a super even interprime number. For example 42 is a super even interprime number and 39 is a super odd interprime number.

If the interprime number is palindromic in nature, then it is a palindromic interprime number. For example 99 is a palindromic interprime number.

If the sphenic number is repunit in nature, then it is a repunit interprime number. For example 111 is a repunit interprime number.

Types of prime numbers: -

(a) Co prime or relatively prime numbers: -

Two numbers 'a' and 'b' are said to be co prime or relatively prime numbers if they share no common factor between them.

For example 10 and 21 are co prime numbers because they don't have a common factor. But 25 and 35 are not co prime numbers because they have a common factor 5.

Gauss Theorem of co primes: -If a, b and c are three natural numbers such that 'c' divides 'ab' then 'c' and 'a' are co primes, and then c divides b. Let a=3, b=10 and c=5. Now 3 and 5 are co prime and the product ab, 30 can be divided by c that is 5.

Note: - Since 1 is a factor of all numbers, the number one cannot be considered as a common factor between two numbers.

Co prime numbers need not be prime numbers. For example in (10, 21) co prime pair both are composite numbers but in (10, 11) co prime pair 10 is a composite number while 11 is a prime number. In case of (11, 17) co prime pair both are prime numbers.

(b) Parent prime: -

A parent prime is defined as one for which the sum of the squares of its digits is also a prime. The sum is therefore the "child" prime. For example 23 is a parent prime because $2^2 + 3^2 = 4 + 9 = 13$. That is, 23 is the parent, (or progenitor) of a prime child, namely 13. But, 13 is not a prime who can be a parent (i.e. have a child), because $1^2 + 3^2 = 1 + 9 = 10$, a composite number.

Another example of an ancestral line of parents and children is 191 to 83 to 73. Here 191 is parent to 83 and grandparent to 73; but 73 is "childless", as the sum (58) of the squares of its two digits (7 and 3) is a composite number. Likewise the longest ancestral lines having 5 generations is 1499 to 179 to 131 to 11 to 2.

(c) Twin primes: -

Prime numbers that differ by 2 are called as twin primes. For example (3,5), (5,7), (11,13), (17,19), (29,31), (41,43), (59,61), (71,73)… are some of the examples of twin primes.

All twin prime pairs with the exception of the first twin prime pair (3, 5), the sum of the primes in the pairs is always is divisible by 12. For example (5, 7) has the sum 12, (11, 13) has 24 and so on.

For the number 810, is a value of n (n=810) for which n-1 and n+1 are twin primes, and so are 2n-1 and 2n+1.

If the prime numbers in the twin primes contain all of their digits as odd, then it is a super odd twin primes. For example (71,73) is a super odd twin prime and (59,61) is not a super odd twin primes.

(d) Cousin primes: -

Prime numbers that differ by 4 are called as cousin primes. For example (3, 7), (7, 11), (13, 17), (19, 23), (37, 41), (43, 47), (67, 71), (79, 83), (97, 101), (103, 107), (109, 113), (127, 131), (163, 167), (193, 197), (223, 227), (229, 233), (277, 281)…

If the prime numbers in the cousin primes contain all of their digits as odd, then it is a super odd cousin primes. For example (67,71) is a super odd cousin prime and (103,107) is not a super odd cousin primes.

(e) Prime triplets: -

Prime numbers of the form $(p, p+2, p+6)$ or $(p, p+4, p+6)$ are called as prime triplets. Some of them are listed as follows, (5, 7, 11), (7, 11, 13), (11, 13, 17), (13,

17, 19), (17, 19, 23), (37, 41, 43), (41, 43, 47), (67, 71, 73), (97, 101, 103), (101, 103, 107), (103, 107, 109), (107, 109, 113), (191, 193, 197), (193, 197, 199), (223, 227, 229), (227, 229, 233), (277, 281, 283), (307, 311, 313), (311, 313, 317), (347, 349, 353)...

If the prime numbers in the prime triplets contain all of their digits as odd, then it is a super odd prime triplet. For example (311,313,317) is a super odd prime triplet and (307,311,313) is not a super odd prime triplet.

(f) Prime quadruplets: -

Prime numbers of the form (p, $p+2$, $p+6$, $p+8$) are called as prime quadruplets. For example (5, 7, 11, 13), (11, 13, 17, 19), (101, 103, 107, 109), (191, 193, 197, 199), (821, 823, 827, 829), (1481, 1483, 1487, 1489), (1871, 1873, 1877, 1879), (2081, 2083, 2087, 2089), (3251, 3253, 3257, 3259), (3461, 3463,3467, 3469), (5651, 5653, 5657, 5659), (9431, 9433, 9437, 9439)...

If the prime numbers in the prime quadruplets contain all of their digits as odd, then it is a super odd prime quadruplet. For example (11,13,17,19) is a super odd prime quadruplets and (101,103,107,109) is not a super odd prime quadruplets.

(g) Reflux prime: -

It means the reflux number of the prime numbers is also a prime number. That is 13, 31; 17, 71; 37, 73; 79, 97; 1193, 3911.

All the reflux prime numbers are odd and further they are super odd reflux prime numbers.

(h) Safe prime: -

It is a prime number of the form $2p + 1$, where p is also a prime. (Conversely, the prime p is called as a Sophie Germain prime.) The safe prime sequence is 3, 5, 11, 23, 47, 59, 83, 107, 167, 179, 227, 263, 347, 359, 383, 467, 479, 503, 563, 587, 719, 839, 863, 887, 983, 1019, 1187, 1283, 1307, 1319, 1367, 1439, 1487, 1523, 1619, 1823, 1907...

Palindromic safe prime number is a safe prime number which is palindromic in nature. For example 11, 383... are palindromic safe primes.

All safe prime numbers are odd numbers. If the odd safe prime number contains all its digits as odd, then it is called as super odd safe prime number. For example 359 is a super odd safe prime number and 383 is not a super odd safe prime number.

(i) Circular prime: -

If cyclic numbers of a number remains prime number then the number is called as circular prime number. For example the number 197 has the cyclic numbers 971 and 719 are prime numbers. The circular prime number sequence is 11, 13, 17, 37, 79, 113, 137, 197...

137 is a special circular prime because
its cyclic numbers 371 and 713 are prime,
its each individual number 1, 3 and 7 are prime,
its transpose numbers that is 731, 713, 173, 317 and 371 are prime numbers,
its reflux number 731 is a prime,
its each two digit number arranged in any of the way will be a prime that is
13, 37, 71, 31, 73 and 17 are prime.

Another example for circular prime number is 19937. The cyclic number of this number 99371, 93719, 37199, 71993 and 19937 are prime numbers.

All circular prime numbers are odd numbers and the odd circular prime number contains all its digits as odd and so they are called as super odd circular prime numbers. For example 197 is a super odd circular prime number.

(j) Repunit prime number: -

The repunit numbers which are prime are called as repunit primes. The repunit prime sequence is 11, 1111111111111111111, 11111111111111111111111...

It is to be noted that repunit numbers can only be prime if the number of 1's is prime.

Note: -All repunit primes are circular primes but the reverse is not true.

All repunit primes are palindromic primes but the reverse is not true.

Repdigit primes are not possible.

(k) Palindromic prime number: -

A palindromic prime is a prime which is a palindrome. For example 2, 3, 5, 7, 11, 101, 131, 151, 181, 191, 313, 353, 373...

There are 20 palindromic primes below 10,000.

In base 2, all Mersenne primes are palindromic primes.

RSP Palindromes:

16 is a square number, and its reverse, 61 is a prime number. This is called as Reversible-Square-to-Prime Palindrome, or RSP Palindrome, for short. Here is a chart of all numbers less than 100 (with one exception) that produce RSP Pals.

n	Square	Prime	Palindrome	Prime Factors
4	16	61	1661	11 x 151
14	196	691	196691	11 x 17881
19	361	163	361163	11 x 32833
28	784	487	784487	11 x 71317
32	1024	4201	10244201	11 x 127 x 7333
37	1369	9631	13699631	11 x 1245421
38	1444	4441	14444441	11 x 17 x 77243
41	1681	1861	16811861	11^3 x 17 x 743
62	3844	4483	3844483	11^2 x 7 x 45389
85	7225	5227	72255227	11 x 6568657
89	7921	1297	79211297	11 x 127 x 56701
95	9025	5209	90255209	11 x 79 x 283 x 367
97	9409	9049	94099049	11 x 23^2 x 103 x 157

As to the exception referred to above $40^2 = 1600$. The reverse of 1600 is either 0061, or 61 if the leading zeros are suppressed. This gives us 16000061 and 160061 as two more RSP Pals for this range.

This type of phenomenon is present in higher power numbers also. For example $5^3 = 125$ and its reverse 521 is prime. Hence 125521 is a Reversible Cube to Prime Palindrome.

(l) Gaussian prime: -

Prime numbers of the form '4n+3' where 'n' is a whole number are called as Gaussian primes. The Gaussian prime number sequence is 3, 7, 11, 19, 23, 31, 43, 47, 59, 67, 71, 79, 83...

All Gaussian prime numbers are odd numbers. If the odd Gaussian prime number contains all its digits as odd, then it is called as super odd Gaussian prime number. For example 59 is a super odd Gaussian prime number and 83 is not a super odd Gaussian prime number.

(m) Fermat prime: -

Prime numbers of the form $2^{2^n} + 1$ are called as Fermat primes. The sequence of Fermat primes is 3, 5, 17, 257, 65537...

All Fermat prime numbers are odd numbers. If the odd Fermat prime number contains all its digits as odd, then it is called as super odd Fermat prime number. For example 17 is a super odd Fermat prime number and 257 is not a super odd Fermat prime number.

(n) Carol primes: -

Prime numbers of the form $(2^n - 1)^2 - 2$ where n>2 are called as Carol primes. The Carol prime sequence is 7, 47, 223, 3967, 16127, 1046527, 1073676287...

All Carol prime numbers are odd numbers.

(o) Kynea primes: -

Prime numbers of the form $(2^n+1)^2 - 2$ where 'n' is a whole number. The Kynea prime sequence is 2, 7, 23, 79, 1087, 66047, 263167...

All Kynea prime numbers are odd numbers. If the odd Kynea prime number contains all its digits as odd, then it is called as super odd Kynea prime number. For example 79 is a super odd Kynea prime number and 1087 is not a super odd Kynea prime number.

(p) Wilson primes: -

Primes p for which p^2 divides $(p - 1)! + 1$ are Wilson primes and the sequence is 5, 13, 563...

All Wilson prime numbers are odd numbers. If the odd Wilson prime number contains all its digits as odd, then it is called as super odd Wilson prime number. For example 13 is a super odd Wilson prime number and 563 is not a super odd Wilson prime number.

(q) Woodall primes: -

Prime numbers that can be expressed as $n \cdot 2^n - 1$ where n>2 are Woodall primes and the sequence of it is as follows.

7, 23, 383, 32212254719, 2833419889721787128217599, 195845982777569926302400511, 4776913109852041418248056622882488319...

All Woodall prime numbers are odd numbers.

(r) Thabit prime numbers: -

Prime numbers of the form $(3 \cdot 2^n) - 1$ are called as Thabit prime numbers. The sequence is as follows 2, 5, 11, 23, 47, 191, 383, 6143, 786461, 51539607551, 824633720831, 26388279066623, 1080863910566891903...

Palindromic Thabit prime numbers are the Thabit prime numbers which are palindromic in nature. For example 11, 191, 383...

Except 2, all Thabit prime numbers are odd numbers. If the odd Thabit prime number contains all its digits as odd, then it is called as super odd star prime number. For example 47 is a super odd Thabit prime number and 6143 is not a super odd Thabit prime number.

(s) Star primes: -

Prime numbers of the form $6n (n - 1) + 1$ are called as star primes. The sequence of star primes are 1, 13, 37, 73, 181, 337, 433, 541, 661, 937, 1093, 2053, 2281, 2521, 3037, 3313, 5581, 5953, 6337, 6733, 7561, 7993, 8893, 10333, 10837, 11353, 12421, 12973, 13537, 15913, 18481...

The star primes which are palindrome in nature are called as palindromic star prime numbers. For example 181 is a palindromic star prime number.

All star prime numbers are odd numbers. If the odd star prime number contains all its digits as odd, then it is called as super odd star prime number. For example 11353 is a super odd star prime number and 8893 is not a super odd star prime number.

(t) Factorial primes: -

Prime numbers of the form $n! - 1$ or $n! + 1$ where 'n' is a real number are called as factorial prime numbers. For example 2, 3, 5, 7, 11, 23, 719, 5039, 39916801, 479001599... are factorial primes.

Except 2, all factorial prime numbers are odd numbers. If the odd factorial prime number contains all its digits as odd, then it is called as super odd factorial prime number. For example 719 is a super odd factorial prime number and 5039 is not a super odd factorial prime number.

The factorial prime which is palindromic in nature is called as palindromic factorial prime. For example 11 is a palindromic factorial prime.

(u) Isolated primes: -

Primes p such that neither $p - 2$ nor $p + 2$ is prime are called as isolated primes. For example 2, 23, 37, 47, 53, 67, 79, 83, 89, 97, 113, 127, 131, 157, 163...

Except 2, all isolated prime numbers are odd numbers. If the odd isolated prime number contains all its digits as odd, then it is called as super odd isolated prime number. For example 37 is a super odd isolated prime number and 127 is not a super odd isolated prime number.

The isolated prime which is palindromic in nature is called as palindromic isolated prime. For example 131 is a palindromic isolated prime.

(v) Partition prime: -

Partition numbers that are prime are called as partition prime numbers. For example 2, 3, 5, 7, 11, 101, 17977, 10619863, 6620830889...

Palindromic partition prime numbers are the partition prime numbers which are palindromic in nature. For example 11 and 101 are palindromic partition prime numbers.

All partition prime numbers are odd numbers. If the odd partition prime number contains all its digits as odd, then it is called as super odd partition prime number. For example 17977 is a super odd partition prime number and 10619863 is not a super odd partition prime number.

(w) Wagstaff prime number: -

Prime number 'p' of the form $p = [(2^q + 1)/3]$ where q is another prime is a Wagstaff prime number. For example $43 = [(2^7 + 1)/3]$. Wagstaff primes have applications in cryptology. The Wagstaff primes are 3, 11, 43, 683, 2731, 43691, 174763, 2796203, 715827883, 2932031007403...

The Wagstaff prime which is palindromic in nature is called as palindromic Wagstaff prime. For example 11 is a palindromic Wagstaff prime.

All Wagstaff prime numbers are odd numbers.

(x) Wieferich prime: -

The prime number 'p' for which 'p^2' divides $3^{p-1} - 1$ is called as Wieferich prime. Since the base is 3 to which the power is raised this prime number is called as Wieferich base 3 prime. For example 11, 1006003...

Similarly for Wieferich base 5 prime 'p', 'p^2' divides $5^{p-1} - 1$ and so on.

The following table gives the Wieferich primes of different bases with examples.

No.	Wieferich prime (p)	Condition	Sequence
1.	Base 3	P^2 divides $3^{p-1} - 1$	11, 1006003…
2.	Base 5	P^2 divides $5^{p-1} - 1$	2, 20771, 40487, 53471161, 1645333507, 6692367337, 188748146801…
3.	Base 6	P^2 divides $6^{p-1} - 1$	66161, 534851, 3152573…
4.	Base 7	P^2 divides $7^{p-1} - 1$	5, 491531…
5.	Base 10	P^2 divides $10^{p-1} - 1$	3, 487, 56598313…
6.	Base 11	P^2 divides $11^{p-1} - 1$	71…
7.	Base 12	P^2 divides $12^{p-1} - 1$	2693, 123653…
8.	Base 13	P^2 divides $13^{p-1} - 1$	863, 1747591…
9.	Base 17	P^2 divides $17^{p-1} - 1$	3, 46021, 48947…
10.	Base 19	P^2 divides $19^{p-1} - 1$	3, 7, 13, 43, 137, 63061489…

Some special prime number additions: -

The sum of following five prime numbers again gives a prime number.

$$5 + 7 + 11 + 13 + 17 = 53$$
$$7 + 11 + 13 + 17 + 19 = 67$$
$$11 + 13 + 17 + 19 + 23 = 83$$
$$13 + 17 + 19 + 23 + 29 = 101$$

The sum of following seven prime numbers again gives a prime number.

$$17 + 19 + 23 + 29 + 31 + 37 + 41 = 197$$
$$19 + 23 + 29 + 31 + 37 + 41 + 43 = 223$$
$$23 + 29 + 31 + 37 + 41 + 43 + 47 = 251$$

$$29 + 31 + 37 + 41 + 43 + 47 + 53 = 281$$
$$31 + 37 + 41 + 43 + 47 + 53 + 59 = 311$$

Non commutative property: -
  Division is non commutative. For example $2/5 \neq 5/2$

Division with odd and even numbers: -

Division	Odd number	Even number
Odd number	odd number	even number
Even number	fraction	odd or even number

From the table it can be seen that 10/2=5   16/4=4
From the tables it can be seen that the operations with odd number and even number for addition and subtraction is same. But the operation with odd number and even number for multiplication and division is not same. Division obeys uniquely.

Division with positive and negative numbers: -

Division	Positive number	Negative number
Positive number	+	-
Negative number	-	+

For example $-10/(-5) = 2$

Division Table

Division	1	2	3	4	5	6	7	8	9	10
1	1	2	3	4	5	6	7	8	9	10
2	1/2	1	3/2	2	5/2	3	7/2	4	9/2	5
3	1/3	2/3	1	4/3	5/3	2	7/3	8/3	3	10/3
4	1/4	2/4	3/4	1	5/4	6/4	7/4	2	9/4	10/4
5	1/5	2/5	3/5	4/5	1	6/5	7/5	8/5	9/5	2
6	1/6	2/6	3/6	4/6	5/6	1	7/6	8/6	9/6	10/6
7	1/7	2/7	3/7	4/7	5/7	6/7	1	8/7	9/7	10/7
8	1/8	2/8	3/8	4/8	5/8	6/8	7/8	1	9/8	10/8
9	1/9	2/9	3/9	4/9	5/9	6/9	7/9	8/9	1	10/9
10	1/10	2/10	3/10	4/10	5/10	6/10	7/10	8/10	9/10	1

Divisibility tests of numbers
  Divisibility by 2:- A number is divisible by 2, if and only if the last digit is an even number or zero. For example 220 is divisible by 2 but not 221.
  Divisibility by 3:-
    (a)If the summation of digits in a number is divisible by 3 or multiples of 3 mean that the number also will be divided by 3. For example 729.
    (b)Also the numbers formed by all the digits from 0 to 9 each once are divisible by 3. For example (9013245768/3) = 3004415256.
  Divisibility by 4:- When a number ends with its last two digits as zero or if the last two digits is divisible by 4 means the number is divisible by 4. For example 2748, 2000
  Divisibility by 5:- If a number ends with 5 or zero then it must be divided by 5 without any remainder. For example 6025.
  Divisibility by 6:- If a number is divisible by six means that the last digit will be even and the sum of the digits is divisible by 3. That is a number which is divisible by 2 and 3 is also divisible by their multiplication product. For example 726.
  Divisibility by 7:- Divide the digits of the number into three digit blocks, starting from the right. Compute the alternating three digit sum (the right most block receiving

a positive sign). Now the number is divisible by 7 if and only if the alternating sum is divisible by 7. For example 371469. Split the number into 371/469. Now 469-371=98 which is divisible by 7 and so the number 371469 is divisible by 7.

Divisibility by 8:- When a number's last three digits is divisible by 8 or if the last three digits in the right are '000' means the number is divisible by 8. For example 2432, 7176.

Divisibility by 9:- If the summation of digits in a number is divisible by 9 means that the number also will be divided by 9. For example 927.

Note: - If a number is divisible by 9 means then the number can also be divided by 3, but the numbers which are divisible by 3 may not be divided by 9. Because 3 is the factor of 9. From this we can arrive at the fact that a factor of a number x can divide all the numbers which are divisible by x or factors of x.

For example 18/3=6   18/9=2

21/3=7   21/9=fraction

Divisibility by 10:- If a number is divisible by 10 means that it will have 0 in the unit decimal place. For example 910.

Divisibility by 11:-

(a) When a number's difference between the converse digits or alternate digits become zero or 11 or multiples of 11 means that the number is divisible by 11. For example 4268 and the converse digits are 4, 6 and 2, 8. Now (4+6)-(2+8) =0
The number 4268 is divisible by 11.
Another example 87635064. Now (8+6+5+6)-(7+3+0+4) =25-14=11

(b) If two numbers have the same digits, but every digit which occurs in an even place in one number is in an odd place in the other, then the sum of the numbers is divisible by 11. For example the resultant of 710534 + 405317 is divisible by 11.

(c) Any number of the form ABC…ZZ…CBA is divisible by 11.
For example (12300321/11) = 1118211.

Note: - The largest possible number containing any nine of the ten digits including zero, which is divisible by 11 without any remainder is 987,652,431.

Divisibility by 12:- Check for divisibility by both 3 and 4
For example 4944 can be divided by both 3 and 4 and so it can be divided by 12 also.

Divisibility by 13:- Delete the last digit from the given number. Then subtract 9 times the deleted digit from the remaining number. If what is left is divisible by 13, then so is the original number. For example, 5928-deleting the last digit number we will get 592 and when this number is subtracted by 9 times 8 we get 520. Now the number 520 is divisible by 13 and so the original number is divisible by 13.

In the same way we can continue the divisibility test for higher numbers.

Divisibility by 14: - Check for divisibility by both 2 and 7.

Divisibility by 15: - Check for divisibility by both 3 and 5.

Divisibility by 16: - Check for divisibility by both 2 and 8. It is to be noted that if a number's last digit is divisible by 2 or if the last digit is 0 means the number is divisible by 2. Similarly if a number's last digit is divisible by 4 or if the last digit is 00 means the number is divisible by 4, if a number's last digit is divisible by 8 or if the last digit is 000 means the number is divisible by 8, if a number's last digit is divisible by 16 or if the last digit is 0000 means the number is divisible by 16 and so on.

For some of the bigger numbers and prime numbers the divisibility tests are as follows.

Divisibility by 17: - Take the last digit of the number and multiply it by 5. Subtract this 5th multiple from the original number leaving the last digit. If the resultant obtained after subtraction is divisible by 17, it means the original number is divisible by 17. For example 1105. The last digit of this number is 5 and its 5th multiple is 25. Subtract this from the original number then we get 110-25=85 and the resultant is divisible by 17 and so the original number is divisible by 17.

Divisibility by 18: -A number is divisible by 18 when it is divisible both by 2 and 9. In other words if the sum of the digits of an even number is divisible by 18 means then the number is divisible by 18 also.

Divisibility by 19:- A number is divisible by 19 if and only if the number of tens (not the tenth digit of the number but the total number of integral tens in the whole number) added to twice the number of units is divisible by 19.
For example 475. Here $[47 + (5 \times 2)] = 57$, the resultant 57 is divisible by 19 and so for 475. Another example 4704590 Here $[470459+0] = 470459$ is divisible by 19 and so the number 4704590 can be divided by 19.

Divisibility by 20: -A number is divisible by 20 when it ends in zero and the second last digit is even.

Divisibility by 21: -A number is divisible by 21 when it is divisible both by 3 and 7.

Divisibility by 24: -A number is divisible by 24 when it is divisible both by 3 and 8.

Divisibility by 25:- When a number ends with its last two digits as zero or if the last two digits are divisible by 25 means the number is divisible by 25. For example 6075.

Divisibility by 30: -A number is divisible by 30 when it is divisible by both 3 and 10.

Divisibility by 37:- A number is divisible by 37 if and only if the resultant of the three digit blocks, starting from right is divisible by 37. For example 47109362. Split the number into 47/109/362 and now 362+109+47=518 which is divisible by 37 and so the number 47109362 is divisible by 37.

If a number of the form abc is divisible by 37 means its cyclic numbers bca and cab are also divisible by 37. For example 259 is divisible by 37 and so for 592 and 925.

A 6-digit number in which the first 3 digits are consecutively-increasing numbers; and the last 3 digits, consecutively-decreasing numbers, is divisible by 37. For instance, the numbers 123987, 234765 and 567543 are all divisible by 37.

Divisibility by 40: - When a number is divisible by both 5 and 8, then the number is divisible by 40.

Divisibility by 50:- When a number ends with its last two digits as zero or 50 then the number is divisible by 50. For example 1550 is divisible by 50.

Divisibility by 80: - When a number is divisible by both 5 and 16, then the number is divisible by 80.

Divisibility by 101:- Break the number into two digit blocks starting from the right. Compute its alternating two digit sum starting from right. Then the sum is divisible by 101 if and if only the alternating sum is divisible by 101.
For example 46763. The number is split into 4/67/63 and the alternate two digit sum yields 63-67+4=0 and is divisible by 101 and so the number 46763 is divisible by 101.

Divisibility test for higher numbers: - To find the divisibility test of higher numbers, multiplication rule can be used for multiple divisors. For example, to check the divisibility of a number by 57, check whether the number is divisible by 19 and 3 since $57 = 19 \times 3$.

Squareful number: -

A number is said to be "Squareful" if it contains at least one square in its factorization. The squareful number sequence is 4, 8, 9, 12, 16, 18, 24, 25… It is to be noted that all square numbers are squareful numbers but all squareful numbers are not square numbers. For example 18 is a squareful number but not a square number.

Special division property of 2519 and 2520: -

The specialty of 2519 is

> when the number is divided by 1, the remainder is 0
> when the number is divided by 2, the remainder is 1
> when the number is divided by 3, the remainder is 2
> when the number is divided by 4, the remainder is 3
> when the number is divided by 5, the remainder is 4
> when the number is divided by 6, the remainder is 5
> when the number is divided by 7, the remainder is 6
> when the number is divided by 8, the remainder is 7
> when the number is divided by 9, the remainder is 8
> when the number is divided by 10, the remainder is 9

The number 2520 can be divided by 1, 2, 3, 4, 5, 6, 7, 8, 9 and 10 without any remainder.

Special divisions: -

The number 381,654,729 is a special number in which the number formed by the first two digits from left side onwards is divisible by 2, the number formed by the first three digits is divisible by 3 and so on until the entire nine digit number is divisible by 9. Likewise the number 3816547290 is the only base 10 number that uses each of the 10 digits exactly once with the leftmost k digits are evenly divisible by k.

That is 3 is evenly divisible by 1

> 38 is evenly divisible by 2
>
> 381 is evenly divisible by 3
>
> ……………………………
>
> 3816547290 is evenly divisible by 10.

The number 27,216 is divisible by 2, 7, 6, 27, 72, 21, 16, 162 and 216. All these numbers are in the given number.

The number 58 when divided by 3, 4, 5 and 6 gives a remainder 1, 2, 3 and 4 respectively.

The number 123654 has a special division property in which the number 123654 is divisible by 6

> 12345 is divisible by 5
>
> 1234 is divisible by 4
>
> 123 is divisible by 3
>
> and 12 is divisible by 2.

If the number of digits in a number is six and the number is divisible by 7, then the new number formed by shifting the unit digit to the extreme left is also divisible by 7. For example 251,734 is a six digit number and is divisible by 7. So the number 425,173 is also divisible by 7. That is 425,173÷7 = 60,739

The number 56 is a special number

$$(111+1)/(1+1) = 56$$
$$(222+2)/(2+2) = 56$$
$$(333+3)/(3+3) = 56$$
$$...................................$$
$$(999+9)/(9+9) = 56$$

Using the numbers from 1 to 9 only once, the following special divisions can be done as follows.

$$\frac{6729}{13458} = \frac{1}{2}, \frac{5832}{17496} = \frac{1}{3}, \frac{4392}{17568} = \frac{1}{4}$$

$$\frac{2769}{13845} = \frac{1}{5}, \frac{2943}{17658} = \frac{1}{6}, \frac{2394}{16758} = \frac{1}{7}$$

$$\frac{3187}{25496} = \frac{1}{8} \; and \; \frac{6381}{57429} = \frac{1}{9}$$

Using the numbers from 0 to 9 only once, the following special divisions can be done as follows. 6918 = 20754 / 3, and each digit is contained in the equation exactly once.

Division miracle with three digits number: -

Write any three digit number and write it twice. For example 123 is written twice 123123. Divide this number by 7 followed by 11 and 13. There will be no remainder and the resultant is the starter itself. 123123/7 = 17589; 17589/11 = 1599; 1599/13 = 123.

Also it is seen that the product of any two of the above numbers (7, 11 and 13) followed by the third number division also gives the same number started.
123123/91 = 1353; 1353/11 = 123
Also if we divide the number by 1001 we will get the same number started.
123123/ (7×11×13) = 123123/1001 = 123
The reverse is true. That is to get a three digit number twice multiply the number by 11 and the product by 91 or simply multiply by 1001. For example 666 and to write it twice 666 × 1001 = 666,666 (or)
666 × 11 = 7326 followed by 7326 × 91 = 666666.

Note: - The above rule is applicable for every number of the form ABCABC. Similar type of rule is applicable for every number of the form ABABAB divisible by 3, 7, 13 and 37. For example to get a number of the form ABABAB multiply any two digit number by 10101.

Since we are getting 10101 with 3×7×13×37, any two digit number AB multiplied by 10101 or with 3×7×13×37 will give the result of the form ABABAB.

Similarly we are getting 10101 with 259×39 and so any two digit number AB multiplied by 10101 or with 259×39 will give the result of the form ABABAB.

Further we are getting 100001 with 11×9091 and so any five digit number ABCDE multiplied by 100001 or with 11×9091 will give the result of the form ABCDEABCDE.

Abundant number or Excessive number or Supersaturated number: -

The composite number 'n' is abundant if the sum of all its positive divisors except itself is more than 'n'. These numbers have abundant factors above perfection and so they are called as abundant numbers. For example 12 whose factors are 1, 2, 3, 4, 6 and the sum of factors (16) is greater than the original number. So 12 is an abundant number. The abundant number series is 12, 18, 20, 24, 30, 36, 40, 42, 48, 54, 60, 100...

There are 2487 abundant numbers below 10,000.

945 is the smallest odd abundant number.

Perfect number or Saturated number: -

The composite number 'n' is called as saturated number if the sum of all its positive divisors except itself is equal to 'n'. For example 28 whose factors are 1, 2, 4, 7, 14 and the sum of factors (28) is equal to the original number. So 28 is a saturated number. The saturated number is also known as perfect number.

So a perfect number is defined as the number whose summations of the factors or divisors (excluding the number itself) gives the same number or it is the number whose summations of the factors including the number itself is twice the original number. For example 6 and the summation of the factors of this number 1+2+3=6.

28 and the summation of the factors of this number 1+2+4+7+14=28

496 and the summation of the factors of this number

1+2+4+8+16+31+62+124+248 = 496

8128 and the summation of the factors of this number

1+2+4+8+16+32+64+127+254+508+1016+2032+4064 = 8128

33,550,336 and the summation of the factors of this number 1 + 2 + 4 + 8 + 16 + 32 + 64 + 128 + 262112 + 256 + 131056 + 512 + 65528 + 1024 + 32764 + 2048 + 4096 + 8191 + 16382 + 524224 + 1048448 + 2096896 + 4193792 + 8387584 + 16775168 = 33550336.

Other examples are 8589869056, 137438691328, 2305843008139952128, 2658455991569831744654692615953842176 and

191561942608236107294793378084303638130997321548169216.

If $(2^k - 1)$ is a prime number, then $(2^{k-1})(2^k - 1)$ is a perfect number. For example if k=3, then $2^k - 1$ is 7 and we get the perfect number 28.

Note: - For 'n' digit number only one perfect number is present.

There are only about a dozen perfect numbers up to 10^{160}.

All the perfect numbers are even numbers.

Further if the even Niven number contains all its digits as even, then it is a super even harshad number. For example 48 is a super even Niven number and 45 is not a super even Niven number.

All the perfect numbers are triangular numbers.

All the perfect numbers are hexagonal numbers.

All even perfect numbers end in 6 or 8.

Like the prime numbers, perfect numbers are randomly present and are infinite.

Perfect numbers are between deficient and abundant numbers.

One interesting property is that the sum of the reciprocals of all the divisors (including the number itself) of a perfect number is 2. For example, take the case of the perfect number 28:

$$\frac{1}{1} + \frac{1}{2} + \frac{1}{4} + \frac{1}{7} + \frac{1}{14} + \frac{1}{28} = 2$$

The perfect number can be written in binary with (n) ones and (n-1) zeros

$$6_{10}=110_2$$
$$28_{10}=11100_2$$
$$496_{10}=111110000_2$$
$$8128_{10}=1111111000000_2$$
$$33550336_{10}=1111111111111000000000000_2$$
$$8589869056_{10}=1111111111111111110000000000000000_2$$
$$137438691328_{10} = 11111111111111111111000000000000000000_2$$

………………………………………………………………………..

Deficient number or Unsaturated number: -

The composite number 'n' is deficient if the sum of all its positive divisors except itself is less than 'n'. For example 4, 8, 9, 10, 14, 15, 16, 21, 22, 25, 26, 27, 32, 33, 34, 35, 38, 39, 44, 45, 46, 49, 50, 128… are deficient numbers. There are 7508 deficient numbers below 10,000.

The deficient number which is odd is called as odd deficient number and the deficient number which is even is called as even deficient number. For example 9 is an odd deficient number and 10 is an even deficient number.

Further if the odd deficient number contains all of its digits as odd, then it is a super odd deficient number and if the even deficient number contains all of its digits as even, then it is a super even deficient number. For example 39 is a super odd deficient number and 46 is a super even deficient number.

The deficient number which is prime is called as prime deficient number and the deficient number which is composite is called as composite deficient number. For example 19 is prime deficient number and 21 is composite deficient number. It is to be noted that all prime numbers are deficient numbers but the reverse is not true.

The deficient number which is palindromic in nature is called as palindromic deficient number. For example 22 is a palindromic deficient number.

It is to be noted that all the abundant numbers, saturated numbers and deficient numbers are composite numbers.

Vampire number or Vampire Friedman number: -

The number 'n' is called a vampire number if there exists a factorization of n, using n's digits. For example 126, which can be factorized as 21×6=126 and the factors contain the digits as that of original number. The vampire number sequence 153, 688, 1206, 1255, 1260, 1395, 1435, 1503, 1530, 1827…

There are 15 vampire numbers below 10,000. The vampire numbers are special type of Friedman numbers which involve only multiplication. So all the vampire numbers are Friedman numbers but the reverse is not true.

A prime vampire number, is a vampire number whose factors are its prime factors. The sequence of prime vampire numbers is 117067, 124483, 146137, 371893, 536539…

Niven number: -

A Niven number is any whole number that is divisible by the sum of its digits. That is the number digit of the number is a divisor of the number. For example 126 is a Niven number because its sum of the digits 9 (1+2+6) is a divisor of the number 126. The sequence of Niven number is 12, 18, 20, 21, 24, 27, 30, 36, 40, 42, 45, 48, 50…

The Niven number which is odd is called as odd Niven number and the Niven number which is even is called as even Niven number. For example 24 is an even Niven number and 21 is an odd Niven number.

Further if the even Niven number contains all its digits as even, then it is a super even harshad number. For example 48 is a super even Niven number and 45 is not a super even Niven number.

The square numbers in the Niven number series are called as square Niven numbers. For example 36 is a square Niven number.

Harshad number: -

A harshad number is defined as the number whose summation of the digits is one of the divisors of the number or simply it is a number which is divisible by the sum of its own digits. For example 1458 whose summation of the digits is 1+4+5+8=18 and 18 is a divisor of the number 1458.

Harshad number series is 12, 20, 24, 27, 30, 36, 40, 42, 45, 48, 50…

Note: - All the n! is a harshad number except for n = 432.

The harshad number which is odd is called as odd harshad number and the harshad number which is even is called as even harshad number. For example 45 is an odd harshad number and 48 is an even harshad number.

Further if the even harshad numbers contains all its digits as even, then it is a super even harshad number. For example 42 is a super even harshad number and 36 is not a super even harshad number.

BODMAS: -

Suppose if the sum involves several operations the order of preference to do the operations is given by the Rule of BODMAS. The expansion is Bracket, Of, Division, Multiplication, Addition and Subtraction. (Note: - Of means multiplication)

For example 75% of 220×5(7+3)/10-100=75% of 220×5(10)/10-100

$$= (75/100) \times 220 \times 5 \times (10/10) - 100$$
$$= 165 \times 5 \times (10/10) - 100$$
$$= 165 \times 5 - 100$$
$$= 825 - 100$$
$$= 725$$

Specialty of 13:-

1, 3, 4, 9, 10 and 12 when divided by 13, the resultant form a group of numbers.

1/13 = 0.076923	076923
3/13 = 0.230769	230769
4/13 = 0.307692	307692
9/13 = 0.692307	692307
10/13=0.769230	760230
12/13=0.923076	923076

In all the above six fractions 0, 7, 6, 9, 2 and 3 have come in the same order. In the above resultant numbers after 6 nine has come and behind six is seven in a cyclic manner.

Another set of fractions 2/13 = 0.153846   153846
5/13 = 0.384615   384615
6/13 = 0.461538   461538

$$7/13 = 0.538461 \quad 538461$$
$$8/13 = 0.615384 \quad 615384$$
$$11/13 = 0.846153 \quad 846153$$

In these fractions 1, 5, 3, 8, 4 and 6 have come in the same order. In all the above resultant numbers after 3, eight has come and behind 3 is five in a cyclic manner.

On multiplying the number 76923 by successive multiples of 13, a curious numerical pattern is obtained as follows.

$$76923 \times 13 = 0999999$$
$$76923 \times 26 = 1999998$$
$$76923 \times 39 = 2999997$$

$$76923 \times 52 = 3999996$$
$$76923 \times 65 = 4999995$$
$$76923 \times 78 = 5999994$$

$$76923 \times 91 = 6999993$$
$$76923 \times 104 = 7999992$$
$$76923 \times 117 = 8999991$$

$$76923 \times 130 = 9999990$$
$$76923 \times 143 = 10999989$$
$$76923 \times 156 = 11999988$$

………………………...

Palindromic equalities involving 13:

13 x 62 = 26 x 31

13 x 93 = 39 x 31

13 x 13 = 169 and 961 = 31 x 31

Rules of fractions:-

(a) $(a\pm b)/c = (a/c) \pm (b/c)$

For example consider a=6; b=9 and c=3 then $(6+9)/3 = (6/3) + (9/3)$
$$15/3 = 2 + 3$$
$$5 = 5$$

(b) $a/(b+c) \neq (a/b) + (a/c)$

For example consider a=6; b=9 and c=3 then $6/(9+3) \neq (6/9) + (6/3)$
$$6/12 \neq 2/3 + 2$$
$$½ \neq 8/3$$

(c) If $a/b = c/d$ then $(a+b)/b = (c+d)/d$

For example consider a=2; b=3; c=4 and d=6 then $2/3 = 4/6$
$$\text{also } (2+3)/3 = (4+6)/6$$
$$5/3 = 10/6$$

(d) If $a/b = c/d$ then $(a-b)/b = (c-d)/d$

For example consider a=2; b=3; c=4 and d=6 then $2/3 = 4/6$
$$\text{also } (2-3)/3 = (4-6)/6$$
$$(-1)/3 = (-2)/6$$

(e) If $a/b = c/d$ then $(a+b)/(c+d) = b/d$

For example consider a=2; b=3; c=4 and d=6 then $2/3 = 4/6$
$$\text{also } (2+3)/(4+6) = 3/6$$
$$5/10 = ½$$

(f) If a is less than b then a/b is less than 1.

For example consider a=2 and b=3 then 2/3 is less than 1.

(g) (b-c)/a = -(c-b)/a

For example consider a=2; b=3 and c=4

(3-4)/2 = -(4-3)/2

(-1)/2 = (-1)/2

(h) a/(b-c) = -a(c-b)

For example consider a=2; b=3 and c=4

2/(3-4) = -2(4-3)

   -2 = -2

(i) (a±b)/ (c±d) = (b±a)/(d±c)

For example consider a=2; b=3; c=4 and d=5

(2+3)/(4+5) = (3+2)/(5+4)

         5/9 = 5/9

(j) (a±b)/(c±d) = (a – or + b)/(c – or + d)

For example consider a=2; b=3; c=4 and d=5

(2-3) / (4-5) = (3-2) / (5-4)

   (-1)/(-1) = 1/1

          1 = 1

Note: - In all the above examples a, b, c and d are real numbers.

(k) Multiplication of the numerator and denominator by the same quantity will not change the value of the fraction.

For example 2/5 when multiplied by 3 both the numerator and denominator will give 6/15 but the value remains the same.

(l) If the fraction is proper fraction means, the product obtained by dividing the square of the numerator with the square of the denominator will be a square number. For example in the above case 2/5 can be written as 4/25 which will yield 0.25 and is a square number. The same rule can be extended for cube, square root and cube root. That is the product obtained by dividing the cube of the numerator with the cube of the denominator will be a cube. For example 8/125 will give 0.064 and is a cube number.

But the rule cannot be applied for improper fractions.

(m) The value of the fraction increases with the increase of numerator and decreases with the increase of denominator.

For example 2/5= 0.4; 3/5=0.6; But 2/6=0.33

(n) To find the smaller fraction cross multiply the fraction

For example 2/3, 5/9, ½, 11/18

(18)2/3, 5/9(15)   5/9 is smaller than 2/3

(18)1/2, 11/18(22)   ½ is smaller than 11/18

Now   (10)5/9, ½ (9)   ½ is smallest among these fractions.

(o) Even numbers of fractions alone obey the cross multiplication rule.
For example ½ = 2/4 = 4/8 = 8/16. These fractions are collectively called as equivalent fractions.

$$1×4×4×16 = 2×2×8×8$$

$$256 = 256$$

Odd numbers of fractions do not obey the cross multiplication rule.

For example ½ = 2/4 = 4/8

$$1×4×4 = 2×2×8$$

$$16 \neq 32$$

(p) Fractions are neither even nor odd numbers.

Mods:-

If a system comes repetitively like 7 days a week, 24 hours per day then they all obey mod system. For example if a system comes repetitively after every 7 times then 8 can be indicated as follows. 8=1(mod 7)

If a=b(mod m) and c=d(mod m) then a+c = b+d(mod m)

For example 8=1(mod 7) and 9=2(mod 7) then 8+9 = 1+2(mod 7)

That is 17 = 3(mod 7)

If a=b (mod m) and c=d (mod m) then a-c = b-d (mod m) provided that a>c

For example 9=2(mod 7) and 8=1(mod 7) then 9-8 = 2-1(mod 7)

That is 1 = 1(mod 7)

If a=b (mod m) and c=d (mod m) then a×c = b×d (mod m)

For example 8=1(mod 7) and 9=2(mod 7) then 8×9 = 1×2(mod 7)

72 = 2(mod 7)

If a=b (mod m) and c=d (mod m) then a/c = b/d (mod m)

For example 8=1(mod 7) and 16=2(mod 7) then 8/16 = 1/2(mod 7)

1/2 = 1/2(mod 7)

Ratio: -

Ratio is the comparison of the two things in terms of fractions. For example if x and y earn Rs.2000/- and Rs.3000/- respectively then y is earning 3/2 times of x. This can be represented as a ratio and the symbol is ":" pronounced as "is to". The above relationship is represented in terms of ratio as x: y :: 2:3 Here the first term, 2 is called as precedent or antecedent and the second term, 3 is called as consequent.

It is to be noted that the product of terminal numbers in the ratio will be equal to the product of middle numbers. This is called as the cross multiplication rule of ratios.

For example 2:3:: 2000:3000

2×3000 = 3×2000

6000 = 6000

The cross multiplication rule of ratios is obeyed by even number of ratios only.

For example 2:3:: 4:6:: 8:12:: 16:24

2×6×8×24 = 3×4×12×16

2304 = 2304

For odd number of ratios the cross multiplication rule is not obeyed.

For example 2:3:: 4:6:: 8:12

2×6×8 ≠ 3×4×12

96 ≠ 144

Sum of the numbers in a ratio must be a factor of the measurement. For example if a carton box containing a dozen mirror is dropped the ratio of broken mirrors to unbroken mirrors may be 2:1, 3:1, 5:1 and 7:5 but not 3:2 because in the former cases the sum of the ratios 2+1, 3+1, 5+1 and 7+5 are factors of 12. But the sum of 3+2 is not a factor of 12.

The difference between ratios and fractions is that fractions represent part per whole and ratios represent part per part, that is ¼ means out of 4 (whole part) 1 part is represented but 1:4 means out of 5 (whole part) 1 part is related to 4 parts.

Proportionality: -

Proportionality is also a type of ratio which involves comparison between two things. Otherwise proportionality is a uniform ratio where if one property increases or decreases with respect to another property is known as proportionality.

There are two types of proportionality namely direct proportionality and inverse proportionality.

Direct proportionality: - If one property increases with respect to other property or decreases with other property, it is known as direct proportionality.

For example working days and wages.

Inverse proportionality: - If one property increases while other property decreases or vice versa, it is known as inverse proportionality.

For example number of workers and time of completion of work.

In direct proportion the multiplication factor will be inversed while in the inverse proportion it will come as such.

For example if 'x' is getting Rs.100/- per day means what he will get for 15 days? This is a direct proportion $(15/1) \times 100 = 1500$/-

If 10 people can complete a construction work in 15 days means, in how many days 5 people will be able to complete the work? This is an inverse proportion.

$(10/5) \times 15 = 30$. The work will be completed in 30 days.

Percentage: -

Percentage is also a type of comparison of two things with respect to 100 or simply it is a ratio to 100. This is a direct proportion. This is denoted by '%'. Percentage can also be called as parts per hundred.

For example if the price of an edible oil is increased to Rs.200/- from Rs.150/- then what is the price increase in terms of percentage can be calculated as follows.

For 150 the increase is 200 then for 100(percentage) what is the increase

$(100/150) \times 200 = (400/3) = 133.3\%$

To convert fractions or decimals to percentage multiply by 100.

For example 4/3 can be represented in terms of percentage as follows.

$(4/3) \times 100 = 133.3\%$

To convert percentage into fractions divide it by 100.

For example 25% can be represented in terms of fraction as follows.

$(25/100) = 0.25$

If A is R% more than B, then B is less than A by $[\{R/(100+R)\} \times 100]$ %. Similarly if A is R% less than B, then B is more than A by $[\{R/(100-R)\} \times 100]\%$

For example if A is getting 200% more than B, then B is less than A by 16.6%

If A is getting 20% less than B, then B is more than A by 25%

Like percentages, per thousands (parts per thousand or ppt), per millions (parts per million or ppm), per billion (parts per billion) and so on can be expressed. The rules applicable for percentages are applicable for per thousands, per millions, per billions and so on.

It is quite obvious that to convert percentages into ppm multiply by 10,000.

Arithmetic mean: -

It is the resultant obtained by the division of the summation of the numbers divided by the number of things. It is also called as arithmetic average. For example the arithmetic mean of 4, 6 and 9 is $[(4+6+9)/3] = 6.3$

There is only one arithmetic mean or average for a list of data available.

The arithmetic mean can be added, subtracted, multiplied and divided.

Amicable numbers: -

Two distinct positive integers (m, n) are said to be amicable pair when each integer is the sum of the proper divisors of the other. Proper divisors mean all divisors of a number except the number itself. The proper divisors of 220 are 1, 2, 4, 5, 10, 11, 20, 22, 44, 55 and 110. Their sum is 284. The proper divisors of 284 are 1, 2, 4, 71 and 142. Their sum is 220. Therefore (220, 284) is a pair of amicable numbers. Some other examples of amicable pairs are (1184, 1210), (2620, 2924), (5020, 5564), (6232, 6368), (17296, 18416) and (9363584, 9437056).

There exist pairs of odd amicable numbers and even amicable numbers, like (12,285; 14,595) and (2620; 2924), but it is not known whether there exists any pair with one of the numbers odd and one even.

One of the method to find out amicable number pair is if 'n' is a positive integer such that $3 \times 2^n - 1$, $3 \times 2^{n+1} - 1$, and $3^2 \times 2^{2n+1} - 1$ are all prime, then $2^{n+1}(3 \times 2^n - 1)(3 \times 2^{n+1} - 1)$ and $2^{n+1}(3^2 \times 2^{2n+1} - 1)$ form an amicable pair. For example if n=1 we get 220 and 284 which are amicable numbers.

Three distinct positive integers (m, n and p) are said to be amicable triples when each integer is the sum of the proper divisors of the other. For example (1980, 2016, 2556), (9180, 9504, 11556), (21668, 22200, 27312) and (103340640, 123228768, 124015008) are some of the amicable triples.

Demlo number: -

Consider the number 136653. If you add the first two and last two digits that is (13+53) will give 66 which is the digits in between the numbers. This type of phenomenon is called as demlo number. It is to be noted that Demlo number is not possible in the case of single digit and double digit numbers. Obviously the Demlo numbers start from three digit number onwards only.

The demlo number series is 121, 253, 473.....
A demlo number has three parts. Addition of the first and last parts gives the digit or digits of the middle part.

A demlo number which is palindromic in nature is called as palindromic demlo number. For example 121 is a palindromic demlo number.

A demlo number which is odd is called as odd demlo number and the demlo number which is even is called as even demlo number. For example 473 is an odd demlo number and 462 is an even demlo number.

A test for demlo number is it is divisible by 11. So it is obvious that all demlo numbers are composite numbers. For example in the above case (136653/11) = 12423.

Demlo number is obtained by adding a two digit number two times, three digit number four times, four digit number by six times and five digit number eight times as follows.

For example for 42 the demlo number is 42 for 234 it is 234 for 1012 it is 1012

42	234	1012
462	234	1012
	234	1012
	259974	1012
		1012
		112444332

for 20150 (five digit number) the demlo number is 20150

20150
20150
20150
20150
20150
20150
<u>20150</u>
<u>223888886650</u>

Likewise we can proceed further for six digit number ten times and so on. For example the Demlo number for the six digit number 123456 is 123456

123456
123456
123456
123456
123456
123456
123456
123456
<u>123456</u>
<u>137173333319616</u>

Now in all the above cases the additions of the first and last parts give the digits of the middle part.

The Angel number – 421: -

Take any real number. If it is an even number, divide it by 2. If it is odd number, multiply it by 3 and add 1 to it. If we repeat the process the repetition will end up with a loop of 4, 2 and 1. The number formed by the loop value 421 is called as angel number.

For example 7. It is an odd number, multiply it by 3 and add 1 to it $(7 \times 3)+1=22$

22 is an even number, divide it by 2. $(22/2)=11$

11 is an odd number, multiply it by 3 and add 1 to it $(11 \times 3)+1=34$

34 is an even number, divide it by 2. $(34/2)=17$

17 is an odd number, multiply it by 3 and add 1 to it $(17 \times 3)+1=52$

52 is an even number, divide it by 2. $(52/2)=26$

26 is an even number, divide it by 2. $(26/2)=13$

13 is an odd number, multiply it by 3 and add 1 to it $(13 \times 3)+1=40$

40 is an even number, divide it by 2. $(40/2)=20$

20 is an even number, divide it by 2. $(20/2)=10$

10 is an even number, divide it by 2. $(10/2)=5$

5 is an odd number, multiply it by 3 and add 1 to it $(5 \times 3)+1=16$

16 is an even number, divide it by 2. $(16/2)=8$

8 is an even number, divide it by 2. $(8/2)=4$

4 is an even number, divide it by 2. $(4/2)=2$

2 is an even number, divide it by 2. $(2/2)=1$

7. MIRROR IMAGE AND PALINDROME NUMBERS

When a number is written in the reverse order, the new number obtained is called as mirror image number or reflex number. For example if the number 176 is placed before a mirror the image obtained is 671 and is its reflux number.

Palindrome numbers: -

When a number written in the reverse order is equal to the original number itself then it is called as a palindrome number or symmetrical number. It obeys the reflux property or simply it is equal to the mirror image of the original number. For example 101 when written in the reverse order is 101 and it is a palindrome number.

The palindrome numbers sequence is 11, 22, 33, 44, 55, 66, 77, 88, 99, 101, 111, 121, 131, 141, 151, 161, 171, 181, 191…

It is to be noted that all palindrome numbers are mirror image numbers but the reverse is not true. That is all palindrome numbers are mirror image numbers but all mirror image numbers are not palindrome numbers.

There are 198 palindromic numbers below 10,000.

The palindromic even numbers are the even numbers which are palindromic in nature. For example 242, 414, 662266, 808… are palindromic even numbers. Further if the palindromic even number contains only even digits then it is a super even palindromic number. For example 242 is a super even palindromic number and 414 is not a super even palindromic number.

The palindromic odd numbers are the odd numbers which are palindromic in nature. For example 1441, 121, 303, 525, 747, 99199… are palindromic odd numbers. Further if the palindromic even number contains only odd digits then it is a super odd palindromic number. For example 99199 is a super odd palindromic number and 747 is not a super odd palindromic number.

The palindromic primes are 11, 101, 131, 151…

The palindromic square numbers are 121, 484, 676, 10201, 12321…

The alternating number which is palindromic in nature is called as palindromic alternating number. For example 545 is a palindromic alternating number.

The palindromic composite numbers are the composite numbers which are palindromic in nature. For example 22, 33, 44, 55, 66, 77… are palindromic composite numbers.

Palindromic pronic number is the pronic number which is palindromic in nature. For example 272, 6006, 289982… are palindromic pronic numbers.

The property of palindrome numbers or symmetry numbers is possible in other number systems also. For example $33_{10} = 100001_2$ and $99_{10} = 1100011_2$. This is can be called as equivalency in palindrome numbers.

All numbers in base 10 with one digit are palindromic. So by including zero we have 10 single digit palindrome numbers.

The number of palindromic numbers with two digits is nine. They are 11, 22, 33, 44, 55, 66, 77, 88 and 99.

There are 90 palindromic numbers with three digits. (Using the rule of product: 9 choices for the first digit – which determines the third digit as well – multiplied by 10 choices for the second digit). That is {101, 111, 121, 131, 141, 151, 161, 171, 181, 191… 909, 919, 929, 939, 949, 959, 969, 979, 989, and 999}

90 palindrome numbers with four digits: (Again, 9 choices for the first digit multiplied by ten choices for the second digit. The other two digits are determined by

the choice of the first two). That is {1001, 1111, 1221, 1331, 1441, 1551, 1661, 1771, 1881, 1991... 9009, 9119, 9229, 9339, 9449, 9559, 9669, 9779, 9889 and 9999}

There are 199 palindrome numbers below 10^4. Below 10^5 there are 1099 palindromic numbers and for other exponents of 10^n where n>5 we have the total number of palindrome numbers as 1999, 10999, 19999, 109999, 199999, 1099999... respectively.

The number of palindromic numbers formed with the number 7537252 by changing its digits is 6. There are two twos, two fives two sevens and only one 3. So 3 must be in the middle of the palindromic numbers. 2, 5 and 7 can be arranged in 3! ways (Note: -The symbol '!' represents factorial). So we get a total of six palindromic numbers 7523257, 2753572, 5723275, 7253527, 2573752 and 5273725.

The palindromic numbers are of two types, one containing odd number of digits like 171, 82828... and the other containing even number of digits like 99, 8448...

Binary palindrome numbers: -

Like in the decimal system, palindromic numbers can be considered in other numbering systems also. For example, the binary palindromic numbers are: 11, 101, 111, 1001, 1111, 10001, 10101, 11011, 11111, 100001 ...

It is to be noted that all the binary palindromic numbers are odd numbers since the leading zeros are not counted as part of a palindrome. This also means that any number in a base copalindromic with binary number system should be an odd number.

Classification of palindrome numbers

Numbers

Mirror image
122 and 221

Palindrome
131 and 131

Odd palindrome
121 and 121

Even palindrome
212 and 212

Even repdigit
2222 and 2222

Odd repunit
1111 and 1111

Odd repdigit
33 and 33

Composite even
repdigit

Prime odd
repunit Eg. 11

Composite odd
repunit Eg. 111

Composite odd
repdigit Eg. 77

It is to be noted that there is no even repunit number possible, since all repunit numbers are odd numbers. Also there is no prime even repdigit number, as all even repdigit numbers are composite in nature. Further there is no prime odd repdigit number as all repdigit numbers are composite numbers.

Specialty of 1089: -

The multiplication of 1089 with 9 is the reverse of the number that is a palindrome number and similarly the multiplication of 10989 with 9 is the reverse of the number and this can be extended to 109989, 1099989…

That is 1089×9=9801

   10989×9=98901

  109989×9=989901

  1099989×9=9899901

....

Also it is to be noted that 1089 when divided by 9 is 121 and is a palindrome number. This can be extended to 10989, 109989, 1099989… which will give 1221, 12221, 122221… respectively.

The number 1089 gives palindrome number as follows

   1089×1=1089 and its palindrome number 9801=1089×9

   1089×2=2178 and its palindrome number 8712=1089×8

   1089×3=3267 and its palindrome number 7623=1089×7

   1089×4=4356 and its palindrome number 6534=1089×6

   1089×5=5445 and 5445 is a palindrome number.

It is to be noted that the first digit of the resultant is equal to the multiplier itself. Further in the resultants, the first digits run from 1 to 9, the second digits run from 0 to 8, third digits run from 8 to 0 and the last digits run from 9 to 1.

The multiplication of 1089 and its palindrome is a square number. For example $1089×9801=10673289=3267^2$. Another example is $169 × 961 = 403^2$.

The smallest number that is reversed on multiplication is 1089 followed by the second smallest number that is reversed on multiplication is 2178. That is 1089×9=9801 and 2178×4=8712.

Further the fun triangle made with 2178 by multiplication with 4 is as follows

That is 2178×4=8712

   21978×4=87912

  219978×4=879912

  2199978×4=87999912

....

Some characteristic features of palindrome numbers: -

The sum of the digits of the palindrome number is equal to the sum of the digits in the original number. For example 1551 and its sum of the digits that the digital root is 3. The palindrome of the above number is again 1551 and its digital root is 3.

A test for palindrome number is if we split the number into two parts the summation of the digits of the left side is equal to the summation of the digits of the right hand side. For example take 251152. If we split into 251 and 152, the summation of the digits of the left and right hand side is 8. Another example is 123454321 and if we split it into two parts 1234/5/4321, the summation of the digits of the left and right hand side is 10.

Any number when written in the reverse order and the reverse number or reflux number is written adjacent to the original number then the resultant obtained will be a palindrome number. For example 786 and its reflux number is 687. When the reverse number is written adjacent to the original number then we get 786687 and this is a palindrome number.

Any n^{th} digit number when subtracted from its reflux number gives the resultant equal to 9 or multiples of 9. For example 3142 and its reflux number is 2413. If we subtract the smaller number from the bigger number that is 3142 – 2143 = 729. Now the resultant is the divisor of 9.

The numbers 11, 111, 1111… and the multiples of these numbers from 1 to 9 give perfect palindrome numbers. For example 111 × 3 = 333. Now $/\dfrac{333}{333}/$ where /---/ represents a mirror and 333 if written in the reverse order is 333.

The total number of palindrome numbers will be more in pental system when compared with decimal system. The total number of palindrome numbers will be highest in binary system as the number of repeating digits are 0 and 1 only.

All the Repunit numbers are palindrome numbers but the converse is not true. Similarly all the palindromic numbers are undulating numbers but the reverse is not true.

The number containing 'n' repunits separated by (n-2) zeros between the repunits is divisible by the number formed by 'n' repunits. For example 1001001001 can be divided by 1111 and 20202 can be divided by 222.

The palindromic number 3113 when split into parts, the left side part is mirror image of the right side part. When left side part is added to right side part will give a repunit number. That is 31+13=44. Other examples are 3223, 1001, 1111, 1221, 1331, 1441, 1551, 1661, 1771, 1881… Since we get a repunit number from these palindromic numbers, these numbers are also called as prorepunit numbers and prorepdigit number.

Palindrome numbers are classified based upon the property of natural number which is having the palindromicity property. That is palindrome number can be classified into palindromic odd number, palindromic even number, palindromic prime number, palindromic composite number palindromic square number and so on.

Formation of palindrome numbers: -
The product of the following multiplications is same whether read from right or left.

(a) 6x7x6 = 252
(b) 21978×4=87912
(c) 139 × 109 = 15151

(c) 836 × 836 = 698896
(d) 152207 × 73 = 11,111,111
(e) 14287143 ×7 = 100,010,001

(f) 142857143 ×7 = 1,000,000,001
(g) 27994681 × 441 = 12345654321
(h) 333333666667 ×33 = 11000011000011
(i) 11011011 × 91 = 1,002,002,001

The multiplication of 12345679 with multiples of 9 will give the palindrome numbers as follows.

(a) 12345679× 9 =111111111
(b) 12345679×18=222222222

(c) 12345679×27=333333333

(d) 12345679×36=444444444

(e) 12345679×45=555555555

(f) 12345679×54=666666666

(g) 12345679×63=777777777

(h) 12345679×72=888888888

(i) 12345679×81=999999999

The resultant of the following additions will give the palindrome number. It is to be noted that the addendum and adduct of the following additions are mirror image to each other.

(a) 13+31 = 44

(b) 47+74 = 121

(c) 605+506 = 1111

(d) 1353+3531 = 4884

Products of the following consecutive numbers give palindrome numbers.

(a) 16×17=272

(b) 77x78 = 6006

(c) 77x78x79 = 474474

(d) 538×539=289982

(e) 1621×1622=2629262

(f) 2457×2458=6039306

Lychrel number: -

A Lychrel number is a natural number which cannot form a palindrome through the iterative process of repeatedly reversing its base 10 digits and adding the resulting numbers. The repeated process of doing this iterative process lead to palindrome number and the number of iterations depend upon the number. For example 56 becomes palindromic after one iteration: 56+65 = 121.

57 becomes palindromic after two iterations: 57+75 = 132 and 132+231 = 363.

59 becomes a palindrome after 3 iterations: 59+95=154, 154+451=605 and 605+506=1111.

89 takes an unusually large 24 iterations to reach the palindrome 8813200023188 and so on.

The first known number starting from 0 that does not apparently form a palindrome is a three digit number that is 196 and is the smallest Lychrel number. The Lychrel number sequence is 196, 295, 394, 493, 592, 689, 691, 788, 790, 879, 887, 978, 986, 1495, 1497, 1585, 1587, 1675, 1677, 1765, 1767, 1855, 1857, 1945, 1947, 1997…

Palindrome special of zero and eight: -

The number 8 and its derivatives like 88, 888… are also palindrome numbers and these numbers have special characteristics in that they remain unchanged when they are written upside down. A tetradic (or four-way) number is a number that remains unchanged when flipped back to front, mirrored up-down, or flipped up-down. Since the only numbers that remain unchanged which turned up-side-down or mirrored are 0, 1, and 8 (here, the numerals 1 and 8 are assumed to be written as a single stroke and symmetrical pair of loops, respectively), a tetradic number is precisely a palindromic number containing only 0, 1, and 8 as digits. The first few are therefore 1, 8, 11, 88, 101, 111, 181, 808, 818...

The tetradic numbers are of two types, one containing odd number of digits like 101, 80808… and the other containing even number of digits like 88, 8008…

It is to be noted that all tetradic numbers are palindrome numbers but the reverse is not true.

Fun with palindrome numbers: -

The number 696 (n) is a palindrome number and its n (n+8), that is 489984 is also a palindrome number.

Both 2009 and its palindrome number 9002 are multiples of 7.

The resultant formed from the addition of the following numbers is palindrome of the resultant formed from the multiplication. This is called as Weird property.

$$9 + 9 = 18 \text{ and its palindrome number } 81 = 9 \times 9$$

$$24 + 3 = 27 \text{ and its palindrome number } 72 = 24 \times 3$$

Weird triangle

$$47 + 2 = 49 \text{ and reversely } 94 = 47 \times 2$$

$$497 + 2 = 499 \text{ and reversely } 994 = 497 \times 2$$

$$4997 + 2 = 4999 \text{ and reversely } 9994 = 4997 \times 2$$

………………………………………………………

The numbers 46 and 96 are peculiar; their product does not change if the digits are interchanged. That is $46 \times 96 = 64 \times 69 = 4416$. This property is called as palindromic equality. In a similar manner, we have only the following thirteen equations.

(1) $12 \times 42 = 21 \times 24 = 504$

(2) $12 \times 63 = 21 \times 36 = 756$

(3) $12 \times 84 = 21 \times 48 = 1008$

(4) $13 \times 62 = 31 \times 26 = 806$

(5) $13 \times 93 = 31 \times 39 = 1209$

(6) $14 \times 82 = 41 \times 28 = 1148$

(7) $23 \times 64 = 32 \times 46 = 1472$

(8) $23 \times 96 = 32 \times 69 = 2208$

(9) $24 \times 63 = 42 \times 36 = 1512$

(10) $24 \times 84 = 42 \times 48 = 2016$

(11) $26 \times 93 = 62 \times 39 = 2418$

(12) $34 \times 86 = 43 \times 68 = 2924$

(13) $36 \times 84 = 63 \times 48 = 3024$

Some more fun is, in the following multiplications the multiplicand and the multipliers are mirror image numbers.

$$144 \times 441 = 252 \times 252$$

$$156 \times 651 = 273 \times 372$$

$$168 \times 862 = 294 \times 492$$

$$276 \times 672 = 384 \times 483$$

$$13344 \times 44331 = 23352 \times 25332$$

$$13356 \times 65331 = 23373 \times 37332$$

$$13368 \times 86331 = 23394 \times 49332$$
$$1224 \times 4221 = 2142 \times 2412$$
$$1236 \times 6321 = 2163 \times 3612$$

$$1248 \times 8421 = 2184 \times 4812$$
$$1584 \times 4851 = 2772 \times 2772$$
$$1596 \times 6951 = 2793 \times 3972$$

Special palindrome addition

Using only prime numbers, the resultant obtained is also prime number and is equal to the reverse of each individual sums as follows.

$7 + 11 + 13 =$	31	13
$17 + 19 + 23 =$	59	95
$29 + 31 + 37 =$	97	79
$41 + 43 + 47 =$	131	131
$53 + 59 + 61 =$	173	371
$61 + 71 + 73 =$	211	112
$79 + 83 + 89 =$	251	152
	953	953

Again the resultant 953 is a prime number.

Special palindromic division

$$147741 \text{ divided by } 37 = 3993$$
$$852258 \text{ divided by } 37 = 23034$$

Special palindromic powers

$$68^2 + 89^2 + 96^2 = 86^2 + 98^2 + 69^2$$

A comparative study of divisibility and the reflux property of numbers: -

(a) When a number is divisible by 3 means the mirror image of that number is also divisible by 3. For example 225 and the mirror image of that number is 522. Both the numbers are divisible by 3.

Note: - The above rule is applicable for the multiplication operation also. That is if a number is multiplied by 3, the reflux number of the product obtained is divisible by three. For example $123 \times 3 = 369$ and the reflux number of the product 963 is divisible by 3.

The same rule is applicable for 9, 11, 33 and 99 also.

(b) When a number is divisible by 9 means the mirror image of that number is also divisible by 9. For example 558 and the mirror image of that number is 855. Both the numbers are divisible by 9.

(c) When a number is divisible by 33 means the mirror image of that number is also divisible by 33. For example 330.

(d) When a number is divisible by 99 means the mirror image of that number is also divisible by 99. For example 990.

(e) When a number is divisible by 11 means the mirror image of that number is also divisible by 11. For example 220, 8162, 1322, 209…

$209 \div 11 = 19$ and $902 \div 11 = 82$

One more special with this palindromicity is, if any of the digits of the multiplicand is not 9 means the resultant obtained after multiplying 11 is reverse number of the resultant obtained when the multiplicand is reversed.

$8162 \times 11 = 89782$ and $2618 \times 11 = 28798$

The square of the palindrome number of the following
 (a) $12^2 = 144$ and $441 = 21^2$ and again 12 is the mirror image of 21.
 (b) $13^2 = 169$ and $961 = 31^2$ and again 13 is the mirror image of 31.
 (c) $20^2 = 400$ and $004 = 02^2$ and again 20 is the mirror image of 02.

 (d) $30^2 = 900$ and $009 = 03^2$ and again 30 is the mirror image of 03.
 (e) $102^2 = 10404$ and $40401 = 201^2$, again 102 is the mirror image of 201.
 (f) $103^2 = 10609$ and $90601 = 301^2$, again 103 is the mirror image of 301.

 (g) $110^2 = 12100$ and $00121 = 011^2$, again 110 is the mirror image of 011.
 (h) $112^2 = 12544$ and $44521 = 211^2$, again 112 is the mirror image of 211.
 (i) $113^2 = 12769$ and $96721 = 311^2$, again 113 is the mirror image of 311.

 (j) $120^2 = 14400$ and $00441 = 021^2$, again 120 is the mirror image of 021.
 (k) $122^2 = 14884$ and $48841 = 221^2$, again 122 is the mirror image of 221.
 (l) $130^2 = 16900$ and $00961 = 031^2$, again 130 is the mirror image of 031.

 (m) $200^2 = 40000$ and $00004 = 002^2$, again 200 is the mirror image of 002.
 (n) $201^2 = 40401$ and $10404 = 102^2$, again 201 is the mirror image of 102.
 (o) $210^2 = 44100$ and $00144 = 012^2$, again 210 is the mirror image of 012.

 (p) $211^2 = 44521$ and $12544 = 112^2$, again 211 is the mirror image of 112.
 (q) $220^2 = 48400$ and $00484 = 022^2$, again 220 is the mirror image of 022.
 (r) $221^2 = 48841$ and $14884 = 122^2$, again 221 is the mirror image of 122.

 (s) $1012^2 = 1024144$ and $2101^2 = 4414201$
 (t) $1112^2 = 1236544$ and $2111^2 = 4456321$
 (u) $1212^2 = 1468944$ and $2121^2 = 4498641$
 (v) $2012^2 = 4048144$ and $2102^2 = 4418404$
 Likewise we can extend for higher power palindrome numbers.
For example $20^3 = 800$ and $008 = 02^3$, again 20 is the mirror image of 02.

 It is to be noted that in the above cases, the numbers we have taken initially are not palindrome numbers. But in the following cases the numbers we have taken initially are palindrome numbers and the square of the palindrome number of the following is the same number itself.
 (a) $11^2 = 121$ and the mirror image of the resultant is $121 = 11^2$.
 (b) $22^2 = 484$ and the mirror image of $484 = 22^2$.
 (c) $101^2 = 10201$ and the mirror image of $10201 = 101^2$.

 (d) $111^2 = 12321$ and the mirror image of $12321 = 111^2$.
 (e) $121^2 = 14641$ and the mirror image of $14641 = 121^2$.
 (f) $202^2 = 40804$ and the mirror image of $40804 = 202^2$.

 (g) $212^2 = 44944$ and the mirror image of $44944 = 212^2$.
 (h) $1001^2 = 1002001$ and the mirror image of $1002001 = 1001^2$.
 (i) $1111^2 = 1234321$ and the mirror image of $1234321 = 1111^2$.

(j) $2002^2 = 4008004$ and the mirror image of $4008004 = 2002^2$.

(k) $11111^2 = 123454321$ and the mirror image of $123454321 = 11111^2$.

Curiously in the following equation we get

$21^2 - 12^2 = 441 - 144 = 297$ is mirror image of $792 = 31^2 - 13^2 = 961 - 169$

For example $11^3 = 1331$ and the mirror image of the resultant is $1331 = 11^3$. Similarly for $101^3 = 1030301$

$1001^3 = 1003003001$

$10001^3 = 1000300030001$

.... and

$111^3 = 1367631$

$10101^3 = 1030607060301$

$1001001^3 = 1003006007006003001$

....

Likewise we can extend for higher power palindrome numbers.

The only known palindromic cube whose root is not palindromic is $2201^3 = 10662526601$. The following squares 121, 484, 676, and 69696 are palindromic as well as undulating in natures.

Interesting Palindromic Triangular Numbers

539593131395935 consists only of the odd digits 1, 3, 5 and 9.

8208268228678028 consists only the even digits 0, 2, 6 and 8.

Palindromic Squares of Palindromes

$$10001^2 = 100020001$$
$$11011^2 = 121242121$$
$$11111^2 = 123454321$$
$$11211^2 = 125686521$$

Fascinating Palindromes

Start palindrome	Divided by	Gives this palindrome	Divided by	Gives this palindrome
121	11	11	11	1
1234321	11	112211	11	10201
12345654321	11	1122332211	11	102030201
123456787654321	11	11223344332211	11	1020304030201

Curious Palindromes and Prime Factorizations

Palindrome number	Prime factorization	Addition of prime factorization	Remarks
9	3×3	$3+3$	6
989	23×43	$23+43$	66
98789	223×443	$223+443$	666
9876789	2223×4443	$2223+4443$	6666
987656789	22223×44443	$22223+44443$	66666
98765456789	222223×444443	$222223+444443$	666666
9876543456789	2222223×4444443	$2222223+4444443$	6666666
....	....	....	....

Prime Numbers among the Palindromes

There are 15 palindromic primes with three digits are present. That is 101, 131, 151, 181, 191, 313, 353, 373, 383, 727, 757, 787, 797, 919 and 929.

11 is the only palindromic prime number in two digit numbers. There is no other palindromic prime number comprising even number of digits since they are divisible by 11. Conversely every palindromic prime must contain an odd number of digits with one exception that 11 is a palindromic prime having even number of digits.

There are no palindromic primes with 4 digits because they all have the factor 11. For example 4554 has the factor 11.

There are 93 primes with 5 digits.
There is no palindromic prime number with 6 digits because they all have the factor 11.

There are 668 palindromic primes with 7 digits. For example 1878781, 1879781, 1880881 and 1881881 are some of the seven digit palindromic prime number.

The palindromic prime number 1888081808881 reads the same upside down or when viewed in a mirror.

The smallest palindromic prime number containing all the digits from 0 to 9 is 1023456987896543201.

Interesting 9-Digit Palindromic Primes

The following palindromic primes have special property of first and last digit same keeping the middle digits only different.

188888881
199999991
355555553

The following palindromic primes have special property of having the middle digit same and keeping the other digits same.

111181111
111191111
777767777

The following palindromic primes are smoothly undulating primes

323232323
727272727
919191919

The following palindromic primes are positive smooth seesaw number or peak prime numbers.

123494321
345676543
345686543

The following palindromic primes are negative smooth seesaw number or valley prime numbers.

765404567
987101789
987646789

Palindromic magic squares

The palindromic number magic square is formed from the basic magic square as follows

6	1	8
7	5	3
2	9	4

The 2 digit palindromic number magic square can be derived as

66	11	88
77	55	33
22	99	44

Here the magic square addition sum is 165. Similarly the 3 digit palindromic number magic square can be derived as follows. Here the magic square addition sum is 1665, irrespective of whether the palindromic numbers are added row wise or column wise or diagonal wise.

444	999	222
333	555	777
888	111	666

We can extend the magic square to all the palindromic numbers whose end digits are from 1 to 9, as follows. Here the magic square addition sum is 1605.

636	131	838
737	535	333
232	939	434

Likewise in the following magic square, the magic square addition sum is 1968.

646	696	626
636	656	676
686	616	666

Further the palindromic number magic squares can be extended to higher digit palindromic numbers as follows.

43434	93939	23232
33333	53535	73737
83838	13131	63636

Here 160605 is the magic square addition sum.

8. ARITHMETIC OF ZERO AND INFINITY

Zero is really a typical number without which it is very difficult to express the bigger numbers. This is denoted as '0' and can also be called as cipher.

General properties of zero: -

Zero is an even number.

Zero is a neutral number in the number system. That is zero is neither a positive integer nor a negative integer. That is there is no +0 or -0.

Zero is the additive identity and it can also be considered as subtractive identity.

$x + 0 = x$ where x is a real number

$x - 0 = x$ where x is a real number

$0 - x = -x$ where x is a real number

$0 + x = x$ (provided $x \neq 0$)

$x - x = 0$ where x is a real number

$x \times 0 = 0$ (provided $x \neq 0$)

$0 + 0 = 0$

$0 - 0 = 0$

Dividing by zero is undefined. But $0 \div x = 0$ because $0 = x \times 0$

Factorial zero is 1.

$0^n = 0$, where 'n' is a positive number. That is 0 is a special power number.

Zero has no face value but it has the place value.

Place value of zero: -

In any number, the position of each digit or numerical from the right shows how many ones, tens, hundreds, thousands and so on are there in the number. For example in the number 1,002 the place value of each digit is as follows.

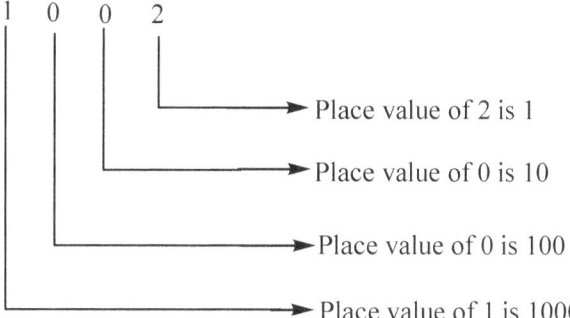

Zero written in the left side of a number will not increase the value of the number. But if the zero is written between the digits of a number or after the number from left side onwards will increase the value of the number.

For example 123 and 0123, 00123, 000123… (Equation 1) are same as that of 123. But 1023, 10023, 100023… (Equation 2) has the greater value than their corresponding counterparts of equation 1 and 1203, 12003, 120003… (Equation 3) has still greater value corresponding to their counterparts of equations 1 and 2. Similarly 1230, 12300, 123000… (Equation 4) has still greater value when compared to the values of equations 1, 2 and 3.

Similar arguments can be extended to infinity also.

From the above statements we can say that unlike other integers which have the discrete values, 0 and ∞ have not discrete value at all. That is zero and infinity has a set of values with each individual integer themselves. That means 1 can be represented as 01, 001, 0001… which implies that integer 1 has zero in itself or the other way that is zero has 1 itself then we can represent 1 as 1zero (don't confuse with the number 10). The same argument can be extended to the integer 2 and so on. Then we will have 2zero, 3zero... Similarly if we extrapolate the argument for ∞ we will get 1∞, 2∞, 3∞...

In a broader sense we can state a little bit confusing but still with authenticity that zero has infinite sets of discrete values and infinity has infinite sets of discrete values.

From the above utopian statement we can derive one more value that is 0∞ which give the actual relationship between 0 and ∞.

If we relate zero to the birth of a species that is as the origin of species, then infinity can be related to the death of the species.

Classification of zero: -

There are three classes of zeros. They are leading zeros, captive zeros and trailing zeros.

Leading zeros are zeros that precede all of the non zero digits. They never count as significant figures. For example 0.0025 the three zeros indicate the position of the decimal point. The number has only two significant figures 2 and 5.

Captive zeros are zeros that fall between non zero digits. They always count as significant figures. For example 1.008 has four significant figures.

Trailing zeros are zeros at right end of the number. They are significant only if the number is written with a decimal point. For example 100 has only one significant figure, but written as 100. , it has three significant figures.

Scientific notation number: -

It is the representation of any number as the powers of 10. Or it is the representation of larger number concisely as the powers of 10.

For example $0.00034 = 3.4 \times 10^{-4}$ and $26000000 = 26 \times 10^{6}$

Significant figures:

All non zero digits are significant. For example 187, 0.162 each has three significant figures.

Zeros to the left of the first non zero digit in the number are not significant. For example 0.002, 0.0004 each has one significant figure. But 0.0020 has two significant figures. Zero between non zero digits are significant. For example 120003, 0.00130042 each has six significant figures.

In the case of 4.0 two significant figures are there because this number signifies the exact value which lies between 4.1 to 4.9 and so zero is considered as a significant figure along with four. Similarly the number 4.00 has three significant figures.

Nomenclature of bigger numbers: -

Zero is an important number in representing bigger numbers.

S.No.	Number	Name
1.	10^{-18}	Ioto
2.	10^{-15}	Femto
3.	10^{-12}	Pico
4.	10^{-9}	Nano
5.	10^{-6}	Micro
6.	10^{-3}	Milli
7.	10^{-2}	Centi
8.	10^{-1}	Deci
9.	10^{1}	Ten

10.	10^2	Hundred
11.	10^3	Thousand or kilo
12.	10^4	Ten thousand
13.	10^5	Lakh
14.	10^6	Ten Lakh or Million or mega
15.	10^7	Crore or Ten million
16.	10^8	Hundred million
17.	10^9	Billion or giga
18.	10^{10}	Ten billion
19.	10^{11}	Hundred billions
20.	10^{12}	Trillion
21.	10^{13}	Ten trillions
22.	10^{14}	Hundred trillions
23.	10^{15}	quadrillion
24.	10^{16}	Ten quadrillions
25.	10^{17}	Hundred quadrillions
26.	10^{18}	Quintillion
27.	10^{19}	Ten quintillions
28.	10^{20}	Hundred quintillions
29.	10^{21}	Sextillion
30.	10^{22}	Ten sextillions
31.	10^{23}	Hundred sextillions
32.	10^{24}	Septillion
33.	10^{25}	Ten septillion
34.	10^{26}	Hundred septillion
35.	10^{27}	Octillion
36.	10^{28}	Ten octillions
37.	10^{29}	Hundred octillions
38.	10^{30}	Nonillion
39.	10^{33}	Decillion
40.	10^{36}	Undecillion
41.	10^{39}	Duodecillion
42.	10^{42}	Tredecillion
43.	10^{45}	Quatuordecillion
44.	10^{48}	Quindecillion
45.	10^{51}	Sexdecillion
46.	10^{54}	Septendecillion
47.	10^{57}	Octodecillion
48.	10^{60}	Novemdecillion
49.	10^{63}	Vigintillionn
50.	10^{66}	Unvigintillion
51.	10^{69}	Duovigintillion
52.	10^{72}	Trevigintillion
53.	10^{75}	Quattorvigintillion
54.	10^{78}	Quinvigintillion
55.	10^{81}	Sexavigintillion
56.	10^{84}	Septevigintillion
57.	10^{87}	Octavigintillion

58.	10^{90}	Novemvigintillion
59.	10^{93}	Trigintillion
60.	10^{96}	Untrigintillion
61.	10^{99}	Duotrigintillion
62.	10^{100}	Centillion or Googol

Note: - (a) The superscript on zero indicates the number of zeros. For example 10^{12} indicates the number 1 followed by 12 zeros. That is 1,000,000,000,000.

(b)The kilo, mega and giga like numbers in decimal value is different from that of kilo, mega and giga like numbers represented in binary usage. The following table represents a list of comparative usage of bigger numbers in decimal and binary number system.

No.	Name	Decimal value	Binary usage	Binary equivalent
1.	Kilo	10^3	Kilobyte (KB)	$2^{10}=1024$
2.	Mega	10^6	Megabyte (MB)	$2^{20}=1,048,576$
3.	Giga	10^9	Gigabyte (GB)	$2^{30}=1,073,741,824$
4.	Tera	10^{12}	Terabyte (TB)	$2^{40}=1,099,511,627,776$
5.	Peta	10^{15}	Petabyte (PB)	$2^{50}=1,125,899,906,842,624$
6.	Exa	10^{18}	Exabyte (EB)	$2^{60}=1,152,921,504,606,846,976$
7.	Zetta	10^{21}	Zettabyte (ZB)	$2^{70}=1,180,591,620,717,411,303,424$
8.	Yotta	10^{24}	Yottabyte (YB)	$2^{80}=1,208,925,819,614,629,174,706,176$

Mathematical fallacy: -

The equation 1 = 2 can be proved as follows.

Multiply both sides with zero and is $1 \times 0 = 2 \times 0 = 0$

Cancel the common number zero we get 1=2

This means at zero state all numbers are equal.

INFINITY

Infinity is the imaginary number and is considered to be the last number of the whole number series. That is, both zero and the infinity are imaginary numbers and are at the extreme end of the whole number system. It is denoted by the symbol '∞'.

General properties of infinity: -

Infinity is a neutral number in the number system. That is like zero, infinity is neither a positive integer nor a negative integer. That is there is no -∞ or +∞.

$0 + ∞ = ∞$ because 0 is an additive identity.

$0 - ∞ = ∞$

$0 × ∞ = 0$

$0 ÷ ∞ = 0$

$∞ - 0 = ∞$

$x + ∞ = ∞$

$x - ∞ = ∞$

$x × ∞ = ∞$

$x ÷ ∞ = ∞$

$∞ + ∞ = ∞$

$∞ - ∞ = ∞$

$∞ ÷ ∞ = 1$

$x^∞ = ∞$ provided that $x ≠ 0$

$∞! = ∞$

Infinity has no place value. But it has the face value.

Between the whole numbers 1 to 2 infinity numbers are there like 1.1, 1.11, 1.111…1.2, 1.22…1.21, 1.2121…But if we consider only the whole numbers there is no numbers between 1 and 2, because 1.0, 1.00, 1.000…is equal to 1 only.

Mathematical fallacy: -

The equation 1 = 2 can be proved as follows.

Multiply both sides with infinity and is $1 × ∞ = 2 × ∞$

Cancel the common number infinity we get 1 = 2

This means at infinity state all numbers are equal.

Note: -Since at both '0'state and '∞' state all numbers are equal, it doesn't mean that 0 is equal to ∞. This can be confirmed by the following arguments.

Comparison of zero and infinity: -

Case (a) If we consider the rational number system as a big utopian circle, then 0 and ∞ will be the opposite members in the system.

0

-1   1

-2   2

-3   3

…… ∞……..

Case (b) If we consider the whole number system as a big utopian circle starting from 0 to ∞, then 0 and ∞ will be the neighbors in the system.

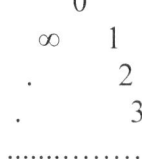

0

∞   1

.   2

.   3

……………

Case (c) Geometrical approach of 0 and ∞:- If we consider the arithmetic numbers in geometrical view, 1 can be represented as .(dot), 2 is represented as straight line, 3 as triangle, 4 as square, 5 as pentagon, 6 as hexagon, 7 as heptagon and so on. If we extrapolate the same concept for infinity we will get a circle and the imaginary origin of the circle is zero and the number of points present in the circumference of the circle is infinity.

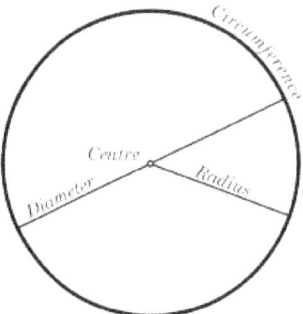

Case (d) Probability theory of 0 and ∞:-The probability of occurring an event for zero time is 100% and the probability of occurring an event for an infinity time is 0%.

Case (e) Logical approach of 0 and ∞:-The difference between zero and infinity is there are multiples of infinity states are possible, but in zero state there is no multiplicity. That is if we multiply any number by infinity the resultant will be infinity, but logically there is a difference between 2∞ and 3∞ though the answer is same. But the multiples of zero are zero, as any number multiplied by zero becomes zero.

Case (f) Astronomical approach of 0 and ∞:- During new moon day, zero photon of light is transmitted from moon and infinity photon of light is transmitted from moon during full moon day.

Case (g) Theosophical approach of 0 and ∞:- Hypothetically 0 and ∞ are polymorphs of God. Zero and infinity are God's own super natural numbers. Skeptically at 0 and ∞ states all numbers are equal, but in reality it is not possible.

As such 0 has no value but if it is placed next to a number the value of the number increases. If the number is associated with infinity means it attains the eternal infinity state. Simply zero is nothing, but it is everything and infinity is of everything with its existence nothing.

Likewise each human being is a number of his own kind and must associate with God to add value in their life and attain the eternal state and they should not ask about the existence of Almighty like zero and infinity.

9. SEQUENCES AND SERIES

A sequence or series of numbers is nothing but arrangement of numbers having common relationship or common difference among themselves.
For example 0, 2, 4, 6, 8, 10, 12, 14, 16, 18…

Fibonacci series: -

The series 1, 1, 2, 3, 5, 8, 13, 21, 34, 55, 89, 144, 233, 377, 610, 987, 1597, 2584… In this series every term except the first and second term is the sum of two previous terms. That is 3+2=5 and so on. This series is called as Fibonacci series. The 'n'th term in this series is usually denoted by F_n. Thus F_6 of Fibonacci series is 8. There are 19 Fibonacci numbers below 10,000.

The Fibonacci numbers which are even are called as even Fibonacci numbers and the Fibonacci numbers which are odd are called as odd Fibonacci numbers. For example 2, 8, 34, 144, 610… are even Fibonacci numbers and 1, 3, 5, 13, 21… are odd Fibonacci numbers.

In this sequence if 'n' is divisible by 5, then F_n is divisible by 5. That is F_{10} = 55. Here 10 is divisible by 5 and so the 10th term 55 is divisible by 5.

If F_n is the 'n'th Fibonacci number and if a = F_{2n-1} and b = F_{2n+1} then $a^2 + 1$ is divisible by b and $b^2 + 1$ is divisible by a.
For example let n = 6 then a = F_{12-1} = F_{11} = 89 and b = F_{2n+1} = F_{13} = 233. Now $a^2 + 1$ = 7922 and $b^2 + 1$ = 54290. Then 7922 is divisible by 233 and 54290 is divisible by 89.

Take any four consecutive numbers in the Fibonacci sequence. The product of first and fourth numbers and the twice the product of the middle numbers form the sides a right triangle. For example 5, 8, 13 and 21 are the four consecutive numbers in the Fibonacci sequence. Now 5×21=105; 8×13×2=208. So the 105 and 208 are the sides of a right triangle. That is $105^2 + 208^2 = 11025 + 43264 = 54289 = 233^2$.

The product of F_{2n}, F_{2n+2}, and F_{2n+4} of three consecutive Fibonacci numbers is equal to the product of three consecutive natural numbers.
For example if n=2 then $F_4 \times F_6 \times F_8$ = 3×8×21 = 504 = 7×8×9
    if n=3 then $F_6 \times F_8 \times F_{10}$ = 8×21×55 = 9240 = 20×21×22.

The prime numbers in the Fibonacci sequence are called as Fibonacci prime numbers. For example 2, 3, 5, 13, 89, 233, 1597, 28657, 514229, 433494437, 2971215073, 99194853094755497, 1066340417491710595814572169, 19134702400093278081449423917… are prime Fibonacci numbers. The composite numbers in the Fibonacci sequence is called as composite Fibonacci numbers. For example 21 is a composite Fibonacci number.

The repunit number in the Fibonacci sequence is called as repunit Fibonacci number. For example 55 is a repunit Fibonacci number.

Golden ratio: The decimal number rounded to 1.61 is called as golden ratio. The ratio of consecutive Fibonacci numbers tends to be the golden ratio. For example, 21/13 = 1.61538…; 144/89 = 1.617977…; 89/55=1.618181…

Lucas series is obtained by the Fibonacci rule of adding the latest two to get the next is kept, but here we start from 2 and 1 (in this Lucas series) instead of 0 and 1 for the (ordinary) Fibonacci numbers. The numbers obtained are called as Lucas numbers.
The Lucas series 2, 1, 3, 4, 7, 11, 18…

In fact, for every series formed by adding the latest two values to get the next, and no matter what two values we start with we will always end up having terms whose ratio is Phi=1·6180339… that is the golden ratio.

The prime numbers in the Lucas sequence is called as Lucas primes. Thus the Lucas prime sequence is 2, 3, 7, 11, 29, 47, 199, 521, 2207, 3571, 9349, 3010349…

Pascal series: -

The following series which starts from 1 and ends with 1 with specified difference is known as Pascal's series. That is 1 9 36 84 126 84 36 9 1

The triangle formed by using the Pascal's series as depicted below is known as Pascal's triangle.

```
                1
              1   1
            1   2   1
          1   3   3   1
        1   4   6   4   1
      1   5  10  10   5   1
    1   6  15  20  15   6   1
  1   7  21  35  35  21   7   1
1   8  28  56  70  56  28   8   1
1  9  36  84  126  126  84  36  9  1
```
....

Diagrammatically it can be represented as follows.

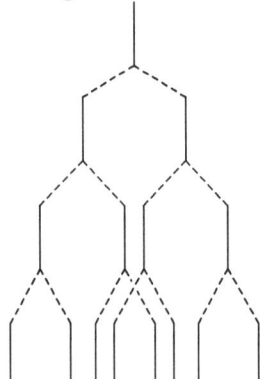

....

Note: - Each line is split into two lines and the total number of lines or the intensity of lines represents Pascal series. It is to be noted that all natural numbers are Pascal numbers.

The coefficients of the nth power of the binomial $(a+b)^n$ will give Pascal's triangle.

That is $(a+b)^1 = (a+b)$ $=a+b$ 1 1

$(a+b)^2 = (a+b)(a+b)$ $=a^2+2ab+b^2$ 1 2 1

$(a+b)^3 = (a+b)(a+b)(a+b)=a^3+3a^2b+3ab^2+b^3$ 1 3 3 1

....

The summation of the two consecutive numbers in the series will be equal to the following number in the next series. That is it will form a triangle. For example 3+3=6.

The first angular number series in the Pascal's triangle is uniformly 1.

The second angular number series is a real number series. That is 1, 2, 3, 4, 5… or in the other way as 1, (1+1), (1+1+1), (1+1+1+1), (1+1+1+1+1)…

The third angular number series is nothing but the triangular number series. That is 1, 3, 6, 10, 15… or 1,(1+2), (1+2+3), (1+2+3+4), (1+2+3+4+5)…One more method of writing the same series is 1̲, (1+2̲), (3+3̲), (6+4̲), (10+5̲)…

The fourth angular number series is tetrahedral number series. That is 1, 4, 10, 20, 35... are tetrahedral numbers.

The sum of the numbers in any row is equal to 2^n, where n is the number of the row. For example: $2^0 = 1$

$$2^1 = 1+1=2$$
$$2^2 = 1+2+1=4$$
$$2^3 = 1+3+3+1=8$$
$$2^4 = 1+4+6+4+1=16$$

....

The triangle is symmetric in nature. That is if we split the triangle into two halves the left side part is symmetrical or mirror image to the right side part of the triangle.

If the 1st element in a row is a prime number (remember, the 0th element of every row is 1), all the numbers in that row (excluding the 1's) are divisible by it. For example, in row 7 (1 7 21 35 35 21 7 1) 7, 21, and 35 are all divisible by 7.

The first 5 powers of 11 are in the top of the triangle. $11^0=1$, $11^1=11$, $11^2=121$, $11^3=1331$ and $11^4=14641$. When these numbers are stacked in a pyramid it will form the top part of the triangle.

If you start at any given 1 and go diagonally down and then make a one step left you will find the answer to the numbers that you just followed. For example:
$1 + 2 + 3 + 4 + 5 = 15$, or try any other number $1 + 8 + 36 + 120 + 330 = 495$.

The odd numbers in the Pascal triangle are called as odd Pascal numbers and the even numbers in the Pascal triangle are called as even Pascal numbers. For example 6 is an even Pascal number and 15 is an odd Pascal number.

The repunit numbers in the Pascal triangle are called as repunit Pascal numbers. For example 11 is a repunit Pascal number.

The palindromic number in the Pascal triangle is called as palindromic Pascal number. For example 252 is a palindromic Pascal number.

The prime numbers in the Pascal's triangle are called as prime Pascal numbers. That is 2, 3, 5, 7, 11, 13, 17, 19... are prime Pascal numbers and the composite numbers in the Pascal triangle are called as composite Pascal numbers.

The power numbers in the Pascal triangle are called as power Pascal numbers. For example 9, 36, 8... are some of the power Pascal numbers.

Pascal's triangle is useful in expressing the splitting pattern of hydrogen atoms in a compound in the NMR spectrum.

Arithmetic progression:-

It is a symmetrical sequence having the property that each term after the first is obtained by adding a fixed number to the preceding term.
For example 100, 95, 90, 85, 80, 75...

In the above sequence if we add (-5) to the preceding term, we get the following number in the series and this number is called as the common difference. The common difference is denoted by the symbol 'd'.

There are two types of arithmetic progressive series called as finite arithmetic progressive series and infinite arithmetic progressive series.
For example 100, 95, 90, 85, 80... is called as infinite arithmetic progressive series and 100, 95, 90, 85, 80. is called as finite arithmetic progressive series.

The 'n'th number in the arithmetic progression is n = a+ (n-1) d
where n = 'n'th number in the series
a = first number in the series
d = common difference in the series

This formula is applicable to both finite arithmetic progressive series and infinite arithmetic progressive series.

For example using the above formula the 7th term of the series 100, 95, 90, 85, 80... can be calculated as follows. Here a=100, n=7 and d=5

Applying to the above formula 7th term = 100 + [(7-1) × (-5)]

$$7^{th} \text{ term} = 100 + [6 × (-5)]$$
$$7^{th} \text{ term} = 100 + [-30]$$
$$7^{th} \text{ term} = 70$$

The sum (S) of 'n' terms of arithmetic series is

S = {n/2[2a + (n-1) d]}   where S = Sum of 'n' terms of arithmetic series

n = 'n'th number in the series

a = first number in the series

d = common difference in the series

This formula is applicable to both finite arithmetic progressive series and infinite arithmetic progressive series.

For example using the above formula the sum of the arithmetic series up to 5th term of the series 100, 95, 90, 85, 80... can be calculated as follows. Here a=100, n=5 and d=5

Now the sum of the series up to 5th term S = 5/2[(2×100) + (5-1) (-5)]

$$= 5/2[200 + (-20)]$$
$$= 5/2(180)$$
$$= 450.$$

For finite arithmetic progressive series the sum (S) of 'n' terms of arithmetic series is calculated by the formula

S = [n/2(a+l)]   where S = sum of 'n' terms of arithmetic series

n = 'n' termed arithmetic series

a = first number in the series

l = last number in the series

For example using the above formula the sum of the finite arithmetic series up to 5 terms of the series 100, 95, 90, 85 and 80 can be calculated as follows.

Here a=100, n=5 and l=80

S = 5/2(100+80)

= 5/2(180)

= 450.

A sequence of real numbers is an arithmetic sequence of first order.

1, 2, 3, 4, 5……..   Primary sequence

1, 1, 1, 1………   First difference sequence

Similarly the sequence of square numbers is an arithmetic sequence of second order.

1, 4, 9, 16, 25……Primary sequence

3, 5, 7, 9……..   First difference sequence

2, 2, 2……   Second difference sequence

Likewise the sequence of cubic numbers is an arithmetic sequence of third order and so on. That is 1, 8, 27, 64, 125…Primary sequence

7, 19, 37, 61……First difference sequence

12, 18, 24………Second difference sequence

6, 6…………..   Third difference sequence

A sequence of numbers whose 'k'th difference is constant is called as an arithmetic sequence of 'k'th order.

The sequence of fourth power numbers is an arithmetic sequence of fourth order and the constant can be calculated by multiplying the real numbers from 1 to 4.

That is $1\times2\times3\times4 = 24$ and similarly for the fifth power number series the constant is calculated by multiplying the real numbers from 1 to 5. That is $1\times2\times3\times4\times5 = 120$.

The arithmetic sum of real numbers starting from 1 to n (where n is a real number) is calculated by the formula $S = [n(n+1)/2]$

For example $1+2+3+4+5$.

The sum of numbers up to 5 is $S_5 = [5(5+1)/2]$
$$= [(5\times6)/2]$$
$$= [30/2]$$
$$= 15.$$

The sum of even numbered real number arithmetic series starting from '0' onwards is $[2\times n(n+1)/2]$ that is $n(n+1)$. Similarly for 3rd multiples starting from '1' onwards is $[3\times n(n+1)/2]$ and for 4th multiples $[4\times n(n+1)/2]$ that is $2n(n+1)$ and so on.

The arithmetic sum of square numbers starting from '1' onwards is $[n(n+1)(2n+1)/6]$ where 'n' is a real number.

For example $1^2+2^2+3^2+...$ For example if n=3

The sum of square numbers up to 3 is $\{3(3+1)[(2\times3)+1]/6\}$
$$= \{(3\times4\times7)/6\}$$
$$= 14.$$

The arithmetic sum of cubic numbers starting from '1' onwards is $[n(n+1)/2]^2$ where 'n' is a real number.

For example $1^3+2^3+3^3+...$ Here n=3

The sum of cubic numbers up to 3 is $[3(3+1)/2]^2$
$$= [(3\times4)/2]^2$$
$$= 36.$$

The arithmetic sum of odd number series starting from '1' onwards is $[(n+1)/2]^2$ where 'n' is the last number of series.

For example 1, 3, 5, 7......   Here n=7

The arithmetic sum of odd number series starting from 1 to 7 is $[(7+1)/2]^2 = 16$.

The arithmetic sum of odd number series starting from '1' onwards can also be calculated by another formula n^2 where 'n' is the number of term in the series.

For example 1, 3, 5, 7.........   Here n=4

The arithmetic sum of odd number series starting from 1 to 7 is $4^2 = 16$.

The arithmetic sum of odd numbered squares is $[n(4n^2 - 1)/3]$ where 'n' is the number of the term in the series.

For example $1^2+3^2+5^2+7^2+9^2$ Here n=5

The arithmetic sum of odd numbered squares up to 9 is $5[4(5)^2 - 1]/3= (5\times99)/3 = 165$.

The arithmetic sum of addition of multiples of two from '1' onwards can be calculated by multiplying the last number by 2 and then subtract 1 from the resultant.

For example $1+2+4+8+16+32+...$

Or in other words $1+2 + (2\times2) + (2\times2\times2) + (2\times2\times2\times2) +...$

The sum up to 32 is $(32\times2)-1=63$.

The arithmetic sum of addition of multiples of three from '1' onwards can be calculated by adding the last number with half of the last number minus 1.

For example $1+3+9+27+81+243+...$

Or in other words $1+3 + (3\times3) + (3\times3\times3) + (3\times3\times3\times3) +...$

For example the sum up to 81 is $[81 + (81-1)/2] = 121$.

If an arithmetic progression is added, subtracted, multiplied or divided by any constant number the resultant will also be an arithmetic progressive series.

Geometric progression: -

It is a sequence having the property that each term after the first is obtained by multiplying a fixed number to the preceding term.

For example 1, 2, 4, 8, 16...

The nth term in the geometric progression is $a \times r^{n-1}$ where 'a' is the first term, 'r' is the common ratio and 'n' is the nth term.

Sum of the first 'n' terms of a geometric progression is $[a(r^n - 1)/(r - 1)]$, if r>1 and $[a(1 - r^n)/(1 - r)$, if r<1.

If every term of a geometric progression is raised to the same power, the resulting series is also a geometric progression. The reciprocal of a geometric progression is also a geometric progression.

Harmonic progression: -

If a sequence having the property that each reciprocals of the term form an arithmetic progression, it is known as harmonic progression.

For example ½, 1/4, 1/6, 1/8, 1/10...

If the given sequence of numbers is both in arithmetic progression and harmonic progression then they will be also in geometric progression.

10. POLYGONAL NUMBERS

The real numbers whose units or digits can be laid to form a specified geometrical figure are called as polygonal numbers. These are one type of 2-dimensional figurate numbers. Except 1 and 2, all real numbers when represented in digits can be laid to form a closed geometrical figure.

The number one will form a point and the number two will form a straight line. A closed geometrical figure is obtained from the number three onwards. Based upon the geometrical shape formed, the numbers can be classified as triangular number, square number, pentagonal number and so on.

Since the study of polygonal numbers links between arithmetic and geometry it can also be called as 'Number geometry'.

The study of polygonal numbers is applied in logistics.

Triangular number: -

A triangular number is one whose units or digits can be laid to form an equilateral triangle. The triangular number series is 1, 3, 6, 10, 15, 21, 28, 36, 45, 55, 66, 72, 91, 105, 120, 136, 153, 171, 190, 210, 231, 253, 276, 300, 325, 351, 378, 406, 435, 465, 496, 528, 561, 595, 630, 666…

The 'n'th triangular number can be calculated by using the formula $[n(n+1)/2]$. That is half the sum of any positive integer and its square number is a triangular number. For example if n=10 the 10th triangular number is 55.

The triangular number which is even is called as even triangular number and the triangular number which is odd is called as odd triangular number. For example 15 is an odd triangular number and 10 is an even triangular number.

In the triangular number series if we categorize the numbers as odd (O) number and even (E) number the sequence is O, O, E, E, O, O, E, E… and it repeats further as we move.

Further if the triangular number contains all the digits as even, then it is a super even triangular number and if the triangular number contains all the digits as odd, then it is a super odd triangular number. For example 28 is a super even triangular number and 91 is a super odd triangular number.

The triangular number which is ascending in nature is called as triangular ascending number and the triangular number which is descending in nature is called as triangular descending number. For example 36 is a triangular ascending number and 72 is a triangular descending number.

The triangular numbers are obtained by the consecutive addition of natural numbers as follows.

Addition of natural numbers	Triangular number
1	1
1+2	3
1+2+3	6
1+2+3+4	10
1+2+3+4+5	15
…………………..	……

The triangular numbers obtained like this method in the above case are called as first order triangular numbers.

The second order triangular number is obtained by the consecutive addition of first order triangular numbers as follows.

First order triangular numbers	Second order triangular number
1	1
1 + 3	4
1 + 3 + 6	10
1 + 3 + 6 + 10	20
1 + 3 + 6 + 10 + 15	35
1 + 3 + 6 + 10 + 15 + 21	56
1 + 3 + 6 + 10 + 15 +21 + 28	84
1 + 3 + 6 + 10 + 15 + 21 + 28 + 36	120
1 + 3 + 6 + 10 + 15 + 21 + 28 + 36 + 45	165
1 + 3 + 6 + 10 + 15 + 21 + 28 + 36 + 45 + 55	220
………………………………………………	………

Note: - The numbers 1, 4, 10, 20, 35, 56, 84, 120, 165, 220… are tetrahedral numbers. Similarly the third order triangular number is obtained by the consecutive addition of second order triangular numbers as below.

Second order triangular numbers	Third order triangular number
1	1
1+4	5
1+4+10	15
1+4+10+20	35
1+4+10+20+35	70
…………………..	…..

Likewise we can extend the process to get the fourth order triangular number, fifth order triangular number and so on.

For example the fourth order triangular numbers are 1, 6, 21, 56…

the fifth order triangular numbers are 1, 7, 28, 84…

Relationship between triangular number and square number: - The multiplication of a triangular number by 8 followed by the addition with 1 will give a square number. For example 21 is a triangular number and its multiplication with 8 gives 168. Now the addition of 168 with 1 will give 169 which is nothing but the square of 13. This relationship can be taken as a rule for a triangular number. A number 'x' is a triangular number if and only if "8x + 1" is a square number.

The lowest number which is behaving both as square number and as a triangular number is 36. (Note: - 1 is considered as power number rather than a square number).

No triangular number is a cube or a fourth power number.

Number 3 is the only prime triangular number.

Every positive integer can be written as a sum of three triangular numbers. For example 50 = 28 + 21 + 1.

The triangular number which is palindromic in nature is called as palindromic triangular number. For example 5995 is a palindromic triangular number.

Triangular numbers which are repdigit in nature is called as repdigit triangular numbers. For example 55 and 666 are some of the repdigit triangular number.

The Lucas numbers which can be expressed in triangular shape are called as Lucas triangular numbers. For example 5778.

All hexagonal numbers are triangular numbers just like all the numbers divisible by 6, are divisible by 3.

The power numbers in the triangular number series can be called as power triangular numbers. For example 36 is power triangular number.

The Kaprekar numbers which can be expressed in triangular shape are called as Kaprekar triangular numbers. For example 45, 55, 703…

Centered triangular numbers: -

A centered triangular number is one whose units or digits can be laid to form a centered equilateral triangle. The centered triangular number series is 1, 4, 10, 19, 31, 46, 64, 85, 109, 136, 166, 199, 235, 274, 316, 361, 409, 460, 514, 571, 631, 694, 760, 829, 901, 976, 1054, 1135, 1219, 1306, 1396, 1489, 1585, 1684, 1786, 1891, 1999, 2110, 2224, 2341, 2461, 2584, 2710, 2839, 2971, 98689…

A formula to find out the centered triangular number is $[(3n^2 + 3n + 2) / 2]$ where 'n' is a whole number.

If the centered triangular number is an odd number then it is odd centered triangular number and if the centered triangular is an even number then it even centered triangular number. For example 19 is an odd centered triangular number and 2110 is an even centered triangular number.

Further if the centered triangular number contains all the digits as even, then it is a super even centered triangular number and if the centered triangular number contains all the digits as odd, then it is a super odd centered triangular number. For example 460 is a super even centered triangular number and 1135 is a super odd centered triangular number.

The centered triangular number which is ascending in nature is called as centered triangular ascending number and the centered triangular number which is descending in nature is called as centered triangular descending number. For example 235 is a centered triangular ascending number and 631 is a centered triangular descending number.

A centered triangular prime is a centered triangular number that is prime number. The centered triangular prime number sequence is 1, 19, 31, 109, 199, 309… The centered triangular number which is composite is called as centered triangular composite number. For example 4, 10, 46, 64… are called as centered triangular composite numbers. It is to be noted that a centered triangular prime number is always an odd number.

A square number which is also a centered triangular number is called as centered triangular square number. 4, 64, 361… are centered triangular square numbers.

A centered triangular number which is palindromic in nature is called as palindromic centered triangular number. For example 98689 is a palindromic centered triangular number.

Rectangular number: -

The numbers whose dots form a lattice work of rectangles are called as rectangular numbers. In other words any natural number that can be represented as the product of two numbers (not 1 or the number itself) is a rectangular number. For example 1, 6, 8, 10, 12, 14, 15, 16, 18, 20, 21, 22, 24, 26, 27, 28, 30, 32, 33, 34, 35, 36, 38, 40, 42, 44, 45, 46, 48, 49, 50…

The prime numbers will not become rectangular numbers. For example 7 is not a rectangular number because if it is represented by 1×7 as factor, it will form a straight line only.

The rectangular number which is odd is called as odd rectangular number and the rectangular number which is even is called as even rectangular number. For example 14 is an even rectangular number and 15 is an odd rectangular number.

Note: - square number, cube number and other power numbers are also rectangular numbers with special characteristic geometrical representation.

Square number: -

The numbers whose dots form a lattice work of squares are called as square numbers. If we tilt the lattice work of square, rhombus is formed. So the square numbers can also be called as rhombus numbers. They are discussed in chapter 5.

The rotational symmetry of the polygonal number is equal to the number which forms the polygon. For example the rotational symmetry of triangular number is three, the rotational symmetry of square number is four, the rotational symmetry of pentagonal number is five and so on.

Centered square numbers: -

Centered square numbers are formed by the sum of two consecutive square numbers. The first few centered square numbers are 1, 5, 13, 25, 41, 61, 85, 113, 145, 181, 221, 265, 313, 365, 421, 545, 613, 685, 761, 841, 925, 1013, 1105, 1201, 1301, 1405, 1513, 1625, 1741, 1861, 1985, 2113, 2245, 2381, 2521, 2665, 2813, 2965, 3121, 3281, 3445, 3613, 3785, 3961, 4141, 4325...

Palindromic centered square numbers are the centered square numbers which are palindromic in nature. For example 181, 313, 545...

The centered square number which is ascending in nature is called as centered square ascending number and the centered square number which is descending in nature is called as centered square descending number. For example 145 is a centered square ascending number and 841 is a centered square descending number.

Centered square numbers which are prime are called as centered square prime numbers. It is found by the formula $[n^2 + (n+1)^2]$ where 'n' is a real number. The centered square prime numbers are 5, 25, 41, 61, 85, 113...

The centered square numbers which are composite are called as centered square composite numbers. For example 25, 85, 145, 265... are called as centered square composite numbers.

It is to be noted all the centered square numbers and the centered square prime numbers are odd numbers.

All the centered square numbers as well as centered square prime numbers are odd numbers. Further if the centered square number contains all the digits as odd, then it is a super odd centered square number. For example 1013 is a super odd centered square number. Similarly if the centered square prime number contains all the digits as odd, then it is a super odd centered square prime number. For example 113 is a super odd centered square prime number.

Pentagonal number: -

The numbers whose dots form a lattice work of pentagons are called as pentagonal numbers. The pentagonal number series is 1, 5, 12, 22, 35, 51, 70, 92, 117, 145, 176, 210, 247, 287, 330, 376, 425, 477, 532...
There are 81 pentagonal numbers below 10,000.

The 'n'th pentagonal number can be calculated by using the formula $[n(3n-1)/2]$ or $[(3n^2-n)/2]$ where 'n' is a real number. For example 5th pentagonal number is 35.

The nth pentagonal number is one third of the $(3n-1)^{th}$ triangular number. For example the tenth pentagonal number 145 is one third of the 29th triangular number.

The pentagonal number which is odd is called as odd pentagonal number and the pentagonal number which is even is called as even pentagonal number. For example 92 is an even pentagonal number and 51 is an odd pentagonal number.

Further if the pentagonal number contains all the digits as even, then it is a super even pentagonal number and if the pentagonal number contains all the digits as odd, then it is a super odd pentagonal number. For example 22 is a super even pentagonal number and 117 is a super odd pentagonal number.

The pentagonal number which is repdigit in nature is called as repdigit pentagonal number. For example 22 is a repdigit pentagonal number.

The pentagonal number which is ascending in nature is called as pentagonal ascending number and the pentagonal number which is descending in nature is called as pentagonal descending number. For example 145 is a pentagonal ascending number and 532 is a pentagonal descending number.

The pentagonal number which is also is a prime number is called as pentagonal prime number and the pentagonal number which is also a composite number is called as pentagonal composite number. For example 5 is a pentagonal prime number and 51 is a pentagonal composite number.

A number 'x' is a pentagonal number if and only if "24x + 1" is a square number. For example 22 is a pentagonal number because $[(24 \times 22) + 1] = 529$ is a square number.

A pentagonal number should not be a square number.

The pentagonal numbers which are palindromic are called as palindromic pentagonal number. The sequence of palindromic pentagonal number is 1, 22, 1001, 2882, 12521, 720027…

Centered pentagonal numbers: -

The numbers whose dots form a lattice work of centered pentagons are called as centered pentagonal numbers. The centered pentagonal number sequence is 1, 6, 16, 31, 51, 76, 106, 141, 181, 226, 276, 331, 391, 456, 526, 601, 681, 766, 856, 951, 1051, 1156, 1266, 1381, 1501, 1626, 1756, 1891, 2031, 2176, 2326, 2481, 2641, 2806, 2976...

The centered pentagonal number which is odd is called as odd centered pentagonal number and the centered pentagonal number which is even is called as even centered pentagonal number. For example 226 is an even centered pentagonal number and 181 is an odd centered pentagonal number.

Further if the centered pentagonal number contains all the digits as even, then it is a super even centered pentagonal number and if the centered pentagonal number contains all the digits as odd, then it is a super odd centered pentagonal number. For example 226 is a super even centered pentagonal number and 951 is a super odd centered pentagonal number.

The centered pentagonal number which is ascending in nature is called as centered pentagonal ascending number and the centered pentagonal number which is descending in nature is called as centered pentagonal descending number. For example 456 is a centered pentagonal ascending number and 951 is a centered pentagonal descending number.

The centered pentagonal number which is also a prime number is called as centered pentagonal prime number and the centered pentagonal number which is also a composite number is called as centered pentagonal composite number. For example 31 is a centered pentagonal prime number and 681 is a centered pentagonal composite number.

A palindromic centered pentagonal number is a centered pentagonal number which is palindromic in nature. For example 141, 181...

The centered pentagonal numbers can be found using the following formula $\{[5(n-1)^2 + 5(n-1) + 2]/2\}$ where 'n' is a real number starting from 1 onwards.

The square number which is also a centered pentagonal number is called as centered pentagonal square number. For example 16 is a centered pentagonal square number.

Obviously there will be more triangular numbers that the square numbers. Similarly there will be more square numbers than the pentagonal numbers and so on.

Diagrammatically the triangular number, the square number and the rectangular number can be represented as follows. Note that the dots represent the number.

Triangular numbers

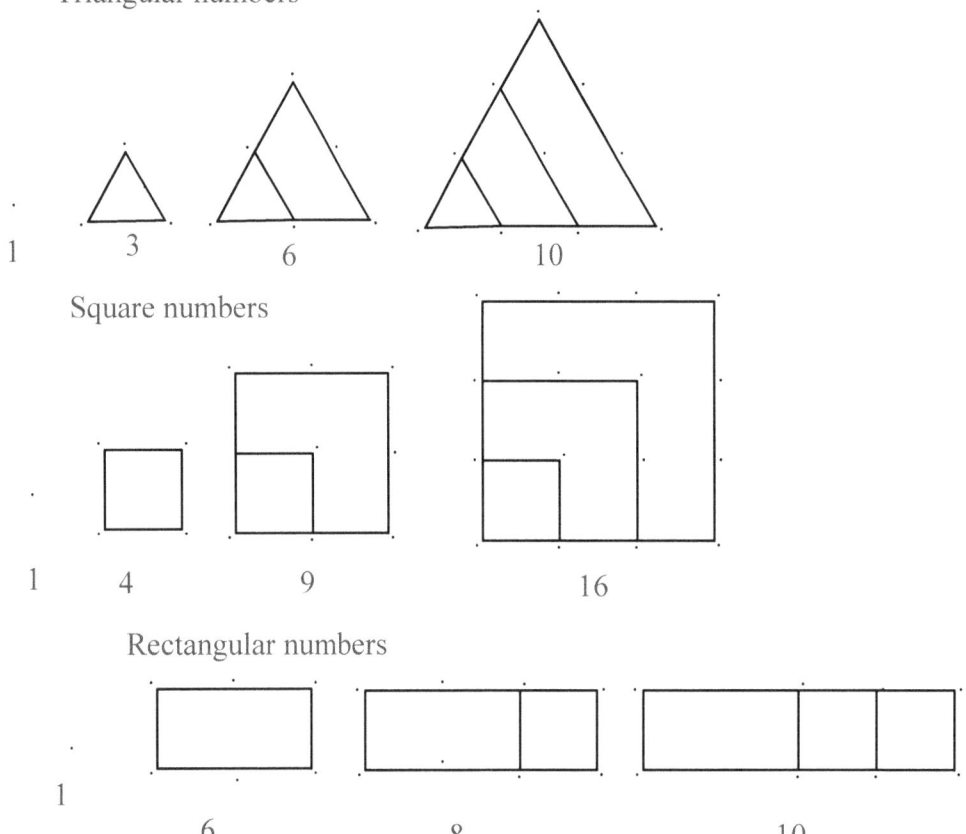

Similarly we can represent the higher polygonal numbers.
Likewise we can draw the centered polygonal numbers as follows.
Centered square numbers

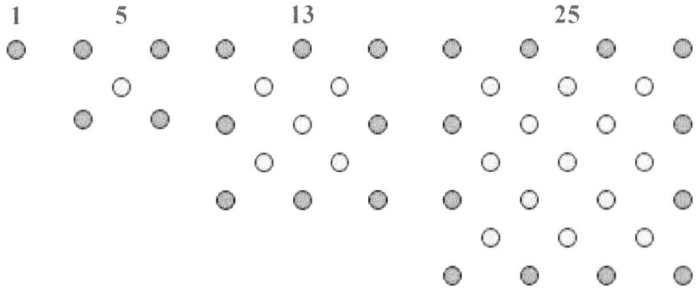

Centered hexagonal numbers

1   7   19   37

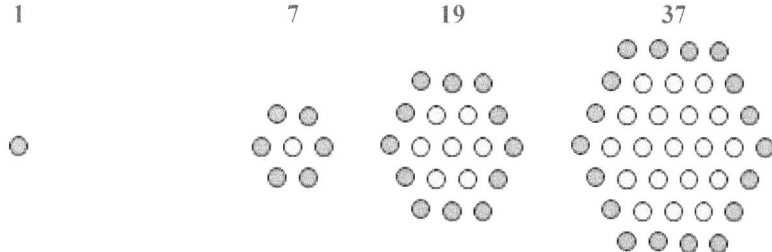

Centered pentatgonal numbers

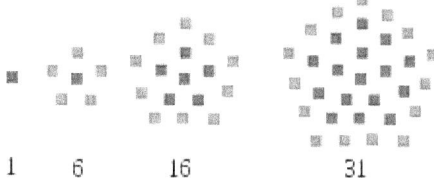

1   6   16   31

Hexagonal number or cornered hexagonal number:-

A hexagonal number is a figurate number, the nth hexagonal number will be the number of points in a hexagon with *n* regularly spaced points on a side. So the hexagonal number can also be called as hexagonal triangular number. For example the hexagonal number sequence is: 1, 6, 15, 28, 45, 66, 91, 120, 153, 190, 231, 276, 325, 378, 435, 496, 561, 630, 703, 780, 861, 946…

It is to be noted that all hexagonal numbers are composite numbers.

The hexagonal number which is repdigit in nature is called as repdigit hexagonal number. For example 66 is a repdigit hexagonal number.

The hexagonal number which is ascending in nature is called as hexagonal ascending number and the hexagonal number which is descending in nature is called as hexagonal descending number. For example 378 is a hexagonal ascending number and 630 is a hexagonal descending number.

The hexagonal number which is odd is called as odd hexagonal number and the hexagonal number which is even is called as even hexagonal number. For example 378 is an even hexagonal number and 153 is an odd hexagonal number.

 Further if the odd hexagonal number contains all of its digits as odd, then it is a super odd hexagonal number and if the even hexagonal number contains all of its digits as even, then it is a super even hexagonal number. For example 153 is a super odd hexagonal number and 28 is a super even hexagonal number.

The formula for the nth hexagonal number is n (2n-1) where 'n' is a real number. Applying the formula the third hexagonal number is 15.

Every hexagonal number is a triangular number, but only every other triangular number or alternate triangular number (the 1st, 3rd, 5th, 7th, etc.) is a hexagonal number.

Like a triangular number, the digital root in base 10 of a hexagonal number can only be 1, 3, 6, or 9.

One can efficiently test whether a positive integer *x* is a hexagonal number by computing the formula n = {[(square root of 8x+1) +1]/4}. If *n* is an integer, then *x* is the *n*th hexagonal number. If *n* is not an integer, then *x* is not a hexagonal number. Applying the formula if x=6 we can get n=2 and then 6 is a second hexagonal number.

Centered hexagonal number: -

A centered hexagonal number, or hex number, is a centered figurate number that represents a hexagon with a dot in the center and all other dots surrounding the center dot in a hexagonal lattice. For example the centered hexagonal number sequence is 1, 7, 19, 37, 61, 91, 127, 169, 217, 271, 331, 397, 469, 547, 631, 721, 817, 919, 1027, 1141, 1261, 1387, 1519, 1657, 1801, 1951, 2107, 2269, 2437, 2611, 2791, 2977, 3169, 3367, 3571, 3781, 3997, 4219, 4447, 4681, 4921, 5167, 5419, 5677, 5941, 6211, 6487...

The nth centered hexagonal number is given by the formula 3n (n-1) +1 where 'n' is a real number. Applying the formula if n=3 we get 19 and 19 is the third centered hexagonal number.

Note:-331 is both a centered pentagonal number as well as a centered hexagonal number.

All centered hexagonal numbers are odd numbers. Further if the odd centered hexagonal number contains all the digits as odd, then it is a super odd centered hexagonal number. For example 1519 is a super odd centered hexagonal number and 817 is not a super odd centered hexagonal number.

The centered hexagonal number which is ascending in nature is called as centered hexagonal ascending number and the centered hexagonal number which is descending in nature is called as centered hexagonal descending number. For example 469 is a centered hexagonal ascending number and 631 is a centered hexagonal descending number.

The centered hexagonal number which is also a prime number is called as centered hexagonal prime number and the centered hexagonal number which is also a composite number is called as centered hexagonal composite number. For example 7, 9, 37... are centered hexagonal prime number and 91, 169... are centered hexagonal composite number.

The centered hexagonal number which is palindromic in nature is called as palindromic centered hexagonal number. For example 919 is a centered hexagonal palindromic number.

The square number which is also a centered hexagonal number is called as centered hexagonal square number. For example 169 is a centered hexagonal square number.

Heptagonal number: -

A heptagonal number is a figurate number, the nth heptagonal number will be the number of points in a heptagon with n regularly spaced points on a side. For example the heptagonal number sequence is: 1, 7, 18, 34, 55, 81, 112, 148, 189, 235, 286, 342, 403, 469, 540, 616, 697, 783, 874, 970, 1071, 1177, 1288, 1404, 1525, 1651, 1782, 1918, 2059, 2205, 2356, 2512, 2673, 2839, 3010, 3186, 3367, 3553, 3744, 3940, 4141, 4347, 4558, 4774, 4995, 5221, 5452, 5688...

The nth heptagonal number is given by the formula $[(5n^2-3n)/2]$. Applying the formula if n=4 we get 34 and 34 is the fourth heptagonal number in the sequence.

The heptagonal number which is odd is called as odd heptagonal number and the heptagonal number which is even is called as even heptagonal number. For example 286 is an even heptagonal number and 4347 is an odd heptagonal number.

Further if the odd heptagonal number contains all of its digits as odd, then it is a super odd heptagonal number and if the even heptagonal number contains all of its digits as even, then it is a super even heptagonal number. For example 1177 is a super odd heptagonal number and 286 is a super even heptagonal number.

The heptagonal number which is prime is called as heptagonal prime number and the heptagonal number which is composite is called as heptagonal composite number. For example 7 is a heptagonal prime number and 81 is a heptagonal composite number.

The heptagonal number which is ascending in nature is called as heptagonal ascending number and the heptagonal number which is descending in nature is called as heptagonal descending number. For example 148 is a heptagonal ascending number and 970 is a heptagonal descending number.

A palindromic heptagonal number is a heptagonal number which is palindromic in nature. For example 55, 616, 4774… are some of the palindromic heptagonal number.

The square number which is also a heptagonal number is called as heptagonal square number. For example 81 is a heptagonal square number.

The centered heptagonal number: -

A centered heptagonal number is a centered figurate number that represents a heptagon with a dot in the center and all other dots surrounding the center dot in successive heptagonal layers. For example the sequence of centered heptagonal number is 1, 8, 22, 43, 71, 106, 148, 197, 253, 316, 386, 463, 547, 638, 736, 841, 953, 1072, 1198, 1331, 1471, 1618, 1772, 1933, 2101, 2276, 2458, 2647, 2843, 3046, 3256, 3473, 3697, 3928, 4166, 4411, 4663, 4922, 5188, 5461, 5741, 6028, 6322, 6623, 6931, 7246...

The centered heptagonal number for n is given by the formula $[(7n^2-7n+2)/2]$ where 'n' is a real number. Applying the formula if n=4 we get 43 and is the fourth centered heptagonal number in the sequence.

The centered heptagonal number which is odd is called as odd centered heptagonal number and the centered heptagonal number which is even is called as even centered heptagonal number. For example 316 is an even centered heptagonal number and 2101 is an odd centered heptagonal number.

Further if the odd centered heptagonal number contains all the digits as odd, then it is a super odd centered heptagonal number. For example 197 is a super odd centered heptagonal number and 386 is not a super odd centered heptagonal number.

The centered heptagonal number which is ascending in nature is called as centered heptagonal ascending number and the centered heptagonal number which is descending in nature is called as centered heptagonal descending number. For example 148 is a centered heptagonal ascending number and 841 is a centered hexagonal descending number.

The centered heptagonal number which is prime is called as centered heptagonal prime number and the centered heptagonal number which is composite is called as centered heptagonal composite number. For example 43 is a centered heptagonal prime number and 106 is a centered heptagonal composite number.

A palindromic centered heptagonal number is a centered heptagonal number which is palindromic in nature. For example 1331 is a palindromic centered heptagonal number.

A repdigit centered heptagonal number is a centered heptagonal number which is repdigit in nature. For example 22 is a repdigit centered heptagonal number.

Octagonal number or cornered octagonal number: -

An octagonal number is a figurate number, the nth octagonal number will be the number of points in an octagon with n regularly spaced points on a side. For example 1, 8, 21, 40, 65, 96, 133, 176, 225, 280, 341, 408, 481, 560, 645, 736, 833, 936, 1045, 1160, 1281, 1408, 1541... are octagonal numbers.

The octagonal number for n is given by the formula $3n^2 - 2n$, where $n > 0$.

The octagonal numbers which are odd are called as odd octagonal numbers and the octagonal numbers which are even are called as even octagonal numbers. For example 21 is an odd octagonal number and 40 is an even octagonal number.

Further if the odd octagonal number contains all of its digits as odd, then it is a super odd octagonal number and if the even octagonal number contains all of its digits as even, then it is a super even octagonal number. For example 133 is a super odd octagonal number and 408 is a super even octagonal number.

Centered octagonal number: -

Centered octagonal number is a centered figurate number that represents an octagon with a dot in the center and all other dots surrounding the center dot in successive octagonal layers. For example the octagonal numbers are 1, 9, 25, 49, 81, 121, 169, 225, 289, 361, 441, 529, 625, 729, 841, 961, 1089, 1225, 1369, 1521, 1681, 1849, 2025, 2209, 2401, 2601, 2809, 3025, 3249, 3481, 3721, 3969, 4225, 4489, 4761, 5041, 5329, 5625...

The centered octagonal number is calculated by the formula $(2n-1)^2$ where 'n' is a real number.

All centered octagonal numbers are odd numbers.

The centered octagonal number which is ascending in nature is called as centered octagonal ascending number and the centered octagonal number which is descending in nature is called as centered octagonal descending number. For example 289 is a centered octagonal ascending number and 961 is a centered octagonal descending number.

In base 10, all centered octagonal numbers are odd numbers and the one's digits follow the pattern 1-9-5-9-1.

An odd number is a centered octagonal number if and only if it is a perfect square.

The square number which is also a centered octagonal number is called as centered octagonal square number. For example 169 is a centered octagonal square number.

A palindromic centered octagonal number is a centered octagonal number which is palindromic in nature. For example 121 is a centered octagonal palindromic number.

Nonagonal or enneagonal number: -

A nonagonal number is a figurate number, the nth nonagonal number will be the number of points in a nonagon with n regularly spaced points on a side. The nonagonal number sequence is 1, 9, 24, 46, 75, 111, 154, 204, 261, 325, 396, 474, 559, 651, 750, 856, 969, 1089, 1216, 1350, 1491, 1639, 1794, 1956, 2125, 2301, 2484, 2674, 2871, 3075, 3286, 3504, 3729, 3961, 4200, 4446, 4699, 4959, 5226, 5500, 5781, 6069, 6364, 6666, 6975, 7291, 7614, 7944, 8281, 8625, 8976, 9334, 9699...

The nonagonal number for n is given by the formula $[(7n^2-5n)/2]$. Applying the formula if n=3 we get 24 and is the third nonagonal number in the sequence.

The nonagonal number which is odd is called as odd nonagonal number and the nonagonal number which is even is called as even nonagonal number. For example 2674 is an even nonagonal number and 5781 is an odd nonagonal number.

Further if the odd nonagonal number contains all of its digits as odd, then it is a super odd nonagonal number and if the even nonagonal number contains all of its digits as even, then it is a super even nonagonal number. For example 75 is a super odd nonagonal number and 2484 is a super even nonagonal number.

A palindromic nonagonal number is a nonagonal number which is palindromic in nature. For example 474 and 969 are some of the examples for palindromic nonagonal number.

The nonagonal number which is repdigit in nature is called as repdigit nonagonal number. For example 6666 is a repdigit nonagonal number.

The palindromic nonagonal number which is odd is called as odd palindromic nonagonal number and the palindromic number which is even is called as even palindromic nonagonal number. For example 969 is an odd palindromic nonagonal number and 474 is an even palindromic nonagonal number.

The nonagonal number which is repunit in nature is called as repunit nonagonal number. For example 111 is a repunit nonagonal number.

The nonagonal number which is repdigit in nature is called as repdigit nonagonal number. For example 6666 is a repdigit nonagonal number.

Centered nonagonal number: -

It is a centered figurate number that represents a nonagon with a dot in the center and all other dots surrounding the center dot in successive nonagonal layers. The centered nonagonal number sequence is 1, 10, 28, 55, 91, 136, 190, 253, 325, 406, 496, 595, 703, 820, 946, 1081, 1225, 1378, 1540, 1711, 1891, 2080, 2278, 2485, 2701, 2926, 3160, 3403, 3655, 3916, 4186, 4465, 4753, 5050, 5356, 5671, 5995, 6328, 6670, 7021, 7381, 7750, 8128, 8515, 8911, 9316.....

The centered nonagonal number for n is given by the formula $[(3n-1)(3n-2)/2]$. Applying the formula if n=4 we get 55 and is the fourth nonagonal number in the sequence.

The centered nonagonal number which is ascending in nature is called as centered nonagonal ascending number and the centered nonagonal number which is descending in nature is called as centered nonagonal descending number. For example 136 is a centered nonagonal ascending number and 820 is a centered nonagonal descending number.

The centered nonagonal number which is odd is called as odd centered nonagonal number and the centered nonagonal number which is even is called as even centered nonagonal number. For example 496 is an even centered nonagonal number and 325 is an odd centered nonagonal number.

Further if the odd centered nonagonal number contains all of its digits as odd, then it is a super odd centered nonagonal number and if the even centered nonagonal number contains all of its digits as even, then it is a super even centered nonagonal number. For example 595 is a super odd centered nonagonal number and 820 is a super even centered nonagonal number.

A palindromic centered nonagonal number is a centered nonagonal number which is palindromic in nature. For example 55, 595…

A centered nonagonal number which is repdigit in nature is called as repdigit centered nonagonal number. For example 55 is a repdigit centered nonagonal number.

Decagonal number: -

A decagonal number is a figurate number, the nth decagonal number will be the number of points in a decagon with n regularly spaced points on a side. The n-th decagonal number is given by the formula $4n^2 - 3n$ where 'n' is a real number. The decagonal numbers are: 1, 10, 27, 52, 85, 126, 175, 232, 297, 370, 451, 540, 637, 742, 855, 976, 1105, 1242, 1387, 1540, 1701, 1870, 2047, 2232, 2425, 2626, 2835, 3052, 3277, 3510, 3751, 4000, 4257, 4522, 4795, 5076, 5365, 5662, 5967, 6280, 6601, 6930, 7267, 7612, 7965, 8326...

The n-th decagonal number can also be calculated by adding the square of n to thrice the (n—1)-th pronic number or, to put it algebraically, as $[n^2 + 3(n^2 - n)]$ where 'n' is a real number.

The decagonal number which is ascending in nature is called as decagonal ascending number and the decagonal number which is descending in nature is called as decagonal descending number. For example 126 is a decagonal ascending number and 540 is a decagonal descending number.

The square number which is also a decagonal number is called as decagonal square number. For example 4000 is a decagonal square number.

A decagonal number which is palindromic in nature is called as palindromic decagonal number. For example 232 is a palindromic decagonal number.

The decagonal number which is odd is called as odd decagonal number and the decagonal number which is even is called as even decagonal number. For example 126 is an even decagonal number and 175 is an odd decagonal number.

Further if the odd decagonal number contains all of its digits as odd, then it is a super odd decagonal number and if the even decagonal number contains all of its digits as even, then it is a super even decagonal number. For example 175 is a super odd decagonal number and 540 is a super even decagonal number.

Centered decagonal number: -

It is a centered figurate number that represents a decagon with a dot in the center and all other dots surrounding the center dot in successive decagonal layers. The centered decagonal number for n is given by the formula $5(n^2 - n) + 1$.

The centered decagonal numbers are 1, 11, 31, 61, 101, 151, 211, 281, 361, 451, 551, 661, 781, 911, 1051, 1202, 1361, 1531, 1711, 1901, 2101, 2311, 2531, 2761, 3001, 3251, 3511, 3781, 4061, 4351, 4651, 4961, 5281, 5611, 5951, 6301, 6661, 7031, 7411, 7801, 8201, 8611, 9031, 9461, 9901...

A Palindromic centered decagonal number is the centered decagonal number which is palindromic in nature. For example 11, 101, 151...

It is to be noted that all the centered decagonal numbers and palindromic centered decagonal numbers are odd numbers. Further if the odd centered decagonal number contains all the digits as odd, then it is a super odd centered decagonal number. For example 1711 is a super odd centered decagonal number and 1361 is not a super odd centered decagonal number.

Like any other centered k-gonal number, the nth centered decagonal number can reckoned by multiplying the (n - 1)th triangular number by k, 10 in this case, then adding 1. As a consequence of performing the calculation in base 10, the centered decagonal numbers can be obtained by simply adding a 1 to the right of each triangular number. Therefore, all centered decagonal numbers are odd and in base 10 always end in 1.

The square number which is also a centered decagonal number is called as centered decagonal square number. For example 361 is a centered decagonal square number.

A centered decagonal prime is a centered decagonal number that is prime. The centered decagonal primes are found by using the formula $[5(n^2-n)+1]$ where n>1.

The centered decagonal primes are 11, 31, 61, 101, 151, 211, 281, 451, 661, 911, 1051, 1201, 1361, 1531, 1901, 2311, 2531, 3001, 3251, 3511, 4651, 5281...

A Palindromic centered decagonal prime number is the centered decagonal prime number which is palindromic in nature. For example 11, 101, 151...

Hendecagonal number: -

A hendecagonal number is a figurate number, the nth hendecagonal number will be the number of points in a hendecagon with n regularly spaced points on a side. The hendecagonal number sequence is 1, 11, 30, 58, 95, 141, 196, 260, 333, 415, 506, 606...

The nth hendecagonal number is calculated by the formula $\frac{1}{2}(9n^2-7n)$ where 'n' is a real number.

A hendecagonal number which is odd is called as odd hendecagonal number and a hendecagonal number which is even is called as even hendecagonal number. For example 506 is an even hendecagonal number and 95 is an odd hendecagonal number.

Further if the odd hendecagonal number contains all of its digits as odd, then it is a super odd hendecagonal number and if the even hendecagonal number contains all of its digits as even, then it is a super even hendecagonal number. For example 95 is a super odd hendecagonal number and 606 is a super even hendecagonal number.

A Palindromic hendecagonal number is the hendecagonal number which is palindromic in nature. For example 11, 141, 333,606...

A hendecagonal number which is repunit in nature is called as repunit hendecagonal number. For example 11 is a repunit hendecagonal number.

A hendecagonal number which is repdigit in nature is called as repdigit hendecagonal number. For example 333 is a repdigit hendecagonal number.

The square number which is also a hendecagonal number is called as hendecagonal square number. For example 196 is a hendecagonal square number.

A hendecagonal number which is also a prime number is called as hendecagonal prime number and the hendecagonal number which is a composite number is called as hendecagonal composite number. For example 11 is a hendecagonal prime number and 606 is a hendecagonal composite number.

Centered hendecagonal numbers: -

A centered hendecagonal number is a figurate number whose dots will form a lattice work of centered hendecagon. The sequence of centered hendecagonal number is 1, 12, 34, 67, 111, 166, 232, 309, 397, 496, 606, 727, 859, 1002, 1156, 1321, 1497, 1684, 1882, 2091, 2311, 2542, 2784, 3037, 3301, 3576, 3862, 4159, 4467, 4786...

A centered hendecagonal number which is odd is called as odd centered hendecagonal number and a centered hendecagonal number which is even is called as even centered hendecagonal number. For example 496 is an even centered hendecagonal number and 309 is an odd centered hendecagonal number.

Further if the odd centered hendecagonal number contains all of its digits as odd, then it is a super odd centered hendecagonal number and if the even centered hendecagonal number contains all of its digits as even, then it is a super even centered hendecagonal number. For example 397 is a super odd centered hendecagonal number and 606 is a super even centered hendecagonal number.

A Palindromic centered hendecagonal number is the centered hendecagonal number which is palindromic in nature. For example 111, 232, 606, 727...

A palindromic centered hendecagonal number which is odd is called as odd palindromic centered hendecagonal number and a palindromic centered hendecagonal number which is even is called as even palindromic centered hendecagonal number. For example 606 is an even palindromic centered hendecagonal number and 727 is an odd palindromic centered hendecagonal number.

A centered hendecagonal number which is repunit in nature is called as repunit centered hendecagonal number. For example 111 is a repunit centered hendecagonal number.

A centered hendecagonal number which is also a prime number is called as centered hendecagonal prime number and a centered hendecagonal number which is also a composite number is called as centered hendecagonal composite number. For example 67 is a centered hendecagonal prime number and 34 is a centered hendecagonal composite number.

606 is a hendecagonal, a centered hendecagonal as well as a palindromic centered hendecagonal number.

Dodecagonal or duodecagonal number: -

A dodecagonal number is a figurate number, the nth dodecagonal number will be the number of points in a dodecagon with n regularly spaced points on a side. The dodecagonal number sequence is 1, 12, 33, 64, 105, 156, 217, 288, 369, 460, 561, 672, 793, 924, 1065, 1216, 1377, 1548, 1729, 1920, 2121, 2332, 2553, 2784, 3025, 3276, 3537, 3808, 4089, 4380, 4681, 4992, 5313, 5644, 5985, 6336, 6697, 7068, 7449, 7840, 8241, 8652, 9073, 9504, 9945...

The dodecagonal number for 'n' is given by the formula $5n^2 - 4n$, with $n > 0$.

The dodecagonal number which is an odd number is called as dodecagonal odd number and the dodecagonal number which is an even number is called as dodecagonal even number. For example 64 is a dodecagonal even number and 105 is a dodecagonal odd number.

Further if the odd dodecagonal number contains all of its digits as odd, then it is a super odd dodecagonal number and if the even dodecagonal number contains all of its digits as even, then it is a super even dodecagonal number. For example 793 is a super odd dodecagonal number and 288 is a super even dodecagonal number.

The dodecagonal number which is ascending in nature is called as dodecagonal ascending number and the dodecagonal number which is descending in nature is called as dodecagonal descending number. For example 369 is a dodecagonal ascending number and 8652 is a dodecagonal descending number.

The dodecagonal number which is undulating in nature is called as undulating dodecagonal number. For example 2121 is an undulating dodecagonal number.

The dodecagonal number which is repunit in nature is called as repunit dodecagonal number. For example 33 is a repunit dodecagonal number. All dodecagonal numbers are composite numbers.

Centered dodecagonal numbers: -

A centered dodecagonal numbers are those numbers whose lattice points will form a centered dodecagon. The centered dodecagonal number sequence is 1, 13, 37, 73, 121, 181, 253, 337, 433, 541, 661, 793, 937, 1093, 1261, 1441, 1633, 1837, 2053, 2281, 2521, 2773, 3037, 3313, 3601, 3901, 4213, 4537, 4873, 5221, 5581, 5953, 6337, 6733, 7141, 7561, 7993, 8437, 8893, 9361, 9841, 10333, 10837...

All centered dodecagonal numbers are odd numbers. Further if the odd centered dodecagonal number contains all the digits as odd, then it is a super odd centered dodecagonal number. For example 937 is a super odd centered dodecagonal number and 1261 is not a super odd centered dodecagonal number.

The centered dodecagonal number which is ascending in nature is called as centered dodecagonal ascending number and the centered dodecagonal number which is descending in nature is called as centered dodecagonal descending number. For example 37 is a centered dodecagonal ascending number and 73 is a centered dodecagonal descending number.

A centered dodecagonal number which is also a prime number is called as centered dodecagonal prime number and the centered dodecagonal prime number which is also a composite number is called as centered dodecagonal even number. For example 13 is a centered dodecagonal prime number and 121 is a centered dodecagonal composite number.

The square number which is also a centered dodecagonal number is called as centered dodecagonal square number. For example 121 is a centered dodecagonal square number.

A Palindromic centered dodecagonal number is the centered dodecagonal number which is palindromic in nature. For example 121, 181 1441…

Tridecagonal number: -

A tridecagonal number is a figurate number, the nth tridecagonal number will be the number of points in a tridecagon with n regularly spaced points on a side. The tridecagonal number sequence is 1, 13, 36, 70, 115, 171, 238, 316, 405, 505, 616, 738...

A tridecagonal number which is odd is called as odd tridecagonal number and a tridecagonal number which is even is called as even tridecagonal number. For example 36 is an even tridecagonal number and 13 is an odd tridecagonal number.

Further if the odd tridecagonal number contains all the digits as odd, then it is a super odd tridecagonal number. For example 115 is a super odd tridecagonal number and 405 is not a super odd tridecagonal number.

A tridecagonal number which is also a prime number is called as tridecagonal prime number and the tridecagonal number which is also a composite number is called as tridecagonal composite number. For example 13 is a tridecagonal prime number and 70 is a tridecagonal composite number.

The square number which is also a tridecagonal number is called as tridecagonal square number. For example 36 is a centered nonagonal square number.

A palindromic tridecagonal number is a Tridecagonal number which is palindromic in nature. For example 171, 505, 616…

Centered tridecagonal number: -

A centered tridecagonal number is a figurate number whose lattice points will form a centered decagon. The sequence of centered tridecagonal numbers is 1, 14, 40, 79, 131, 196, 274, 365, 469, 586, 716, 859, 1015, 1184, 1366, 1561, 1769, 1990, 2224, 2471, 2731, 3004, 3290, 3589, 3901, 4226, 4564, 4915, 5279, 5656…

A centered tridecagonal number which is also a prime number is called as centered tridecagonal prime number and the centered tridecagonal number which is also a composite number is called as centered tridecagonal composite number. For example 79 is a centered tridecagonal prime number and 40 is a centered tridecagonal composite number.

The centered tridecagonal number which is undulating in nature is called as undulating centered tridecagonal number. For example 5656 is an undulating centered tridecagonal number.

A centered tridecagonal number which is odd is called as odd centered tridecagonal number and a centered tridecagonal number which is even is called as even centered tridecagonal number. For example 274 is an even centered tridecagonal number and 365 is an odd centered tridecagonal number.

Further if the odd tridecagonal number contains all of its digits as odd, then it is a super odd tridecagonal number and if the even tridecagonal number contains all of its digits as even, then it is a super even tridecagonal number. For example 79 is a super odd tridecagonal number and 2224 is a super even tridecagonal number.

The square number which is also a centered tridecagonal number is called as centered tridecagonal square number. For example 196 is a centered nonagonal square number.

A palindromic centered tridecagonal number is a centered tridecagonal number which is palindromic in nature. For example 131 is a centered tridecagonal palindromic number.

Tetradecagonal number: -

A tetradecagonal number is a figurate number, the nth tetradecagonal number will be the number of points in a tetradecagon with n regularly spaced points on a side. The tetradecagonal number sequence is 1, 14, 39, 76, 125, 186, 259, 344, 441, 550, 671, 804...

A tetradecagonal number which is odd is called as odd tetradecagonal number and a tetradecagonal number which is even is called as even tetradecagonal number. For example 76 is an even tetradecagonal number and 259 is an odd tetradecagonal number.

Further if the odd tetradecagonal number contains all of its digits as odd, then it is a super odd tetradecagonal number and if the even tetradecagonal number contains all of its digits as even, then it is a super even tetradecagonal number. For example 39 is a super odd tetradecagonal number and 804 is a super even tetradecagonal number.

The square number which is also a tetradecagonal number is called as tetradecagonal square number. For example 441 is a tetradecagonal square number.

Centered tetradecagonal number: -

A centered tetradecagonal number is a figurate number whose lattice points will form a centered tetradecagon. The sequence of centered tetradecagonal number is 1, 15, 43, 85, 141, 211, 295, 393, 505, 631, 771, 925, 1093, 1275, 1471, 1681, 1905, 2143, 2395, 2661, 2941, 3235, 3543, 3865, 4201, 4551, 4915, 5293, 5685, 6091...

All centered tetradecagonal numbers are odd numbers. Further if the odd centered tetradecagonal number contains all the digits as odd, then it is a super odd centered tetradecagonal number. For example 771 is a super odd centered tetradecagonal number and 295 is not a super odd centered tetradecagonal number.

The centered tetradecagonal number which is also a prime number is called as centered tetradecagonal prime number and the centered tetradecagonal number which is also a composite number is called as centered tetradecagonal composite number. For example 43 is a centered tetradecagonal prime number and 925 is a centered tetradecagonal composite number.

A palindromic centered tetradecagonal number is a centered tetradecagonal number which is palindromic in nature. For example 141, 393, 505... are some of the palindromic centered tetradecagonal numbers.

Pentadecagonal number: -

A pentadecagonal number is a figurate number, the nth pentadecagonal number will be the number of points in a pentadecagon with n regularly spaced points on a side. The pentadecagonal number sequence is 1, 15, 42, 82, 135, 201, 280, 372, 477, 595, 726, 870…

A pentadecagonal number which is odd is called as odd pentadecagonal number and a pentadecagonal number which is even is called as even pentadecagonal number. For example 42 is an even pentadecagonal number and 135 is an odd pentadecagonal number.

Further if the odd pentadecagonal number contains all of its digits as odd, then it is a super odd pentadecagonal number and if the even pentadecagonal number contains all of its digits as even, then it is a super even pentadecagonal number. For example 135 is a super odd pentadecagonal number and 280 is a super even pentadecagonal number.

A palindromic pentadecagonal number is a pentadecagonal number which is palindromic in nature. For example 595 is a pentadecagonal palindromic number.

Centered pentadecagonal number: -

A centered pentadecagonal number is a figurate number whose lattice points will form a centered pentagon. The sequence of centered decagonal number is 1, 16, 46, 91, 151, 226, 316, 421, 541, 676, 826, 991, 1171, 1366, 1576, 1801, 2041, 2296, 2566, 2851, 3151, 3466, 3796, 4141, 4501, 4876, 5266, 5671, 6091, 6526…

A centered pentadecagonal number which is odd is called as odd centered pentadecagonal number and a pentadecagonal number which is even is called as even centered pentadecagonal number. For example 46 is an even centered pentadecagonal number and 151 is an odd centered pentadecagonal number.

Further if the odd centered pentadecagonal number contains all of its digits as odd, then it is a super odd centered pentadecagonal number and if the even centered pentadecagonal number contains all of its digits as even, then it is a super even centered pentadecagonal number. For example 1171 is a super odd centered pentadecagonal number and 226 is a super even centered pentadecagonal number.

The centered pentadecagonal number which is undulating in nature is called as undulating centered pentadecagonal undulating number. For example 4141 is a centered pentadecagonal undulating number.

The centered pentadecagonal number which is also a prime number is called as centered pentadecagonal prime number and the centered pentadecagonal number which is also a composite number is called as centered pentadecagonal composite number. For example 151 is a centered pentadecagonal prime number and 316 is a centered pentadecagonal composite number.

A centered Pentadecagonal palindromic number is a centered Pentadecagonal number which is palindromic in nature. For example 151, 676…

Hexadecagonal number: -

A hexadecagonal number is a figurate number, the nth hexadecagonal number will be the number of points in a hexadecagon with n regularly spaced points on a side. The hexadecagonal number sequence is 1, 16, 45, 88, 145, 216, 301, 400, 513, 640, 781, 936…

A hexadecagonal number which is odd is called as odd hexadecagonal number and a hexadecagonal number which is even is called as even hexadecagonal number.

For example 640 is an even hexadecagonal number and 301 is an odd hexadecagonal number.

Further if the odd hexadecagonal number contains all of its digits as odd, then it is a super odd hexadecagonal number and if the even hexadecagonal number contains all of its digits as even, then it is a super even hexadecagonal number. For example 513 is a super odd hexadecagonal number and 640 is a super even heptadecagonal number.

The hexadecagonal number which is ascending in nature is called as hexadecagonal ascending number and the hexadecagonal number which is descending in nature is called as hexadecagonal descending number. For example 145 is a hexadecagonal ascending number and 640 is a hexadecagonal descending number.

The square number which is also a hexadecagonal number is called as hexadecagonal square number. For example 400 is a hexadecagonal square number.

A palindromic hexadecagonal number is a hexadecagonal number which is palindromic in nature. For example 88 is a hexadecagonal palindromic number.

Centered hexadecagonal number: -

A centered hexadecagonal number is a figurate number whose lattice points will form a centered hexadecagon. The centered hexadecagonal sequence is 1, 17, 49, 97, 161, 241, 337, 449, 577, 721, 881, 1057, 1249, 1457, 1681, 1921, 2177, 2449, 2737, 3041, 3361, 3697, 4049, 4417, 4801, 5201, 5617, 6049, 6497, 6961...

All the centered hexadecagonal numbers are odd numbers. Further if the odd centered hexadecagonal number contains all the digits as odd, then it is a super odd centered hexadecagonal number. For example 97 is a super odd centered hexadecagonal number and 1249 is not a super odd centered hexadecagonal number.

The centered hexadecagonal number which is also a prime number is called as centered hexadecagonal prime number and the centered hexadecagonal number which is also a composite number is called as centered hexadecagonal composite number. For example 17 is a centered hexadecagonal prime number and 49 is a centered hexadecagonal composite number.

The centered hexadecagonal number which is ascending in nature is called as centered hexadecagonal ascending number and the centered hexadecagonal number which is descending in nature is called as centered hexadecagonal descending number. For example 1249 is a centered hexadecagonal ascending number and 721 is a centered hexadecagonal descending number.

A palindromic centered hexadecagonal number is a centered hexadecagonal number which is palindromic in nature. For example 161 is a centered hexadecagonal palindromic number.

Heptadecagonal number: -

A heptadecagonal number is a figurate number, the nth heptadecagonal number will be the number of points in a heptadecagon with n regularly spaced points on a side. The heptadecagonal number sequence is 1, 17, 48, 94, 155, 231, 322, 428, 549, 685, 836, 1002...

A heptadecagonal number which is odd is called as odd heptadecagonal number and a heptadecagonal number which is even is called as even heptadecagonal number. For example 94 is an even heptadecagonal number and 231 is an odd heptadecagonal number.

Further if the odd heptadecagonal number contains all of its digits as odd, then it is a super odd heptadecagonal number and if the even heptadecagonal number contains all of its digits as even, then it is a super even heptadecagonal number. For example 155 is a super odd heptadecagonal number and 428 is a super even heptadecagonal number.

The heptadecagonal number which is also a prime number is called as heptadecagonal prime number and the heptadecagonal number which is also a composite number is called as heptadecagonal composite number. For example 17 is a heptadecagonal prime number and 94 is a heptadecagonal composite number.

Centered heptadecagonal number: -

A centered heptadecagonal number is a figurate number, whose lattice points will form a centered heptadecagon. The centered heptagonal number sequence is 1, 18, 52, 103, 171, 256, 358, 477, 613, 766, 936, 1123, 1327, 1548, 1786, 2041, 2313, 2602, 2908, 3231, 3571, 3928, 4302, 4693, 5101, 5526, 5968, 6427, 6903, 7396…

A centered heptadecagonal number which is odd is called as odd centered heptadecagonal number and a centered heptadecagonal number which is even is called as even centered heptadecagonal number. For example 256 is an even centered heptadecagonal number and 3571 is an odd centered heptadecagonal number.

Further if the odd centered heptadecagonal number contains all the digits as odd, then it is a super odd centered heptadecagonal number. For example 171 is a super odd centered heptadecagonal number and 1123 is not a super odd centered heptadecagonal number.

The centered heptadecagonal number which is also a prime number is called as centered heptadecagonal prime number and the centered heptadecagonal number which is also a composite number is called as centered heptadecagonal composite number. For example 103 is a centered heptadecagonal prime number and 52 is a centered heptadecagonal composite number.

A palindromic centered heptadecagonal number is a centered heptadecagonal number which is palindromic in nature. For example 171 is a centered heptadecagonal palindromic number.

Octadecagonal number: -

An octadecagonal number is a figurate number, the nth octadecagonal number will be the number of points in a octadecagon with n regularly spaced points on a side. The octadecagonal number sequence is 1, 18, 51, 100, 165, 246, 343, 456, 585, 730, 891, 1068…

An octadecagonal number which is odd is called as odd octadecagonal number and an octadecagonal number which is even is called as even octadecagonal number. For example 456 is an even octadecagonal number and 891 is an odd octadecagonal number.

Further if the odd octadecagonal number contains all of its digits as odd, then it is a super odd octadecagonal number and if the even octadecagonal number contains all of its digits as even, then it is a super even octadecagonal number. For example 51 is a super odd octadecagonal number and 246 is a super even octadecagonal number.

The square number which is also an octadecagonal number is called as octadecagonal square number. For example 100 is an octadecagonal square number.

A palindromic octadecagonal number is an octadecagonal number which is palindromic in nature. For example 343, 585… are some of the examples for octadecagonal palindromic number.

Centered octadecagonal number: -

A centered octadecagonal number is a figurate number whose lattice points will form a centered octadecagon. The sequence of centered octadecagon number is 1, 19, 55, 109, 181, 271, 379, 505, 649, 811, 991, 1189, 1405, 1639, 1891, 2161, 2449, 2755, 3079, 3421, 3781, 4159, 4555, 4969, 5401, 5851, 6319, 6805, 7309, 7831…

All centered octadecagonal numbers are odd numbers. Further if the odd centered octadecagonal number contains all the digits as odd, then it is a super odd centered octadecagonal number. For example 379 is a super odd centered octadecagonal number and 649 is not a super odd centered octadecagonal number.

The centered Octadecagonal palindromic number is a centered Octadecagonal number that is palindromic in nature. For example 181, 505…

The centered octadecagonal number which is also a prime number is called as centered octadecagonal prime number and the centered octadecagonal number which is also a composite number is called as centered octadecagonal composite number. For example 19 is a centered octadecagonal prime number and 55 is a centered octadecagonal composite number.

The centered octadecagonal repdigit number is a centered octadecagonal number that is repdigit in nature. For example 55 is a centered octadecagonal repdigit number.

Nonadecagonal number: -

A nonadecagonal number is a figurate number, the nth nonadecagonal number will be the number of points in a nonadecagon with n regularly spaced points on a side. This number sequence is 1, 19, 54, 106, 175, 261, 364, 484, 621, 775, 946, 1134…

A nonadecagonal number which is odd is called as odd nonadecagonal number and a nonadecagonal number which is even is called as even nonadecagonal number. For example 106 is an even nonadecagonal number and 621 is an odd nonadecagonal number.

Further if the odd nonadecagonal number contains all the digits as odd, then it is a super odd nonadecagonal number. For example 175 is a super odd nonadecagonal number and 106 is not a super odd nonadecagonal number.

The nonadecagonal number which is also a prime number is called as nonadecagonal prime number and the nonadecagonal number which is also a composite number is called as nonadecagonal composite number. For example 19 is a nonadecagonal prime number and 54 is a nonadecagonal composite number.

A palindromic nonadecagonal number is a nonadecagonal number which is palindromic in nature. For example 484 is a nonadecagonal palindromic number.

Centered nonadecagonal number: -

A centered nonadecagonal number is a figurate number whose lattice points will form a centered nonadecagon. The sequence of centered nonadecagonal number is 1, 20, 58, 115, 191, 286, 400, 533, 685, 856, 1046, 1255, 1483, 1730, 1996, 2281, 2585, 2908, 3250, 3611, 3991, 4390, 4808, 5245, 5701, 6176, 6670, 7183, 7715, 8266…

The centered nonadecagonal number which is also a prime number is called as centered nonadecagonal prime number and the centered nonadecagonal number which is also a composite number is called as centered nonadecagonal composite number. For example 191 is a centered nonadecagonal prime number and 533 is a centered nonadecagonal composite number.

The square number which is also a centered nonadecagonal number is called as centered nonadecagonal square number. For example 400 is a centered nonagonal square number.

A centered nonadecagonal number which is odd is called as odd centered nonadecagonal number and a centered nonadecagonal number which is even is called as even centered nonadecagonal number. For example 286 is an even centered nonadecagonal number and 533 is an odd centered nonadecagonal number.

Further if the odd centered nonadecagonal number contains all of its digits as odd, then it is a super odd centered nonadecagonal number and if the even centered nonadecagonal number contains all of its digits as even, then it is a super even centered nonadecagonal number. For example 115 is a super odd centered nonadecagonal number and 286 is a super even centered nonadecagonal number.

A palindromic centered nonadecagonal number is a centered nonadecagonal number which is palindromic in nature. For example 191 is a centered nonadecagonal palindromic number.

Icosagonal number: -

An icosagonal number is a figurate number, the nth icosagonal number will be the number of points in an icosagon with n regularly spaced points on a side. The icosagonal number sequence is 1, 20, 57, 112, 185, 276, 385, 512, 657, 820, 1001, 1200…

An icosagonal number which is odd is called as odd icosagonal number and an icosagonal number which is even is called as even icosagonal number. For example 112 is an even icosagonal number and 185 is an odd icosagonal number.

Further if the odd icosagonal number contains all of its digits as odd, then it is a super odd icosagonal number and if the even icosagonal number contains all of its digits as even, then it is a super even icosagonal number. For example 57 is a super odd icosagonal number and 820 is a super even icosagonal number.

A palindromic icosagonal number is an icosagonal number which is palindromic in nature. For example 1001 is an icosagonal palindromic number.

Centered icosagonal number: -

A centered icosagonal number is a figurate number whose lattice points will form a centered icosagon. The sequence of centered icosagonal number is 1, 21, 61, 121, 201, 301, 421, 561, 721, 901, 1101, 1321, 1561, 1821, 2101, 2401, 2721, 3061, 3421, 3801, 4201, 4621, 5061, 5521, 6001, 6501, 7021, 7561, 8121, 8701…

All centered icosagonal numbers are odd numbers.

The centered icosagonal number which is also a prime number is called as centered icosagonal prime number and the centered icosagonal number which is also a composite number is called as centered icosagonal composite number. For example 61 is a centered icosagonal prime number and 21 is a centered icosagonal composite number.

A palindromic centered icosagonal number is a centered icosagonal number which is palindromic in nature. For example 121 is a centered icosagonal palindromic number.

General properties of polygonal numbers: -

Nomenclature of polygons: -

To construct the name of a polygon with more than 20 and less than 100 edges or sides, combine the prefixes as follows: 20 icosi- , 30 triaconta- , 40 tetraconta- , 50 pentaconta- , 60 hexaconta- , 70 heptaconta- , 80 octaconta- , and 90 enneaconta-

Then add for 1 -hena- , 2 -di- , 3 -tri- , 4 -tetra- , 5 -penta- , 6 -hexa- , 7 -hepta- , 8 - octa- and 9 –ennea. And finally add –gon. Thus a 42-sided figure is a tetracontadigon and a 50-sided figure is a pentacontagon.

Relation between the odd and polygonal numbers: -

If the polygonal number is odd like triangular number, pentagonal number, heptagonal number, nonagonal number and so on, and if we categorize the numbers as odd (O) number and even (E) number the sequence is O, O, E, E, O, O, E, E… and it repeats further as we move. For example in the case of icosihenagonal number which is an odd polygonal number, icosihenagonal number sequence is 1, 21, 60, 118, 195, 291, 406, 540, 693, 865, 1056, 1266… and is in the order O, O, E, E, O, O, E, E…

If the polygonal number is even like square number, hexagonal number, octagonal number, decagonal number and so on, and if we categorize the numbers as odd (O) number and even (E) number the sequence is O, E, O, E, O, E… and it repeats further as we move. For example in the case of icosagonal number which is an even polygonal number, icosagonal number sequence is 1, 20, 57, 112, 185, 276, 385, 512, 657, 820, 1001, 1200… and is in the order O, E, O, E, O, E…

Icosihenagonal number: -

An icosihenagonal number is a figurate number, the nth icosihenagonal number will be the number of points in an icosihenagon with n regularly spaced points on a side. The icosihenagonal number sequence is 1, 21, 60, 118, 195, 291, 406, 540, 693, 865, 1056, 1266…

An icosihenagonal number which is odd is called as odd icosihenagonal number and an icosihenagonal number which is even is called as even icosihenagonal number. For example 118 is an even icosihenagonal number and 195 is an odd icosihenagonal number.

Further if the odd icosihenagonal number contains all of its digits as odd, then it is a super odd icosihenagonal number and if the even icosihenagonal number contains all of its digits as even, then it is a super even icosihenagonal number. For example 195 is a super odd icosihenagonal number and 60 is a super even icosihenagonal number.

Centered icosihenagonal number: -

The centered icosihenagonal numbers are those numbers whose lattice points will form a centered icosihenagon. The sequence of centered icosihenagonal number is 1, 22, 64, 127, 211, 316, 442, 589, 757, 946, 1156, 1387, 1639, 1912, 2206, 2521, 2857, 3214, 3592, 3991, 4411, 4852, 5314, 5797, 6301, 6826, 7372, 7939, 8527, 9136…

A centered icosihenagonal number which is odd is called as odd centered icosihenagonal number and a centered icosihenagonal number which is even is called as even centered icosihenagonal number. For example 316 is an even centered icosihenagonal number and 211 is an odd centered icosihenagonal number.

Further if the odd centered icosihenagonal number contains all of its digits as odd, then it is a super odd centered icosihenagonal number and if the even centered icosihenagonal number contains all of its digits as even, then it is a super even centered icosihenagonal number. For example 5797 is a super odd centered icosihenagonal number and 442 is a super even centered icosihenagonal number.

185

The square number which is also a centered icosihenagonal number is called as centered icosihenagonal square number. For example 64 is a centered icosihenagonal square number.

A centered icosihenagonal number which is also a prime number is called as centered icosihenagonal prime number and a centered icosihenagonal number which is also a composite number is called as centered icosihenagonal composite number. For example 127 is a centered icosihenagonal prime number and 3214 is a centered icosihenagonal composite number.

The centered icosihenagonal number which is repdigit in nature is a repdigit centered icosihenagonal number. For example 22 is a repdigit icosihenagonal number.

A palindromic centered icosihenagonal number is a centered icosihenagonal number which is palindromic in nature. For example 757 is a centered icosihenagonal palindromic number.

Icosidigonal number: -

An icosidigonal number is a figurate number, the nth icosidigonal number will be the number of points in an icosidigon with n regularly spaced points on a side. The icosidigonal number sequence is 1, 22, 63, 124, 205, 306, 427, 568, 729, 910, 1111, 1332…

An icosidigonal number which is odd is called as odd icosidigonal number and an icosidigonal number which is even is called as even icosidigonal number. For example 124 is an even icosihenagonal number and 63 is an odd icosihenagonal number.

The icosidigonal number which is ascending in nature is called as icosidigonal ascending number and the icosidigonal number which is descending in nature is called as icosidigonal descending number. For example 124 is a icosidigonal ascending number and 910 is a icosidigonal descending number.

The icosidigonal palindromic number is an icosidigonal number that is palindromic in nature. For example 22 is a palindromic icosidigonal number.

The icosidigonal number which is prime is called as icosidigonal prime number and the icosidigonal number which is composite is called as icosidigonal composite number. For example 63 is an icosidigonal prime number and 22 is an icosidigonal composite number.

The icosidigonal repunit number is an icosidigonal number that is repunit in nature. For example 1111 is an icosidigonal repunit number.

Centered icosidigonal number: -

The centered icosidigonal numbers are those numbers whose lattice points will form a centered icosidigon. The sequence of centered icosidigonal number is 1, 23, 67, 133, 221, 331, 463, 617, 793, 991, 1211, 1453, 1717, 2003, 2311, 2641, 2993, 3367, 3763, 4181, 4621, 5083, 5567, 6073, 6601, 7151, 7723, 8317, 8933, 9571…

A centered icosidigonal number which is also a prime number is called as centered icosidigonal prime number and a centered icosidigonal number which is also a composite number is called as centered icosidigonal composite number. For example 133 is a centered icosidigonal prime number and 67 is a centered icosidigonal composite number.

The centered icosidigonal number which is undulating in nature is called as undulating centered icosidigonal number. For example 1717 is an undulating centered icosidigonal number.

All centered icosidigonal numbers are odd numbers. Further if the odd centered icosidigonal number contains all the digits as odd, then it is a super odd centered icosidigonal number. For example 991 is a super odd centered icosidigonal number and 617 is not a super odd centered icosidigonal number.

Icositrigonal number: -

An icositrigonal number is a figurate number, the nth icositrigonal number will be the number of points in an icositrigon with n regularly spaced points on a side. The icositrigonal number sequence is 1, 23, 66, 130, 215, 321, 448, 596, 765, 955, 1166, 1398…

An icositrigonal number which is odd is called as odd icositrigonal number and an icositrigonal number which is even is called as even icositrigonal number. For example 130 is an even icositrigonal number and 321 is an odd icositrigonal number.

Further if the odd icositrigonal number contains all of its digits as odd, then it is a super odd icositrigonal number and if the even icositrigonal number contains all of its digits as even, then it is a super even icositrigonal number. For example 955 is a super odd icositrigonal number and 448 is a super even icositrigonal number.

The icositrigonal number which is prime is called as icositrigonal prime number and the icositrigonal number which is composite is called as icositrigonal composite number. For example 23 is an icositrigonal prime number and 215 is an icositrigonal composite number.

A palindromic icositrigonal number is an icositrigonal number which is palindromic in nature. For example 66 is an icositrigonal palindromic number.

Centered icositrigonal number: -

The centered icositrigonal numbers are those numbers whose lattice points will form a centered icositrigon. The sequence of centered icositrigonal number is 1, 24, 70, 139, 231, 346, 484, 645, 829, 1036, 1266, 1519, 1795, 2094, 2416, 2761, 3129, 3520, 3934, 4371, 4831, 5314, 5820, 6349, 6901, 7476, 8074, 8695, 9339, 10006…

A centered icositrigonal number which is odd is called as odd centered icositrigonal number and a centered icositrigonal number which is even is called as even centered icositrigonal number. For example 1036 is an even centered icositrigonal number and 231 is an odd centered icositrigonal number.

Further if the odd centered icositrigonal number contains all of its digits as odd, then it is a super odd centered icositrigonal number and if the even centered icositrigonal number contains all of its digits as even, then it is a super even centered icositrigonal number. For example 139 is a super odd centered icositrigonal number and 484 is a super even centered icositrigonal number.

A centered icositrigonal number which is also a prime number is called as centered icositrigonal prime number and a centered icositrigonal number which is also a composite number is called as centered icositrigonal composite number. For example 139 is a centered icositrigonal prime number and 70 is a centered icositrigonal composite number.

A palindromic centered icositrigonal number is a centered icositrigonal number which is palindromic in nature. For example 9339 is a centered icositrigonal palindromic number.

Icositetragonal number: -

An icositetragonal number is a figurate number, the nth icositetragonal number will be the number of points in an icositetragon with n regularly spaced points on a side.

The icositetragonal number sequence is 1, 24, 69, 136, 225, 336, 469, 624, 801, 1000, 1221, 1464...

An icositetragonal number which is odd is called as odd icositetragonal number and an icositetragonal number which is even is called as even icositetragonal number. For example 136 is an even icositetragonal number and 469 is an odd icositetragonal number.

Further if the even icositetragonal number contains all of its digits as even, then it is a super even icositetragonal number. For example 624 is a super even icositetragonal number 1464 is not a super even icositetragonal number.

The square number which is also an icositetragonal number is called as icositetragonal square number. For example 225 is an icositetragonal square number.

A palindromic icositetragonal number is an icositetragonal number which is palindromic in nature. For example 1221 is an icositetragonal palindromic number.

Centered icositetragonal number: -

The centered icositetragonal numbers are those numbers whose lattice points will form a centered icositetragon. The sequence of centered icositetragonal number is 1, 25, 73, 145, 241, 361, 505, 673, 865, 1081, 1321, 1585, 1873, 2185, 2521, 2881, 3265, 3673, 4105, 4561, 5041, 5545, 6073, 6625, 7201, 7801, 8425, 9073, 9745, 10441...

All centered icositetragonal numbers are odd numbers. Further if the odd centered icositetragonal number contains all the digits as odd, then it is a super odd centered icositetragonal number. For example 73 is a super odd centered icositetragonal number and 145 is not a super odd centered icositetragonal number.

A centered icositetragonal number which is also a prime number is called as centered icositetragonal prime number and a centered icositetragonal number which is also a composite number is called as centered icositetragonal composite number. For example 75 is a centered icositetragonal prime number and 145 is a centered icositetragonal composite number.

The centered icositetragonal number which is ascending in nature is called as centered icositetragonal ascending number and the centered icositetragonal number which is descending in nature is called as centered icositetragonal descending number. For example 145 is a centered icositetragonal ascending number and 865 is a centered icositetragonal descending number.

The square number which is also a centered icositetragonal number is called as centered icositetragonal square number. For example 361 is a centered icositetragonal square number.

A palindromic centered icositetragonal number is a centered icositetragonal number which is palindromic in nature. For example 505 is a centered icositetragonal palindromic number.

Icosipentagonal number: -

An icosipentagonal number is a figurate number, the nth icosipentagonal number will be the number of points in an icosipentagon with n regularly spaced points on a side. The icosipentagonal number sequence is 1, 25, 72, 142, 235, 351, 490, 652, 837, 1045, 1276, 1530...

An icosipentagonal number which is odd is called as odd icosipentagonal number and an icosipentagonal number which is even is called as even icosipentagonal number. For example 142 is an even icosipentagonal number and 235 is an odd icosipentagonal number.

Further if the odd icosipentagonal number contains all the digits as odd, then it is a super odd icosipentagonal number. For example 351 is a super odd icosipentagonal number and 235 is not a super odd icosipentagonal number.

Centered icosipentagonal number: -

The centered icosipentagonal numbers are those numbers whose lattice points will form a centered icosipentagon. The sequence of centered icosipentagonal number is 1, 26, 76, 151, 251, 376, 526, 701, 901, 1126…

A centered icosipentagonal number which is odd is called as odd centered icosipentagonal number and a centered icosipentagonal number which is even is called as even centered icosipentagonal number. For example 376 is an even centered icosipentagonal number and 151 is an odd centered icosipentagonal number.

Further if the even centered icosipentagonal number contains all of its digits as even, then it is a super even centered icosipentagonal number. For example 26 is a super even centered icosipentagonal number 526 is not a super even centered icosipentagonal number.

A centered icosipentagonal number which is also a prime number is called as centered icosipentagonal prime number and a centered icosipentagonal number which is also a composite number is called as centered icosipentagonal composite number. For example 701 is a centered icosipentagonal prime number and 376 is a centered icosipentagonal composite number.

A palindromic centered icosipentagonal number is a centered icosipentagonal number which is palindromic in nature. For example 151 is a centered icosipentagonal palindromic number.

Icosihexagonal number: -

An icosihexagonal number is a figurate number, the nth icosihexagonal number will be the number of points in an icosihexagon with n regularly spaced points on a side. The icosihexagonal number sequence is 1, 26, 75, 148, 245, 366, 511, 680, 873, 1090, 1331, 1596…

An icosihexagonal number which is odd is called as odd icosihexagonal number and an icosihexagonal number which is even is called as even icosihexagonal number. For example 148 is an even icosihexagonal number and 245 is an odd icosihexagonal number.

Further if the odd icosihexagonal number contains all of its digits as odd, then it is a super odd icosihexagonal number and if the even icosihexagonal number contains all of its digits as even, then it is a super even icosihexagonal number. For example 511 is a super odd icosihexagonal number and 680 is a super even icosihexagonal number.

A palindromic icosihexagonal number is an icosihexagonal number which is palindromic in nature. For example 1331 is an icosihexagonal palindromic number.

Centered icosihexagonal number: -

The centered icosihexagonal numbers are those numbers whose lattice points will form a centered icosihexagon. The sequence of centered icosihexagonal number is 1, 27, 79, 157, 261, 391, 547, 729, 937, 1171…

All centered icosihexagonal numbers are odd numbers. Further if the odd centered icosihexagonal number contains all the digits as odd, then it is a super odd centered icosihexagonal number. For example 1171 is a super odd centered icosihexagonal number and 547 is not a super odd centered icosihexagonal number.

A centered icosihexagonal number which is also a prime number is called as centered icosihexagonal prime number and a centered icosihexagonal number which is also a composite number is called as centered icosihexagonal composite number. For example 79 is a centered icosihexagonal prime number and 729 is a centered icosihexagonal composite number.

Icosiheptagonal number: -

An icosiheptagonal number is a figurate number, the nth icosiheptagonal number will be the number of points in an icosiheptagon with n regularly spaced points on a side. The icosiheptagonal number sequence is 1, 27, 78, 154, 255, 381, 532, 708, 909, 1135, 1386, 1662...

An icosiheptagonal number which is odd is called as odd icosiheptagonal number and an icosiheptagonal number which is even is called as even icosiheptagonal number. For example 154 is an even icosiheptagonal number and 255 is an odd icosiheptagonal number.

Further if the odd icosiheptagonal number contains all the digits as odd, then it is a super odd icosiheptagonal number. For example 1135 is a super odd icosiheptagonal number and 1386 is not a super odd icosiheptagonal number.

A palindromic icosiheptagonal number is an icosiheptagonal number which is palindromic in nature. For example 909 is an icosiheptagonal palindromic number.

Centered icosiheptagonal number: -

The centered icosiheptagonal numbers are those numbers whose lattice points will form a centered icosiheptagon. The sequence of centered icosiheptagonal number is 1, 28, 82, 163, 271, 406, 568, 757, 973, 1216...

A centered icosiheptagonal number which is odd is called as odd centered icosiheptagonal number and a centered icosiheptagonal number which is even is called as even centered icosiheptagonal number. For example 406 is an even centered icosiheptagonal number and 271 is an odd centered icosiheptagonal number.

Further if the odd centered icosiheptagonal number contains all of its digits as odd, then it is a super odd centered icosiheptagonal number and if the even centered icosiheptagonal number contains all of its digits as even, then it is a super even triacontagonal number. For example 973 is a super odd centered icosiheptagonal number and 82 is a super even centered icosiheptagonal number.

The centered icosiheptagonal number which is ascending in nature is called as centered icosiheptagonal ascending number and the centered icosiheptagonal number which is descending in nature is called as centered icosiheptagonal descending number. For example 568 is a centered icosiheptagonal ascending number and 973 is a centered icosiheptagonal descending number.

A centered icosiheptagonal number which is also a prime number is called as centered icosiheptagonal prime number and a centered icosiheptagonal number which is also a composite number is called as centered icosiheptagonal composite number. For example 163 is a centered icosiheptagonal prime number and 82 is a centered icosiheptagonal composite number.

A palindromic icosiheptagonal number is an icosiheptagonal number which is palindromic in nature. For example 757 is an icosiheptagonal palindromic number.

Icosioctagonal number: -

An icosioctagonal number is a figurate number, the nth icosioctagonal number will be the number of points in an icosioctagon with n regularly spaced points on a side.

The icosioctagonal number sequence is 1, 28, 81, 160, 265, 396, 553, 736, 945, 1180, 1441, 1728...

An icosioctagonal number which is odd is called as odd icosioctagonal number and an icosioctagonal number which is even is called as even icosioctagonal number. For example 160 is an even icosioctagonal number and 265 is an odd icosioctagonal number.

Further if the odd icosioctagonal number contains all the digits as odd, then it is a super odd icosioctagonal number. For example 553 is a super odd icosioctagonal number and 945 is not a super odd icosioctagonal number.

An icosioctagonal number which is palindromic in nature is called as palindromic icosioctagonal number. For example 1441 is a palindromic icosioctagonal number.

The square number which is also an icosioctagonal number is called as icosioctagonal square number. For example 81 is an icosioctagonal square number.

Centered icosioctagonal number: -

The centered icosioctagonal numbers are those numbers whose lattice points will form a centered icosioctagon. The sequence of centered icosioctagonal number is 1, 29, 85, 169, 281, 421, 589, 785, 1009, 1261, 1541, 1849...

A formula for finding out the centered icosioctagonal number is $[14n (n+1) +1]$ where 'n' is a whole number. For example if n=5, the 5th centered icosioctagonal number is 421.

All centered icosioctagonal numbers are odd numbers.

A centered icosioctagonal number which is also a prime number is called as centered icosioctagonal prime number and a centered icosioctagonal number which is also a composite number is called as centered icosioctagonal composite number. For example 29 is a centered icosioctagonal prime number and 85 is a centered icosioctagonal composite number.

The centered icosioctagonal number which is ascending in nature is called as centered icosioctagonal ascending number and the centered icosioctagonal number which is descending in nature is called as centered icosioctagonal descending number. For example 169 is a centered icosioctagonal ascending number and 785 is a centered icosioctagonal descending number.

A centered icosioctagonal number which is also a square number is called as centered icosioctagonal square number. For example 169 is a centered icosioctagonal square number.

Icosinonagonal number: -

An icosinonagonal number is a figurate number, the nth icosinonagonal number will be the number of points in an icosinonagon with n regularly spaced points on a side. The icosinonagonal number sequence is 1, 29, 84, 166, 275, 411, 574, 764, 981, 1225, 1496, 1794...

An icosinonagonal number which is odd is called as odd icosinonagonal number and an icosinonagonal number which is even is called as even icosinonagonal number. For example 166 is an even icosinonagonal number and 275 is an odd icosinonagonal number.

Further if the even icosinonagonal number contains all of its digits as even, then it is a super even icosinonagonal number. For example 84 is a super even icosinonagonal number 166 is not a super even icosinonagonal number.

An icosinonagonal number which is also a prime number is called as icosinonagonal prime number and an icosinonagonal number which is also a composite number is called as icosinonagonal composite number. For example 29 is an icosinonagonal prime number and 84 is an icosinonagonal composite number.

Centered icosinonagonal number: -

The centered icosinonagonal numbers are those numbers whose lattice points will form a centered icosinonagon. The sequence of centered icosinonagonal number is 1, 30, 88, 175, 291, 436, 610, 813, 1045, 1306, 1596, 1915, 2089, 2640, 3046…

A formula for finding out the centered icosinonagonal number is $[29n(n+1)/2]+1$ where 'n' is a whole number. For example if n=2, the 3rd centered icosinonagonal number is 88.

A centered icosinonagonal number which is odd is called as odd centered icosinonagonal number and a centered icosinonagonal number which is even is called as even centered icosinonagonal number. For example 1306 is an even centered icosinonagonal number and 1045 is an odd centered icosinonagonal number.

Further if the odd centered icosinonagonal number contains all of its digits as odd, then it is a super odd centered icosinonagonal number. For example 175 is a super odd centered icosinonagonal number and 291 is not a super odd centered icosinonagonal number.

A palindromic centered icosinonagonal number is a centered icosinonagonal number which is palindromic in nature. For example 88 is a icosinonagonal palindromic number.

Triacontagonal number: -

A triacontagonal number is a figurate number, the nth triacontagonal number will be the number of points in a triacontagonal shape with n regularly spaced points on a side. The triacontagonal number sequence is 1, 30, 87, 172, 285, 426, 595, 792, 1017, 1270, 1551, 1860…

A triacontagonal number which is odd is called as odd triacontagonal number and a triacontagonal number which is even is called as even triacontagonal number. For example 172 is an even triacontagonal number and 285 is an odd triacontagonal number.

Further if the odd triacontagonal number contains all of its digits as odd, then it is a super odd triacontagonal number and if the even triacontagonal number contains all of its digits as even, then it is a super even triacontagonal number. For example 1551 is a super odd triacontagonal number and 426 is a super even triacontagonal number.

A palindromic Triacontagonal number is a Triacontagonal number that is palindromic in nature. For example 595, 1551… are some of the examples of palindromic triacontagonal number.

Centered triacontagonal number: -

The centered triacontagonal numbers are those numbers whose lattice points will form a centered triacontagon. The sequence of centered triacontagonal number is 1, 31, 91, 181, 301, 451, 631, 841, 1081, 1351, 1651, 1981…

A formula for finding out the centered triacontagonal number is $[15n(n+1)]+1$ where 'n' is a whole number. For example if n=2, the 3rd centered triacontagonal number is 91.

All centered triacontagonal numbers are odd numbers. Further if the odd centered triacontagonal number contains all the digits as odd, then it is a super odd centered triacontagonal number. For example 1351 is a super odd centered triacontagonal number and 181 is not a super odd centered triacontagonal number.

A centered triacontagonal number which is prime is called as centered triacontagonal prime number and a centered triacontagonal number which is composite is called as centered triacontagonal composite number. For example 31 is a centered icositriacontagonal number and 91 is a centered icositriacontagonal number.

A palindromic centered triacontagonal number is a centered triacontagonal number that is palindromic in nature. For example 181 is a palindromic centered triacontagonal number.

Further polygonal numbers are
31-gonal Numbers 1, 31, 90, 178, 295, 441, 616, 820, 1053...
32-gonal Numbers 1, 32, 93, 184, 305, 456, 637, 848, 1089...
33-gonal Numbers 1, 33, 96, 190, 315, 471, 658, 876, 1225...
34-gonal Numbers 1, 34, 99, 196, 325, 486, 679, 904, 1161...
35-gonal Numbers 1, 35, 102, 202, 335, 501, 700, 932, 1197...
36-gonal Numbers 1, 36, 105, 208, 345, 516, 721, 960, 1233...
37-gonal Numbers 1, 37, 108, 214, 355, 531, 742, 988, 1269...
38-gonal Numbers 1, 38, 111, 220, 365, 546, 763, 1016, 1305...
39-gonal Numbers 1, 39, 114, 226, 375, 561, 784, 1044, 1341...
40-gonal Numbers 1, 40, 117, 232, 385, 576, 805, 1072, 1377...
41-gonal Numbers 1, 41, 120, 238, 395, 591, 826, 1100, 1413...
42-gonal Numbers 1, 42, 123, 244, 405, 597, 847, 1128, 1449...
43-gonal Numbers 1, 43, 126, 250, 415, 603, 868, 1156, 1485...
44-gonal Numbers 1, 44, 129, 256, 425, 618, 889, 1184, 1521...
45-gonal Numbers 1, 45, 132, 262, 435, 633, 910, 1212, 1557...
46-gonal Numbers 1, 46, 135, 268, 445, 648, 931, 1240, 1593...
47-gonal Numbers 1, 47, 138, 274, 455, 663, 952, 1268, 1629...
48-gonal Numbers 1, 48, 141, 280, 465, 678, 973, 1296, 1665...
49-gonal Numbers 1, 49, 144, 286, 475, 693, 994, 1324, 1701...
50-gonal Numbers 1, 50, 147, 292, 485, 708, 1015, 1352, 1737...
51-gonal Numbers 1, 51, 150, 298, 495, 723, 1036, 1380, 1773...
52-gonal Numbers 1, 52, 153, 304, 505, 738, 1057, 1408, 1809...
53-gonal Numbers 1, 53, 156, 310, 515, 753, 1078, 1436, 1845...
54-gonal Numbers 1, 54, 159, 316, 525, 768, 1099, 1464, 1881...
55-gonal Numbers 1, 55, 162, 322, 535, 783, 1120, 1492, 1917...
56-gonal Numbers 1, 56, 165, 328, 545, 798, 1141, 1520, 1953...
57-gonal Numbers 1, 57, 168, 334, 555, 813, 1162, 1548, 1989...
58-gonal Numbers 1, 58, 171, 340, 656, 828, 1183, 1576, 2025...
59-gonal Numbers 1, 59, 174, 346, 575, 843, 1204, 1604, 2061...
60-gonal Numbers 1, 60, 177, 352, 585, 858, 1225, 1632, 2097...
61-gonal Numbers 1, 61, 180, 358, 595, 873, 1246, 1660, 2133...
62-gonal Numbers 1, 62, 183, 364, 605, 888, 1267, 1688, 2169...
63-gonal Numbers 1, 63, 186, 370, 615, 903, 1288, 1716, 2205...
64-gonal Numbers 1, 64, 189, 376, 625, 918, 1309, 1744, 2241...
65-gonal Numbers 1, 65, 192, 382, 635, 933, 1330, 1772, 2277...
66-gonal Numbers 1, 66, 195, 388, 645, 948, 1351, 1800, 2313...
67-gonal Numbers 1, 67, 198, 394, 655, 963, 1372, 1828, 2349...
68-gonal Numbers 1, 68, 201, 400, 665, 978, 1393, 1856, 2385...
69-gonal Numbers 1, 69, 204, 406, 675, 993, 1414, 1884, 2421...
70-gonal Numbers 1, 70, 207, 412, 685, 1008, 1435, 1912, 2457...
71-gonal Numbers 1, 71, 210, 418, 695, 1023, 1456, 1940, 2493...

72-gonal Numbers 1, 72, 213, 424, 705, 1038, 1477, 1968, 2529…
73-gonal Numbers 1, 73, 216, 430, 715, 1053, 1498, 1966, 2565…
74-gonal Numbers 1, 74, 219, 436, 725, 1068, 1519, 2024…
75-gonal Numbers 1, 75, 222, 442, 735, 1083, 1540, 2052…
76-gonal Numbers 1, 76, 225, 448, 745, 1098, 1561, 2080…
77-gonal Numbers 1, 77, 228, 454, 755, 1113, 1582, 2108…
78-gonal Numbers 1, 78, 231, 460, 765, 1128, 1603, 2136…
79-gonal Numbers 1, 79, 234, 466, 775, 1143, 1624, 2164…
80-gonal Numbers 1, 80, 237, 472, 785, 1158, 1645, 2192…
81-gonal Numbers 1, 81, 240, 478, 795, 1173, 1666, 2220…
82-gonal Numbers 1, 82, 243, 484, 805, 1188, 1687, 2248…
83-gonal Numbers 1, 83, 246, 490, 815, 1203, 1708, 2276…
84-gonal Numbers 1, 84, 249, 496, 825, 1218, 1729, 2304…
85-gonal Numbers 1, 85, 252, 502, 835, 1233, 1750, 2332…
86-gonal Numbers 1, 86, 255, 508, 845, 1248, 1771, 2360…
87-gonal Numbers 1, 87, 258, 514, 855, 1263, 1792, 2388…
88-gonal Numbers 1, 88, 261, 520, 865, 1278, 1813, 2416…
89-gonal Numbers 1, 89, 264, 526, 875, 1293, 1834, 2444…
90-gonal Numbers 1, 90, 267, 532, 885, 1308, 1855, 2472…
91-gonal Numbers 1, 91, 270, 538, 895, 1323, 1876, 2500…
92-gonal Numbers 1, 92, 273, 544, 905, 1338, 1897, 2528…
93-gonal Numbers 1, 93, 276, 550, 915, 1353, 1918, 2556…
94-gonal Numbers 1, 94, 279, 556, 925, 1368, 1939, 2584…
95-gonal Numbers 1, 95, 282, 562, 935, 1383, 1960, 2612…
96-gonal Numbers 1, 96, 285, 568, 945, 1398, 1981, 2640…
97-gonal Numbers 1, 97, 288, 574, 955, 1413, 2002, 2668…
98-gonal Numbers 1, 98, 291, 580, 965, 1428, 2023…
99-gonal Numbers 1, 99, 294, 586, 975, 1443, 2044…
100-gonal Numbers 1, 100, 297, 592, 985, 1458, 2065…
101-gonal Numbers 1, 101, 300, 598, 995, 1473, 2086…
102-gonal Numbers 1, 102, 303, 604, 1005, 1488, 2107…
103-gonal Numbers 1, 103, 306, 610, 1015, 1503, 2128…
104-gonal Numbers 1, 104, 309, 616, 1025, 1518, 2149…
105-gonal Numbers 1, 105, 312, 622, 1035, 1533, 2170…
106-gonal Numbers 1, 106, 315, 628, 1045, 1548, 2191…
107-gonal Numbers 1, 107, 318, 634, 1055, 1563, 2212…
108-gonal Numbers 1, 108, 321, 640, 1065, 1578, 2233…
109-gonal Numbers 1, 109, 324, 646, 1075, 1593, 2254…
110-gonal Numbers 1, 110, 327, 652, 1085, 1608, 2275…
111-gonal Numbers 1, 111, 330, 658, 1095, 1623, 2296…
112-gonal Numbers 1, 112, 333, 664, 1105, 1638, 2317…
113-gonal Numbers 1, 113, 336, 670, 1115, 1653, 2338…
114-gonal Numbers 1, 114, 339, 676, 1125, 1668, 2359…
115-gonal Numbers 1, 115, 342, 682, 1135, 1683, 2380…
116-gonal Numbers 1, 116, 345, 688, 1145, 1698, 2401…
117-gonal Numbers 1, 117, 348, 694, 1155, 1713, 2422…
118-gonal Numbers 1, 118, 351, 700, 1165, 1728, 2443…
119-gonal Numbers 1, 119, 354, 706, 1175, 1743, 2464…
120-gonal Numbers 1, 120, 357, 712, 1185, 1758, 2485…
121-gonal Numbers 1, 121, 360, 718, 1195, 1773, 2506…

122-gonal Numbers 1, 122, 363, 724, 1205, 1788, 2527...
123-gonal Numbers 1, 123, 366, 730, 1215, 1803, 2548...
124-gonal Numbers 1, 124, 369, 736, 1225, 1818, 2569...
125-gonal Numbers 1, 125, 372, 742, 1235, 1833, 2590...
126-gonal Numbers 1, 126, 375, 748, 1245, 1848, 2611...
127-gonal Numbers 1, 127, 378, 754, 1255, 1865, 2632...
128-gonal Numbers 1, 128, 381, 760, 1265, 1878, 2653...
129-gonal Numbers 1, 129, 384, 766, 1275, 1893, 2674...
130-gonal Numbers 1, 130, 387, 772, 1285, 1908, 2695...
131-gonal Numbers 1, 131, 390, 778, 1295, 1923, 2716...
132-gonal Numbers 1, 132, 393, 784, 1305, 1938, 2737...
133-gonal Numbers 1, 133, 396, 790, 1315, 1953, 2758...
134-gonal Numbers 1, 134, 399, 796, 1325, 1968, 2779...
135-gonal Numbers 1, 135, 402, 802, 1335, 1983, 2800...
136-gonal Numbers 1, 136, 405, 808, 1345, 1998, 2821...
137-gonal Numbers 1, 137, 408, 814, 1355, 2013...
138-gonal Numbers 1, 138, 411, 820, 1365, 2028...
139-gonal Numbers 1, 139, 414, 826, 1375, 2043...
140-gonal Numbers 1, 140, 417, 832, 1385, 2058...
141-gonal Numbers 1, 141, 420, 838, 1395, 2073...
142-gonal Numbers 1, 142, 423, 844, 1405, 2088...
143-gonal Numbers 1, 143, 426, 850, 1415, 2103...
144-gonal Numbers 1, 144, 429, 856, 1425, 2118...
145-gonal Numbers 1, 145, 432, 862, 1435, 2133...
146-gonal Numbers 1, 146, 435, 868, 1445, 2148...
147-gonal Numbers 1, 147, 438, 874, 1455, 2163...
148-gonal Numbers 1, 148, 441, 880, 1465, 2178...
149-gonal Numbers 1, 149, 444, 886, 1475, 2193...
150-gonal Numbers 1, 150, 447, 892, 1485, 2208...
151-gonal Numbers 1, 151, 450, 898, 1495, 2223...
152-gonal Numbers 1, 152, 453, 904, 1505, 2238...
153-gonal Numbers 1, 153, 456, 910, 1515, 2253...
154-gonal Numbers 1, 154, 459, 916, 1525, 2268...
155-gonal Numbers 1, 155, 462, 922, 1535, 2283...
156-gonal Numbers 1, 156, 465, 928, 1545, 2298...
157-gonal Numbers 1, 157, 468, 934, 1555, 2313...
158-gonal Numbers 1, 158, 471, 940, 1565, 2328...
159-gonal Numbers 1, 159, 474, 946, 1575, 2343...
160-gonal Numbers 1, 160, 477, 952, 1585, 2358...
161-gonal Numbers 1, 161, 480, 958, 1595, 2373...
162-gonal Numbers 1, 162, 483, 964, 1605, 2388...
163-gonal Numbers 1, 163, 486, 970, 1615, 2403...
164-gonal Numbers 1, 164, 489, 976, 1625, 2418...
165-gonal Numbers 1, 165, 492, 982, 1635, 2433...
166-gonal Numbers 1, 166, 495, 988, 1645, 2448...
167-gonal Numbers 1, 167, 498, 994, 1655, 2463...
168-gonal Numbers 1, 168, 501, 1000, 1665, 2478...
169-gonal Numbers 1, 169, 504, 1006, 1675, 2493...
170-gonal Numbers 1, 170, 507, 1012, 1685, 2508...
171-gonal Numbers 1, 171, 510, 1018, 1695, 2523...

172-gonal Numbers 1, 172, 513, 1024, 1705, 2538...
173-gonal Numbers 1, 173, 516, 1030, 1715, 2553...
174-gonal Numbers 1, 174, 519, 1036, 1725, 2568...
175-gonal Numbers 1, 175, 522, 1042, 1735, 2583...
176-gonal Numbers 1, 176, 525, 1048, 1745, 2598...
177-gonal Numbers 1, 177, 528, 1054, 1755, 2613...
178-gonal Numbers 1, 178, 531, 1060, 1765, 2628...
179-gonal Numbers 1, 179, 534, 1066, 1775, 2643...
180-gonal Numbers 1, 180, 537, 1072, 1785, 2658...
181-gonal Numbers 1, 181, 540, 1078, 1795, 2673...
182-gonal Numbers 1, 182, 543, 1084, 1805, 2688...
183-gonal Numbers 1, 183, 546, 1090, 1815, 2703...
184-gonal Numbers 1, 184, 549, 1096, 1825, 2718...
185-gonal Numbers 1, 185, 552, 1102, 1835, 2733...
186-gonal Numbers 1, 186, 555, 1108, 1845, 2748...
187-gonal Numbers 1, 187, 558, 1114, 1855, 2763...
188-gonal Numbers 1, 188, 561, 1120, 1865, 2778...
189-gonal Numbers 1, 189, 564, 1126, 1875, 2793...
190-gonal Numbers 1, 190, 567, 1132, 1885, 2808...
191-gonal Numbers 1, 191, 570, 1138, 1895, 2823...
192-gonal Numbers 1, 192, 573, 1144, 1905, 2838...
193-gonal Numbers 1, 193, 576, 1150, 1915, 2853...
194-gonal Numbers 1, 194, 579, 1156, 1925, 2868...
195-gonal Numbers 1, 195, 582, 1162, 1935, 2883...
196-gonal Numbers 1, 196, 585, 1168, 1945, 2898...
197-gonal Numbers 1, 197, 588, 1174, 1955, 2913...
198-gonal Numbers 1, 198, 591, 1180, 1965, 2928...
199-gonal Numbers 1, 199, 594, 1186, 1975, 2943...
200-gonal Numbers 1, 200, 597, 1192, 1985, 2958...
201-gonal Numbers 1, 201, 600, 1198, 1995, 2973...
202-gonal Numbers 1, 202, 603, 1204, 2005, 2988...
203-gonal Numbers 1, 203, 606, 1210, 2015...
204-gonal Numbers 1, 204, 609, 1216, 2025...
205-gonal Numbers 1, 205, 612, 1222, 2035...
206-gonal Numbers 1, 206, 615, 1228, 2045...
207-gonal Numbers 1, 207, 618, 1234, 2055...
208-gonal Numbers 1, 208, 621, 1240, 2065...
209-gonal Numbers1, 209, 624, 1246, 2075...
210-gonal Numbers 1, 210, 627, 1252. 2085...
211-gonal Numbers 1, 211, 630, 1258, 2095...
212-gonal Numbers 1, 212, 633, 1264, 2105...
213-gonal Numbers 1, 213, 636, 1270, 2115...
214-gonal Numbers 1, 214, 639, 1276, 2125...
215-gonal Numbers 1, 215, 642, 1282, 2135...
216-gonal Numbers 1, 216, 645, 1288, 2145...
217-gonal Numbers 1, 217, 648, 1294, 2155...
218-gonal Numbers 1, 218, 651, 1300, 2165...
219-gonal Numbers 1, 219, 654, 1306, 2175...
220-gonal Numbers 1, 220, 657, 1312, 2185...
221-gonal Numbers 1, 221, 660, 1318, 2195...

222-gonal Numbers 1, 222, 663, 1324, 2205…
223-gonal Numbers 1, 223, 666, 1330, 2215…
224-gonal Numbers 1, 224, 669, 1336, 2225…
225-gonal Numbers 1, 225, 672, 1342, 2235…
226-gonal Numbers 1, 226, 675, 1348, 2245…
227-gonal Numbers 1, 227, 678, 1354, 2255…
228-gonal Numbers 1, 228, 681, 1360, 2565…
229-gonal Numbers 1, 229, 684, 1366, 2275…
230-gonal Numbers 1, 230, 687, 1372, 2285…
231-gonal Numbers 1, 231, 690, 1378, 2295…
232-gonal Numbers 1, 232, 693, 1384, 2305…
233-gonal Numbers 1, 233, 696, 1390, 2315…
234-gonal Numbers 1, 234, 699, 1396, 2325…
235-gonal Numbers 1, 235, 702, 1402, 2335…
236-gonal Numbers 1, 236, 705, 1408, 2345…
237-gonal Numbers 1, 237, 708, 1414, 2355…
238-gonal Numbers 1, 238, 711, 1420, 2365…
239-gonal Numbers 1, 239, 714, 1426, 2375…
240-gonal Numbers 1, 240, 717, 1432, 2385…
241-gonal Numbers 1, 241, 720, 1438, 2395…
242-gonal Numbers 1, 242, 723, 1444, 2405…
243-gonal Numbers 1, 243, 726, 1450, 2415…
244-gonal Numbers 1, 244, 729, 1456, 2425…
245-gonal Numbers 1, 245, 732, 1462, 2435…
246-gonal Numbers 1, 246, 735, 1468, 2445…
247-gonal Numbers 1, 247, 738, 1474, 2455…
248-gonal Numbers 1, 248, 741, 1480, 2465…
249-gonal Numbers 1, 249, 744, 1486, 2475…
250-gonal Numbers 1, 250, 747, 1492, 2485…
251-gonal Numbers 1, 251, 750, 1498, 2495…
252-gonal Numbers 1, 252, 753, 1504, 2505…
253-gonal Numbers 1, 253, 756, 1510, 2515…
254-gonal Numbers 1, 254, 759, 1516, 2525…
255-gonal Numbers 1, 255, 762, 1522, 2535…
256-gonal Numbers 1, 256, 765, 1528, 2545…
257-gonal Numbers 1, 257, 768, 1534, 2555…
258-gonal Numbers 1, 258, 771, 1540, 2565…
259-gonal Numbers 1, 259, 774, 1546, 2575…
260-gonal Numbers 1, 260, 777, 1552, 2585…
261-gonal Numbers 1, 261, 780, 1558, 2595…
262-gonal Numbers 1, 262, 783, 1564, 2605…
263-gonal Numbers 1, 263, 786, 1570, 2615…
264-gonal Numbers 1, 264, 789, 1576, 2625…
265-gonal Numbers 1, 265, 792, 1582, 2635…
266-gonal Numbers 1, 266, 795, 1588, 2645…
267-gonal Numbers 1, 267, 798, 1594, 2655…
268-gonal Numbers 1, 268, 801, 1600, 2665…
269-gonal Numbers 1, 269, 804, 1606, 2675…
270-gonal Numbers 1, 270, 807, 1612, 2685…
271-gonal Numbers 1, 271, 810, 1618, 2695…

272-gonal Numbers 1, 272, 813, 1624, 2705...
273-gonal Numbers 1, 273, 816, 1630, 2715...
274-gonal Numbers 1, 274, 819, 1636, 2725...
275-gonal Numbers 1, 275, 822, 1642, 2735...
276-gonal Numbers 1, 276, 825, 1648, 2745...
277-gonal Numbers 1, 277, 828, 1654, 2755...
278-gonal Numbers 1, 278, 831, 1660, 2765...
279-gonal Numbers 1, 279, 834, 1666, 2775...
280-gonal Numbers 1, 280, 837, 1672, 2785...
281-gonal Numbers 1, 281, 840, 1678, 2795...
282-gonal Numbers 1, 282, 843, 1684, 2805...
283-gonal Numbers 1, 283, 846, 1690, 2815...
284-gonal Numbers 1, 284, 849, 1696, 2825...
285-gonal Numbers 1, 285, 852, 1702, 2835...
286-gonal Numbers 1, 286, 855, 1708, 2845...
287-gonal Numbers 1, 287, 858, 1714, 2855...
288-gonal Numbers 1, 288, 861, 1720, 2865...
289-gonal Numbers 1, 289, 864, 1726, 2875...
290-gonal Numbers 1, 290, 867, 1732, 2885...
291-gonal Numbers 1, 291, 870, 1738, 2895...
292-gonal Numbers 1, 292, 873, 1744, 2905...
293-gonal Numbers 1, 293, 876, 1750, 2915...
294-gonal Numbers 1, 294, 879, 1756, 2925...
295-gonal Numbers 1, 295, 882, 1762, 2935...
296-gonal Numbers 1, 296, 885, 1768, 2945...
297-gonal Numbers 1, 297, 888, 1774, 2955...
298-gonal Numbers 1, 298, 891, 1780, 2965...
299-gonal Numbers 1, 299, 894, 1786, 2975...
300-gonal Numbers 1, 300, 897, 1792, 2985...
301-gonal Numbers 1, 301, 900, 1798, 2995...
302-gonal Numbers 1, 302, 903, 1804, 3005...
303-gonal Numbers 1, 303, 906, 1810, 3015...
304-gonal Numbers 1, 304, 909, 1816, 3025...
305-gonal Numbers 1, 305, 912, 1822, 3035...
306-gonal Numbers 1, 306, 915, 1828, 3045...
307-gonal Numbers 1, 307, 918, 1834, 3055...
308-gonal Numbers 1, 308, 921, 1840, 3065...
309-gonal Numbers 1, 309, 924, 1846, 3075...
310-gonal Numbers 1, 310, 927, 1852, 3085...
311-gonal Numbers 1, 311, 930, 1858, 3095...
312-gonal Numbers 1, 312, 933, 1864, 3105...
313-gonal Numbers 1, 313, 936, 1870, 3115...
314-gonal Numbers 1, 314, 939, 1876, 3125...
315-gonal Numbers 1, 315, 942, 1882, 3135...
316-gonal Numbers 1, 316, 945, 1888, 3145...
317-gonal Numbers 1, 317, 948, 1894, 3155...
318-gonal Numbers 1, 318, 951, 1900, 3165...
319-gonal Numbers 1, 319, 954, 1906, 3175...
320-gonal Numbers 1, 320, 957, 1912, 3185...
321-gonal Numbers 1, 321, 960, 1918, 3195...

322-gonal Numbers 1, 322, 963, 1924, 3205...
323-gonal Numbers 1, 323, 966, 1930, 3215...
324-gonal Numbers 1, 324, 969, 1936, 3225...
325-gonal Numbers 1, 325, 972, 1942, 3235...
326-gonal Numbers 1, 326, 975, 1948, 3245...
327-gonal Numbers 1, 327, 978, 1954, 3255...
328-gonal Numbers 1, 328, 981, 1960, 3265...
329-gonal Numbers 1, 329, 984, 1966, 3275...
330-gonal Numbers 1, 330, 987, 1972, 3285...
331-gonal Numbers 1, 331, 990, 1978, 3295...
332-gonal Numbers 1, 332, 993, 1984, 3305...
333-gonal Numbers 1, 333, 996, 1990, 3315...
334-gonal Numbers 1, 334, 999, 1996, 3325...
335-gonal Numbers 1, 335, 1002, 2002, 3335...
336-gonal Numbers 1, 336, 1005, 2008, 3345...
337-gonal Numbers 1, 337, 1008, 2014...
Likewise we can extend the polygonal numbers.

Star number: -

A star number is a centered figurate number that represents a centered hexagram.

1   13   37

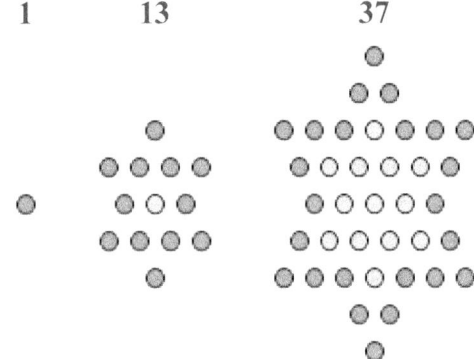

The star number sequence is 1, 13, 37, 73, 121, 181, 253, 337, 433, 541, 661, 793, 937, 1093, 1261, 1441, 1633, 1837, 2053, 2281, 2521, 2773, 3037, 3313, 3601, 3901, 4213, 4537, 4873, 5221, 5581, 5953, 6337, 6733, 7141, 7561, 7993, 8437, 8893, 9361, 9841, 10333, 10837...

The nth star number is given by the formula $6n(n-1)+1$ where 'n' is a real number starting from 1 onwards.

The digital root of a star number is always either 1 or 4.

The star number which is ascending in nature is called as star ascending number and the star number which is descending in nature is called as star descending number. For example 37 is a star ascending number and 9841 is a star descending number.

It is to be noted that all the star numbers are odd numbers. Further if the odd star number contains all the digits as odd, then it is a super odd star number. For example 3313 is a super odd star number and 3037 is not a super odd star number.

The last two digits of a star number in base 10 are always 01, 13, 21, 33, 37, 41, 53, 61, 73, 81, or 93.

The square number which is also a star number is called as star square number. For example 121 is a star square number.

A star prime number is a star number that is prime. The star prime number sequence is 13, 37, 73, 181, 337, 433, 541, 661, 937...

A star composite number is a star number that is composite. For example 121 is a star composite number.

A star palindromic number is a star number that is palindromic. The star palindromic number sequence is 121, 181, 1441...

MULTI-POLYGONAL NUMBERS

Some numbers can be expressed in two and more geometrical shapes and these numbers are generally termed as multi-polygonal numbers. Based on the number of geometrical shapes that can be formed from the numbers, the multi-polygonal numbers are further classified as bipolygonal numbers, tripolygonal numbers and so on.

BIPOLYGONAL NUMBERS: -
(a) Square triangular number: -
Some numbers can be expressed in triangle as well as in square shape. Such numbers are called as triangular square number or square triangular number. The square triangular number series is 1, 36, 1225, 41616, 1413721, 48024900…

A square triangular number which is odd is called as odd square triangular number and a square triangular number which is even is called as even square triangular number. For example 1225 is an odd square triangular number and 36 is an even square triangular number.

(b) Pentagonal triangular number: -
Some numbers can be expressed in triangle as well as in pentagon shape. Such numbers are called as pentagonal triangular number. The pentagonal triangular number sequence is 1, 210, 40755, 7906276, 1533776805…

A pentagonal triangular number which is odd is called as odd pentagonal triangular number and a pentagonal triangular number which is even is called as even pentagonal triangular number. For example 40755 is an odd pentagonal triangular number and 210 is an even pentagonal triangular number.

(c) Heptagonal triangular number: -
Some numbers can be expressed in triangle as well as in heptagon shape. Such numbers are called as heptagonal triangular number. The heptagonal triangular number sequence is 1, 55, 121771, 5720653, 12625478965…

All heptagonal triangular numbers are odd numbers.

(d) Octagonal triangular number: -
Some numbers can be expressed in triangle as well as in octagon shape. Such numbers are called as octagonal triangular numbers. The octagonal triangular number sequence is 1, 21, 11781, 203841, 113123361…

(e) Nonagonal triangular number: -
Some numbers can be expressed in triangle as well as in nonagon shape. Such numbers are called as nonagonal triangular numbers. The nonagonal triangular number sequence is 1, 325, 82621, 20985481, 5330229625, 1353857339341…

(f) Square Pyramidal triangular numbers: -
Some numbers can be expressed in triangle as well as in square pyramid shape. Such numbers are called as square pyramidal triangular numbers. The square pyramidal triangular number sequence is 1, 55, 91, 208335…

The square pyramidal triangular number which is palindromic in nature is called as palindromic square pyramidal triangular number.

(g)Tetrahedral triangular numbers: -

Some numbers can be expressed in triangle as well as in tetrahedron shape. Such numbers are called as tetrahedral triangular numbers. The tetrahedral triangular number sequence is 1, 10, 120, 1540, 7140...

(h) Pentagonal square number: -

Some numbers can be expressed in square as well as in pentagon shape. Such numbers are called as pentagonal square number. The pentagonal square number sequence is 1, 9801, 94109401, 903638458801, 8676736387298001...

Centered pentagonal square number: - Some numbers can be expressed in square as well as in pentagon shape. Such numbers are called as pentagonal square number. The pentagonal square number sequence is 1, 5734...

(i) Hexagonal square number: -

Some numbers can be expressed in square as well as in hexagon shape. Such numbers are called as hexagonal square number. The hexagonal square number sequence is 1, 1225, 1413721, 1631432881...

(j) Heptagonal square number: -

Some numbers can be expressed in square as well as in heptagon shape. Such numbers are called as heptagonal square number. The heptagonal square number sequence is 1, 81, 5929, 2307361, 168662169, 12328771225...

(k) Octagonal square number: -

Some numbers can be expressed in square as well as in octagon shape. Such numbers are called as octagonal square number. The octagonal square number sequence is 1, 225, 43681, 8473921, 1643897025...

(l) Nonagonal square number: -

Some numbers can be expressed in square as well as in nonagon shape. Such numbers are called as nonagonal square number. The nonagonal square number sequence is 1, 9, 1089, 8281, 978121, 7436529...

(m)Square Pyramidal square Numbers: -

Some numbers can be expressed in square as well as in square pyramid shape. Such numbers are called as square pyramidal square numbers. The square pyramidal square number sequence is 1, 4900...

(n)Tetrahedral square Numbers: -

Some numbers can be expressed in tetrahedron as well as in square shape. Such numbers are called as tetrahedral square numbers. The tetrahedral square number sequence is 1, 4, 19600...

(o) Rhombic dodecahedral number: -

Some numbers can be expressed in rhombus as well as in dodecahedron shape. Such numbers are called as rhombic dodecahedral numbers. For example 1, 65, 671, 1105, 1695, 3439, 4641, 6095, 7825, 9855... are some of the examples of rhombic dodecahedral number.

(p) Hexagonal pentagon number: -

Some numbers can be expressed in hexagon as well as in pentagon shape. Such numbers are called as hexagonal pentagon number. The hexagonal pentagon number sequence is 1, 40755, 1533776805...

(q) Heptagonal pentagon number: -

Some numbers can be expressed in pentagon as well as in heptagon shape. Such numbers are called as heptagonal pentagon numbers. The heptagonal pentagon number sequence is 1, 4347, 16701685, 64167869935, 246532939589097...

(r) Octagonal pentagon number: -

Some numbers can be expressed in pentagon as well as in octagon shape. Such numbers are called as octagonal pentagon numbers. The octagonal pentagon number sequence is 1, 176, 1575425, 234631320, 2098015778145...

An octagonal pentagon number which is odd is called as odd octagonal pentagon number and an octagonal pentagon number which is even is called as even octagonal pentagon number. For example 1575425 is an odd octagonal pentagon number and 176 is an even octagonal pentagon number.

(s) Nonagonal pentagon number: -

Some numbers can be expressed in pentagon as well as in nonagon shape. Such numbers are called as nonagonal pentagon numbers. The nonagonal pentagon number sequence is 1, 651, 180868051, 95317119801, 26472137730696901...

(t) Heptagonal hexagon number: -

Some numbers can be expressed in hexagon as well as in heptagon shape. Such numbers are called as heptagonal hexagon numbers. The heptagonal hexagon number sequence is 1, 121771, 12625478965, 1309034909945503...

(u) Octagonal hexagon number: -

Some numbers can be expressed in hexagon as well as in octagon shape. Such numbers are called as octagonal hexagon numbers. The octagonal hexagon number sequence is 1, 11781, 113123361, 1086210502741...

(v) Nonagonal hexagon number: -

Some numbers can be expressed in hexagon as well as in nonagon shape. Such numbers are called as nonagonal hexagon numbers. The nonagonal hexagon number sequence is 1, 325, 5330229625, 1353857339341, 22184715227362706161...

(w) Octagonal heptagon number: -

Some numbers can be expressed in heptagon as well as in octagon shape. Such numbers are called as octagonal heptagon numbers. The octagonal heptagon number sequence is 1, 297045, 69010153345...

(x) Nonagonal heptagon number: -

Some numbers can be expressed in heptagon as well as in nonagon shape. Such numbers are called as nonagonal heptagon numbers. The nonagonal heptagon number sequence is 1, 26884, 542041975, 10928650279834...

The nonagonal heptagon numbers which are odd is called as odd nonagonal heptagon numbers and the nonagonal heptagon numbers which are even is called as even nonagonal heptagon number. For example 26884 is an even nonagonal heptagon number and 542041975 is an odd nonagonal heptagon number.

(y) Nonagonal octagon number: -

Some numbers can be expressed in octagon as well as in nonagon shape. Such numbers are called as nonagonal octagon numbers. The nonagonal octagon number sequence is 1, 631125, 286703855361, 130242107189808901…

(z)Square Pyramidal tetrahedral Numbers: -

Some numbers can be expressed in square pyramid as well as in tetrahedron shape. Such numbers are called as square pyramidal tetrahedral numbers. The only number which is simultaneously tetrahedral and square pyramidal is 1.

TRIPOLYGONAL NUMBERS: -
(a)Hexagonal square triangular numbers: -

Some numbers can be expressed in hexagon, square and triangular shapes and are called as hexagonal square triangular number. For example 1225 is a hexagonal square triangular number.

(b)Hexagonal tridecagonal triangular number: -

Some numbers can be expressed in hexagon, tridecagon and triangular shapes and are called as hexagonal tridecagonal triangular number. For example 4186 is a hexagonal tridecagonal triangular number.

(c)Rhombicubeoctahedral number: -

Some numbers can be expressed in rhombus, cube and octahedral shapes and are called as rhombicubeoctahedral number. For example 8385 is a rhombicubeoctahedral number.

The polygonal numbers discussed above refer to the geometrical shape in two dimensional spaces. The following polygonal numbers discussed like pyramidal number, tetrahedral number, octahedral number… refer to the geometrical shape in three dimensional space (3D-Polygonal numbers).

3D-POLYGONAL NUMBERS

The numbers which can be represented in three dimensional figure are called as 3D-Polygonal numbers and based upon the geometrical shape they are classified as pyramidal number, cube number, tetrahedral number and so on.

PYRAMIDAL NUMBER: -

A pyramidal number is a figurate number that represents a pyramid with a base and a given number of sides. The pyramidal number is named after the geometrical base on which the pyramid is built.

Triangular pyramidal number (three sides): -

This is a figurate number which represents a pyramid with triangle base. The triangular pyramidal number sequence is 1, 4, 10, 20, 35, 56, 84, 120...

It is to be noted that these numbers correspond to placing discrete points in the configuration of a tetrahedron and so triangular pyramidal numbers can also be called as tetrahedral numbers.

Refer the tetrahedral numbers for further details.

Square pyramidal number (4 sides): -

This is a figurate number which represents a pyramid with square base. The square pyramid number sequence is 1, 5, 14, 30, 55, 91, 140, 204, 285, 385, 506, 650, 819...

A formula to find out the nth square pyramid number is $[n (n+1) (2n+1)/6]$ where 'n' is a real number. Applying the formula the fifth square pyramid we will get 55.

A square pyramidal number which is odd is called as odd square pyramidal number and a square pyramidal number which is even is called as even square pyramidal number. For example 140 is an even square pyramidal number and 285 is an odd square pyramidal number.

Further if the odd square pyramidal number contains all of its digits as odd, then it is a super odd square pyramidal number and if the even square pyramidal number contains all of its digits as even, then it is a super even square pyramidal number. For example 204 is a super even square pyramidal number and 91 is a super odd square pyramidal number.

The square pyramidal number which is ascending in nature is called as square pyramidal ascending number and the square pyramidal number which is descending in nature is called as square pyramidal descending number. For example 14 is a square pyramidal ascending number and 140 is a square pyramidal descending number.

The square pyramidal palindromic number is a square pyramidal number which is palindromic in nature. For example 55 is a square pyramidal palindromic number.

The sum of two consecutive square pyramidal numbers is an octahedral number. For example 5 + 14 = 19 and is an octahedral number.

Besides 1, there is only one other number that is both a square and a square pyramid number is 4900, which is both the 70th square number and the 24th square pyramidal number.

Pentagonal pyramidal number (five sides): -

This is a figurate number which represents a pyramid with pentagon base. The pentagonal pyramidal numbers are: 1, 6, 18, 40, 75, 126, 196, 288, 405, 550, 726, 936, 1183, 1470, 1800, 2176, 2601, 3078, 3610, 4200, 4851, 5566, 6348, 7200, 8125, 9126, 10206, 11368, 12615, 13950, 15376, 16896, 18513, 20230, 22050, 23976, 26011, 28158, 30420, 32800, 35301...

A formula to find out the nth pentagonal pyramidal number is $[n^2(n+1)/2]$. So the nth pentagonal pyramidal number is the average of n^2 and n^3.

A pentagonal pyramidal number which is odd is called as odd pentagonal pyramidal number and a pentagonal pyramidal number which is even is called as even pentagonal pyramidal number. For example 228 is an even pentagonal pyramidal number and 405 is an odd pentagonal pyramidal number.

Further if the odd pentagonal pyramidal number contains all of its digits as odd, then it is a super odd pentagonal pyramidal number and if the even pentagonal pyramidal number contains all of its digits as even, then it is a super even pentagonal pyramidal number. For example 288 is a super even pentagonal pyramidal number and 75 is a super odd heptagonal pyramidal number.

The pentagonal pyramidal number which is ascending in nature is called as pentagonal pyramidal ascending number and the pentagonal pyramidal number which is descending in nature is called as pentagonal pyramidal descending number. For example 126 is a pentagonal pyramidal ascending number and 75 is a pentagonal pyramidal descending number.

The square number which is also a pentagonal pyramidal number is called as pentagonal pyramidal square number. For example 196 is a pentagonal pyramidal square number.

The nth pentagonal pyramidal number is also n times the nth triangular number. For example the 5th pentagonal pyramidal number is 75 and is equal to 5 times the 5th triangular number 15 that is $5 \times 15 = 75$.

Hexagonal pyramidal number (six sides): -

This is a figurate number which represents a pyramid with hexagon base. The hexagonal pyramidal numbers are 1, 7, 22, 50, 95, 161, 252, 372, 525, 715, 946, 1222, 1547, 1925, 2360, 2856, 3417, 4047, 4750, 5530, 6391, 7337, 8372, 9500, 10725, 12051, 13482, 15022, 16675, 18445, 20336, 22352, 24497, 26775, 29190, 31746, 34447, 37297, 40300...

A hexagonal pyramidal number which is odd is called as odd hexagonal pyramidal number and a hexagonal pyramidal number which is even is called as even hexagonal pyramidal number. For example 946 is an even hexagonal pyramidal number and 525 is an odd hexagonal pyramidal number.

Further if the odd hexagonal pyramidal number contains all of its digits as odd, then it is a super odd hexagonal pyramidal number. For example 715 is a super odd hexagonal pyramidal number and 6391 is not a super odd hexagonal pyramidal number.

The hexagonal pyramidal palindromic number is a hexagonal pyramidal number which is palindromic in nature. For example 161, 252, 525, 7337... are some of the palindromic hexagonal pyramidal number.

The hexagonal pyramidal repdigit number is a hexagonal pyramidal number which is repdigit in nature. For example 22 is a repdigit hexagonal pyramidal number.

The hexagonal pyramidal number which is prime is called as prime hexagonal pyramidal number and the hexagonal pyramidal which is composite is called as composite hexagonal pyramidal number. For example 372 is a composite hexagonal pyramidal number and 7 is a prime hexagonal pyramidal number.

Heptagonal pyramidal number (seven sides): -

This is a figurate number which represents a pyramid with heptagon base. The heptagonal pyramidal numbers are 1, 8, 26, 60, 115, 196, 308, 456, 645, 880, 1166, 1508, 1911, 2380, 2920, 3536, 4233, 5016, 5890, 6860, 7931, 9108, 10396, 11800, 13325, 14976, 16758, 18676, 20735, 22940, 25296, 27808, 30481, 33320, 36330, 39516, 42883, 46436, 50180, 54120...

A heptagonal pyramidal number which is odd is called as odd heptagonal pyramidal number and a heptagonal pyramidal number which is even is called as even heptagonal pyramidal number. For example 456 is an even heptagonal pyramidal number and 645 is an odd heptagonal pyramidal number.

Further if the odd heptagonal pyramidal number contains all of its digits as odd, then it is a super odd heptagonal pyramidal number and if the even heptagonal pyramidal number contains all of its digits as even, then it is a super even heptagonal pyramidal number. For example 880 is a super even heptagonal pyramidal number and 115 is a super odd heptagonal pyramidal number.

The heptagonal pyramidal number which is ascending in nature is called as heptagonal pyramidal ascending number and the heptagonal pyramidal number which is descending in nature is called as heptagonal pyramidal descending number. For example 456 is a heptagonal pyramidal ascending number and 60 is a heptagonal pyramidal descending number.

The square number which is also a heptagonal pyramidal number is called as heptagonal pyramidal square number. For example 196 is a centered nonagonal square number.

Octagonal pyramidal number (eight sides): -

This is a figurate number which represents a pyramid with octagon base. The octagonal pyramidal numbers are 1, 1045, 4216, 5049, 7030, 9471…

An octagonal pyramidal number which is odd is called as odd octagonal pyramidal number and an octagonal pyramidal number which is even is called as even octagonal pyramidal number. For example 4216 is an even octagonal pyramidal number and 1045 is an odd octagonal pyramidal number.

Enneagonal pyramidal number (nine sides): -

This is a figurate number which represents a pyramid with enneagon (nine sided polygon) base. The enneagonal pyramidal numbers are 1, 885, 2639, 3290, 4040, 5865, 8170, 9520…

An enneagonal pyramidal number which is odd is called as odd enneagonal pyramidal number and an enneagonal pyramidal number which is even is called as even enneagonal pyramidal number. For example 3290 is an even enneagonal pyramidal number and 2639 is an odd decagonal pyramidal number.

Further if the even enneagonal pyramidal number contains all of its digits as even, then it is a super even enneagonal pyramidal number. For example 4040 is a super even enneagonal pyramidal number and 3290 is not a super even enneagonal pyramidal number.

The enneagonal pyramidal number which is undulating in nature is called as undulating enneagonal pyramidal number. For example 4040 is an undulating enneagonal pyramidal number.

Decagonal pyramidal number (ten sides): -

This is a figurate number which represents a pyramid with decagon base. The decagonal pyramidal numbers are 1, 1005, 1375, 5576, 9310...

A decagonal pyramidal number which is odd is called as odd decagonal pyramidal number and a decagonal pyramidal number which is even is called as even decagonal pyramidal number. For example 5576 is an even decagonal pyramidal number and 1375 is an odd decagonal pyramidal number.

Further if the odd decagonal pyramidal number contains all of its digits as odd, then it is a super odd decagonal pyramidal number. For example 1375 is a super odd decagonal pyramidal number and 5576 is not a super odd decagonal pyramidal number.

Hendecagonal pyramidal number (eleven sides): -

This is a figurate number which represents a pyramid with hendecagon base. The hendecagonal pyramidal numbers are 1, 532, 1125, 5153, 6256, 7497…

A hendecagonal pyramidal number which is odd is called as odd hendecagonal pyramidal number and a hendecagonal pyramidal number which is even is called as even hendecagonal pyramidal number. For example 532 is an even hendecagonal pyramidal number and 1125 is an odd hendecagonal pyramidal number.

Further if the odd hendecagonal pyramidal number contains all of its digits as odd, then it is a super odd hendecagonal pyramidal number. For example 5153 is a super odd hendecagonal pyramidal number and 1125 is not a super odd hendecagonal pyramidal number.

Dodecagonal pyramidal number (twelve sides): -

This is a figurate number which represents a pyramid with dodecagon base. The dodecagonal pyramidal numbers are 1, 876, 1245, 2266, 3731, 4600, 5720, 8313, 9861…

A dodecagonal pyramidal number which is odd is called as odd dodecagonal pyramidal number and a dodecagonal pyramidal number which is even is called as even dodecagonal pyramidal number. For example 876 is an even dodecagonal pyramidal number and 1245 is an odd dodecagonal pyramidal number.

Further if the odd dodecagonal pyramidal number contains all of its digits as odd, then it is a super odd dodecagonal pyramidal number and if the even dodecagonal pyramidal number contains all of its digits as even, then it is a super even dodecagonal pyramidal number. For example 4600 is a super even dodecagonal pyramidal number and 3731 is a super odd dodecagonal pyramidal number.

The dodecagonal pyramidal number which is ascending in nature is called as dodecagonal pyramidal ascending number and the dodecagonal pyramidal number which is descending in nature is called as dodecagonal pyramidal descending number. For example 1245 is a dodecagonal pyramidal ascending number and 876 is a dodecagonal pyramidal descending number.

Cube number: -

The numbers whose dots form a lattice work of cubes are called as cube numbers. They are discussed in chapter 5.

Centered cube number: -

The cube numbers whose dots form a lattice work of centered cubes are called as centered cube numbers. The centered cube numbers are 1, 9, 35, 91, 189, 341, 559, 855, 1241, 1729, 2331, 3059, 3925, 4941, 6119, 7471, 9009, 10745, 12691, 14859,

17261, 19909, 22815, 25991, 29449, 33201, 37259, 41635, 46341, 51389, 56791, 62559, 68705, 75241, 82179, 89531, 97309, 105525...

The nth centered cube number is found by using the formula $n^3+(n-1)^3$ where 'n' is a real number.

All centered cube numbers are odd numbers. Further if the odd centered cube number contains all of its digits as odd, then it is a super odd centered cube number. For example 559 is a super odd centered cube number and 51389 is not a super odd centered cube number.

The centered cube number which is ascending in nature is called as centered cube ascending number and the centered cube number which is descending in nature is called as centered cube descending number. For example 35 is a centered cube ascending number and 91 is a centered cube descending number.

A palindromic centered cube number is a centered cube number that is palindromic in nature. For example 9009 is a palindromic centered cube number.

Tetrahedral number: -
The numbers whose dots form a lattice work of tetrahedrons are called as tetrahedral numbers. For example the numbers 1, 4, 10, 20, 35, 56, 84, 120, 165, 220, 969, 1771... are tetrahedral numbers.

A tetrahedral number which is odd is called as odd tetrahedral number and a tetrahedral number which is even is called as even tetrahedral number. For example 120 is an even tetrahedral number and 84 is an odd tetrahedral number.

Further if the odd tetrahedral number contains all of its digits as odd, then it is a super odd tetrahedral number and if the even tetrahedral number contains all of its digits as even, then it is a super even tetrahedral number. For example 220 is a super even tetrahedral number and 35 is a super odd tetrahedral number.

The tetrahedral which is ascending in nature is called as tetrahedral ascending number and the tetrahedral number which is descending in nature is called as tetrahedral descending number. For example 56 is a tetrahedral ascending number and 10 is a tetrahedral descending number.

The tetrahedral numbers which are palindromic in nature is called as tetrahedral palindromic numbers. For example 969 and 1771 are some of the examples for tetrahedral palindrome numbers.

The tetrahedral number which also forms a square is called as tetrahedral square number. For example 36 is a tetrahedral square number.

The number 6545 and its reflux number are tetrahedral numbers. Another example exhibiting the similar property is 5456.

Centered tetrahedral number: -
The numbers whose dots form a lattice work of centered tetrahedrons are called as centered tetrahedral numbers. For example 1, 589, 2059, 3605, 4255, 4979, 5781, 6665, 7635, 8695, 9849... are centered tetrahedral numbers.

All centered tetrahedral numbers are odd numbers.

Octahedral number: -
An octahedral number is a figurate number that represents an octahedron, or two square pyramids placed together, one upside-down underneath the other. For example the numbers 1, 6, 19, 44, 85, 146, 231, 344, 489, 670, 891, 1156, 1469, 1834, 2255, 2736, 3281, 3894, 4579, 5340, 6181, 7106, 8119, 9224, 10425, 11726, 13131, 14644,

16269, 18010, 19871, 21856, 23969, 26214, 28595, 31116, 33781, 36594, 39559, 42680... are some of the octahedral numbers.

An octahedral number which is odd is called as odd octahedral number and an octahedral number which is even is called as even octahedral number. For example 1156 is an even octahedral number and 891 is an odd octahedral number.

Further if the odd octahedral number contains all of its digits as odd, then it is a super odd octahedral number. For example 39559 is a super odd octahedral number and 23969 is not a super odd octahedral number.

The nth octahedral number can be obtained by adding the $(n-1)^{th}$ and n^{th} square pyramidal numbers together, or by using the following formula: $[n(2n^2+1)/3]$.

The octahedral number which is undulating in nature is called as undulating octahedral number. For example 13131 is an undulating octahedral number.

Octahedral palindromic number is an octahedral number that is palindromic. For example 44, 13131... are some of the examples of palindromic octahedral number.

An octahedral number which is prime number is called as prime octahedral number and an octahedral number which is composite number is called as composite octahedral number. For example 19 is a prime octahedral number and 85 is a composite octahedral number.

Centered octahedral number: -

A centered octahedral number is a figurate number that represents a centered octahedron. For example 1, 833, 1159, 2625, 3303, 4089, 6017, 7175... are some of the examples of centered octahedral number.

All centered octahedral numbers are odd numbers. If the odd centered octahedral number contains all of its digits as odd, then it is a super odd centered octahedral number. For example 7175 is a super odd centered octahedral number and 4089 is not a super odd centered octahedral number.

Centered dodecahedral number: -

The numbers whose dots form a lattice work of centered dodecahedrons are called as centered dodecahedral numbers. For example 1, 1661, 2743, 4215, 6137, 8569... are centered dodecahedral numbers.

All centered dodecahedral numbers are odd numbers.

A centered dodecahedral number which is palindromic in nature is called as centered dodecahedral palindromic number. For example 1661 is a centered dodecahedral palindromic number.

Icosahedral number: -

The numbers whose dots form a lattice work of icosahedrons are called as icosahedral numbers. For example the numbers 1, 1128, 1629, 2260, 6384, 7890, 9616... are some of the examples of icosahedral numbers.

An icosahedral number which is odd is called as odd icosahedral number and an icosahedral number which is even is called as even icosahedral number. For example 1128 is an even icosahedral number and 1629 is an odd icosahedral number.

Further if the even icosahedral number contains all of its digits as even, then it is a super even icosahedral number. For example 2260 is a super even icosahedral number and 6384 is not a super even icosahedral number.

Centered icosahedral number: -

The numbers whose dots form a lattice work of centered icosahedrons are called as centered icosahedral numbers. For example 1, 1415, 2057, 2869, 5083, 6525, 8217… are some of the centered icosahedral numbers.

All centered icosahedral numbers are odd numbers.

Diagrammatically the classification of polygonal numbers can be represented as follows.

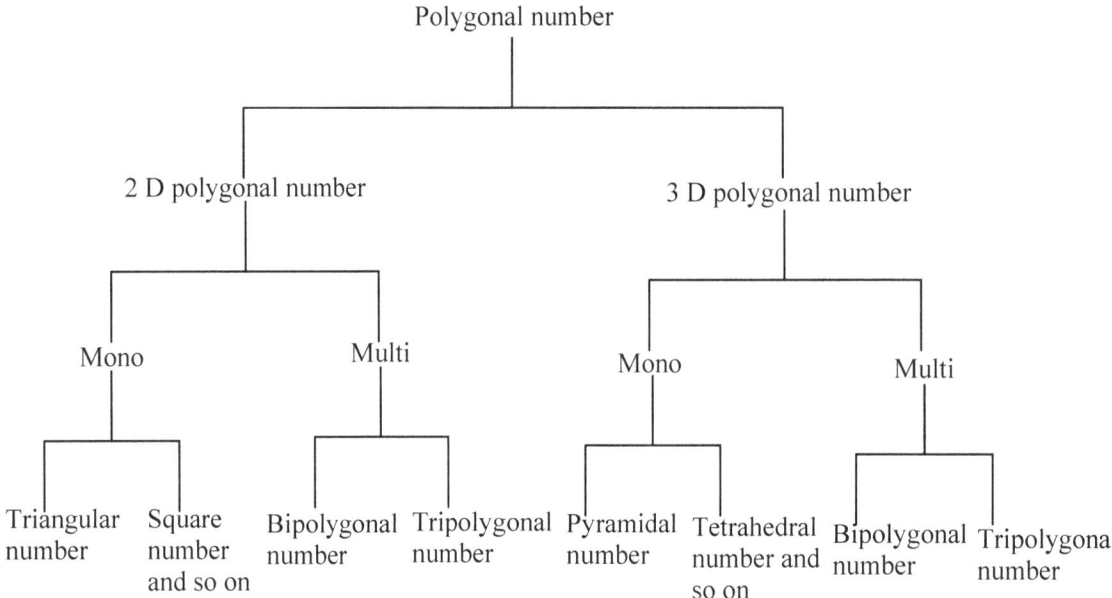

11. MATHEMATICS CAN BE FUN

The series of numbers can be arranged in regular geometrical shapes so that the sum obtained is constant regardless of adding the numbers in any side. The sum obtained is called as magic sum and based upon the geometrical shape formed they are classified as magic triangle, magic square, magic star, magic square border and so on.

Magic triangles
Magic triangles are triangles having the peculiar property that the sum is constant regardless of adding the numbers, forming the triangle in any side.

Magic triangles with consecutive numbers starting from 1 to n (where n>2) can be solved by applying the formula [n(n+1)/2] where 'n' is the end number or highest number in the magic triangle. For example if 'n' is 3, by applying the above formula, end number 'E' is 6. The magic triangle is formed as follows.

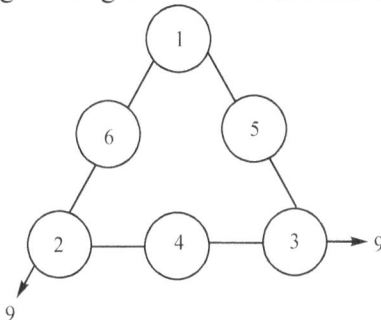

The magic triangle sum 'M' is 9, a constant regardless of adding the numbers in any side.

The order of a magic triangle is determined by the number of entries between any two corners of the triangle. For example in the above case order 'O' = 1. In other words for forming a 1st order magic triangle, 6 consecutive numbers are required.

The relation between the order number and the end number of the triangle border puzzle is

Order number (O)	1	2	3	4	5
End number (E)	6	9	12	15	18

From the table we can arrive E = 3(O+1)

For example for order 6, the end number is 21.

The sum of consecutive numbers that make up the magic triangle denoted as 'S', can be found by using the formula [n(n+1)/2]. For example in the above case, sum 'S' = (6×7)/2 = 21.

Any magic triangle puzzle will give four sums

(I) the sum of consecutive numbers that make up the magic triangle denoted as 'S'

(II) the sum of corner numbers in the magic triangle denoted as 'C'

(III) the sum of non corner numbers in the magic triangle denoted as 'B' and

(IV) the magic triangle sum denoted as 'M'

For example in the magic triangle puzzle of order 1, starting from 1 to 6

(I) S = [(6×7)/2] = 21

(II) C = 1+2+3 = 6

(III) B = 6+4+5 = 15

(IV) M = 1+6+2 = 2+4+3 = 1+5+3 = 9

The relationship between S, C and M is S + C = 3M

C = 3M – S

M = ⅓(S+C)

S + C = 4M for magic square border puzzle and S + C = 5M for magic pentagonal border puzzle and so on.

The relationship between B, S and M is B = 2S – 3M

'M' lies between ⅓S and ⅔S. The 'S' lies between 'C' and 'B'.

For magic triangle puzzle of order 1, there are four solutions available and they are

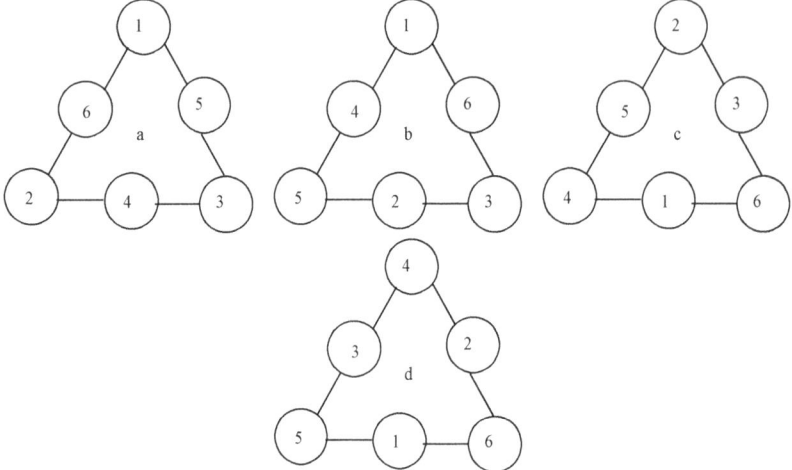

Now for the above magic triangle puzzles a, b, c and d

	a	b	c	d
S	21	21	21	21
C	6	9	12	15
M	9	10	11	12
B	15	12	9	6

There are four magic border sums for magic triangle of order 1 and seven magic border sums for magic triangle of order 2 and so on. The formula is 3n + 1 where 'n' is the order of the magic triangle. For magic triangle of order 3 the magic border sums possible are [(3×3)+1] = 10.

The magic triangles of odd order magic triangles can be solved in this method. For example the magic triangle of order 3 is formed as follows. Here the magic triangle border sum is 28.

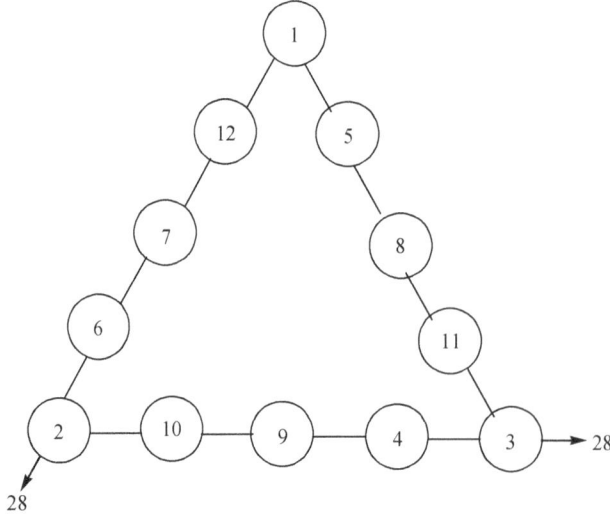

Similarly the higher odd ordered magic triangles are solved by filling up the consecutive numbers in clockwise and anticlockwise directions alternatively. The following magic triangle of order 5 has the magic triangle border sum 59.

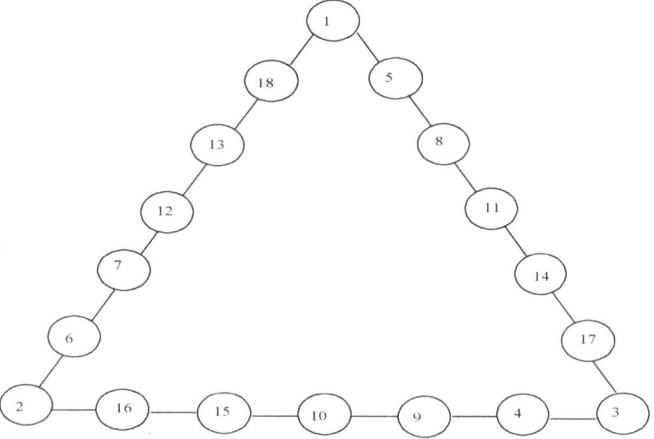

The even ordered magic triangles have irregular arrangements. The following magic triangle of order 2 has the magic triangle border sum 17.

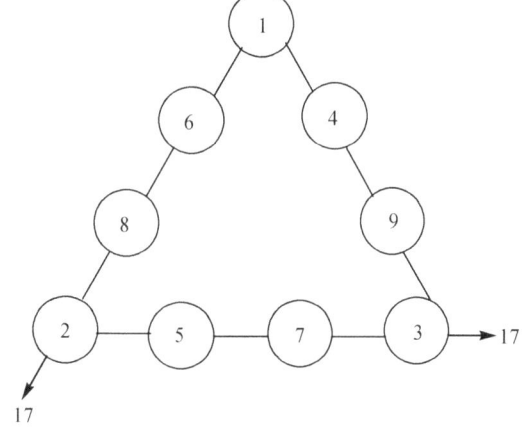

This magic triangle can be written in another way, keeping the magic border sum same is as follows.

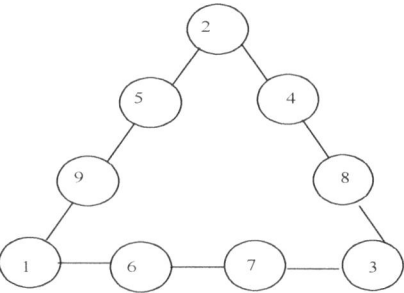

The above magic triangle of order 2 can be arranged in such a manner so that the sum becomes 20.

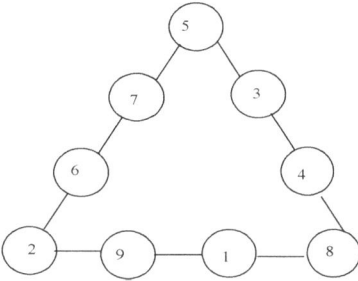

A special triangle is formed from 1 to 25, so that the sum of the individual sub-hexagons will be equal as follows.

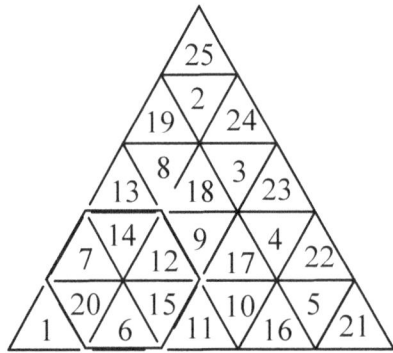

The above triangle is split into sub-hexagons so that the sum is 74 in all the sub-hexagons. For example in the following sub-hexagon 4, sum is 13+8+18+9+12+14=74.

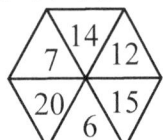

Sub hexagon 1

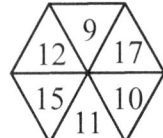

Sub hexagon 2

Sub hexagon 3

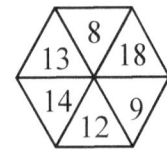

Sub hexagon 4

Sub hexagon 5

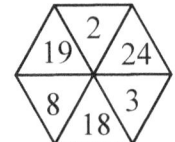

Sub hexagon 6

MAGIC SQUARES

Magic squares are the squares having the peculiar property that the sum is constant regardless of adding the numbers in any side. For example a magic square can be formed by using the numbers from 1 to 9 is as follows.

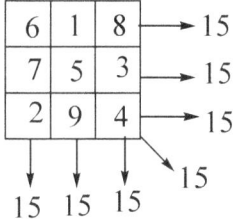

In the above magic square the row sum is 15, the column sum is 15 and the diagonal sum is also 15 regardless of adding the numbers in any side. Here 15 is called as magic square sum.

Magic squares are classified into two types namely odd magic squares and even magic squares based upon the number of squares present in the magic square. For example the above magic square is an odd magic square.

The order of magic square puzzle is equal to the number of rows or number of columns present in the magic square puzzle. For example in the above case the order is 3.

A table of order of magic square starting from 1 onwards against the magic square sum is as follows.

Order of magic square	3	4	5	6	7	8	16	24
Magic square sum	15	34	65	111	175	260	2056	1379

Note: - Magic square of order 1 and 2 is not possible.

A magic square sum remains unchanged when rotated or reflected.

If we delete the number in the central square keeping the outermost border square only, then we will get the magic square border puzzle of the same order. For example in the above case if we remove the number in the central square of 5, then we will get the magic border puzzle of order 3.

8	3	4
1		9
6	7	2

Here except the diagonal sum, the row sum and the column sum is equal. But the numbers are not consecutive as 5 gets left out from 1 to 9.

Thrice the amount of central number of the magic square is equal to the magic square sum. In the above case 5×3=15.

The sum of exact opposite numbers in the magic square is ⅔ of magic square sum. In the above case 9+1=10 which is ⅔ of 15.

A simple way of arranging this type of magic square having nine consecutive numbers is as follows.

6th number	1st number	8th number
7th number	5th number	3rd number
2nd number	9th number	4th number

For example a 3rd order magic square using the consecutive numbers from 20 to 28 is

25	20	27
26	24	22
21	28	23

A third order magic square comprising prime numbers only is as follows.

71	5	101
89	59	29
17	113	47

Here the magic square sum is 177.

A third order magic square comprising Curzon numbers only is as follows.

113	341	65
125	173	221
281	5	233

Here the magic square sum is 519.

Special 4×4 magic square: -

The magic square of 4×4 starting from 1 onwards is as follows. It is the continuous sequence magic square. Here the magic square sum is 34.

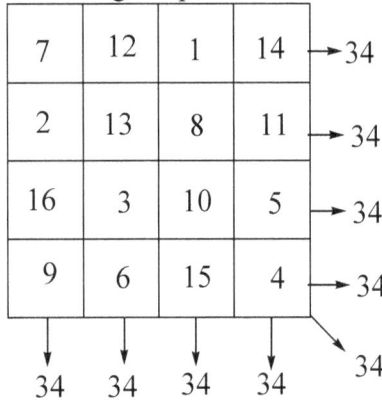

The above magic square is of special interest.

Column-wise addition of the magic square sum is 34.

7 + 2 + 16 + 9 = 34	**7**	12	**1**	14
12 + 13 + 3 + 6 = 34	**2**	13	**8**	11
1 + 8 + 10 + 15 = 34	**16**	3	**10**	5
14 + 11 + 5 + 4 = 34	**9**	6	**15**	4

Row-wise addition of the magic square sum is 34.

7 + 12 + 1 + 14 = 34	**7**	**12**	**1**	**14**
2 + 13 + 8 + 11 = 34	2	13	8	11
16 + 3 + 10 + 5 = 34	**16**	**3**	**10**	**5**
9 + 6 + 15 + 4 = 34	9	6	15	4

Addition of four corner group numbers is 34.

7 + 12 + 2 + 13 = 34	**7**	**12**	1	14
1 + 14 + 8 + 11 = 34	**2**	**13**	8	11
16 + 3 + 9 + 6 = 34	*16*	*3*	*10*	*5*
10 + 5 + 15 + 4 = 34	*9*	*6*	*15*	*4*

Addition of four middle group is numbers is 34.

12 + 1 + 13 + 8 = 34	7	**12**	**1**	14
2 + 13 + 16 + 3 = 34	**2**	**13**	8	11
3 + 10 + 6 + 15 = 34	**16**	**3**	10	5
8 + 11 + 10 + 5 = 34	9	**6**	15	4

Diagonal-wise addition of the magic square sum is 34.

7 + 13 + 10 + 4 = 34	**7**	12	1	*14*
14 + 8 + 3 + 9 = 34	2	**13**	*8*	11
	16	*3*	**10**	5
	9	6	15	**4**

Addition of four corner numbers as well as four inner numbers is 34.

7 + 14 + 4 + 9 = 34	*7*	12	1	*14*
13 + 8 + 3 + 10 = 34	2	**13**	**8**	11
	16	**3**	**10**	5
	9	6	15	*4*

Addition of four opposite numbers both vertically and horizontally is 34.

12 + 6 + 1 + 15 = 34	7	**12**	1	14
2 + 11 + 16 + 5 = 34	*2*	13	8	*11*
	16	3	10	*5*
	9	**6**	**15**	4

Addition of both sets of symmetrically placed opposite numbers is 34.

12 + 2 + 15 + 5 = 34	7	**12**	*1*	14
1 + 11 + 16 + 6 = 34	*2*	13	8	*11*
	16	3	10	*5*
	9	*6*	**15**	4

Diabolic magic square or pandiagonal magic square is defined as the magic square having the magic sum equal in all directions as well as along the broken diagonals. For example in the above magic square, the magic sum 34 is obtained in all directions as well as along the broken diagonals as follows.

The sum of the numbers in four broken diagonals formed by the numbers (7, 11, 10, 6) is 34, (14, 2, 3, 15) is 34, (4, 16, 3, 1) is 34 and (9, 12, 8, 5) is 34.

The sum of the first two rows equals the sum of the bottom two rows and the sum of the squares of the numbers in the first two rows equals the sum of the squares of the numbers in the bottom two rows.

49 + 144 + 1 + 196 = 390	7	12	1	14
4 + 169 + 64 + 121 = 358	2	13	8	11
256 + 9 + 100 + 25 = 390	16	3	10	5
81 + 36 + 225 + 16 = 358	9	6	15	4

The sum of the first two columns equals the sum of the right side two columns and the sum of the squares of the numbers in the first two columns equals the sum of the squares of the numbers in the right side two columns.

49 + 4 + 256 + 81 = 390	7	12	1	14
144 + 169 + 9 + 36 = 358	2	13	8	11
1 + 64 + 100 + 225 = 390	16	3	10	5
196 + 121 + 25 + 16 = 358	9	6	15	4

The same is true for the numbers of the first and third rows (and the sum of their squares) as respects the numbers of the second and fourth rows. Likewise for the columns. The sum of one set of symmetrically placed numbers, their squares and cubes equals the sum of the other set of symmetrically placed numbers, their squares and cubes.

Examples of magic square of order 4×4 with the random sequence of numbers are as follows.

Table 1

44	51	2	7
6	3	48	47
50	45	8	1
4	5	46	49

Table 1:- The sum of the numbers along the row, column or diagonal is 104.

Table 2:- The sum of the numbers along the row, column or diagonal is 76.

Table 2

30	37	2	7
6	3	34	33
36	31	8	1
4	5	32	35

A special 4×4 magic square: -In the following magic square, the magic square sum is constant irrespective keeping the magic square upside down or reflecting the magic square.

96	11	89	68	The magic square sum is 264 irrespective of the addition of numbers along the row, column or diagonal. Each number of the left side magic square is reflex number of right side magic square.	69	11	98	86
88	69	91	16		88	96	19	61
61	86	18	99		16	68	81	99
19	98	66	81		91	89	66	18

Again the magic square sum is not altered even if it is turned upside down.

19	98	66	81	The magic square sum is 264 irrespective of the addition of numbers along the row, column or diagonal. Each number of the left side magic square is turned upside down of the right side magic square.	91	89	66	18
61	86	18	99		16	68	81	99
88	69	91	16		88	96	19	61
96	11	89	68		69	11	98	86

Another example of magic square having the above property is given below. Here the magic square sum is 176.

25	18	51	82
81	52	15	28
12	21	88	55
58	85	22	11

The magic square of 5×5 starting from 1 onwards is as follows. Here the magic square sum is 65.

15	8	1	24	17
16	14	7	5	23
22	20	13	6	4
3	21	19	12	10
9	2	25	18	11

Likewise we can extend the magic squares for 6×6 starting from 1 onwards and so on. One of the examples for the magic square of 7×7 starting from 1 onwards is as follows. Here the magic square sum is 175.

30	39	48	1	10	19	28
30	47	7	9	18	27	29
46	6	8	17	26	35	37
5	14	16	25	34	36	45
13	15	24	33	42	44	4
21	23	32	31	43	3	12
22	31	40	49	2	11	20

Magic square border puzzle: -

The magic square puzzle with consecutive numbers from 1 to 9 is unique giving only one magic square sum whereas the magic square border puzzle with consecutive numbers starting from 1 to 8 of the third order give four magic square border sums as follows.

Square border puzzle A:-

6	4	2
5	XII	7
1	8	3

219

Square border puzzle B:-

8	3	2
4	XIII	6
1	7	5

Square border puzzle C:-

8	2	4
5	XIV	3
1	6	7

Square border puzzle D:-

8	1	6
4	XV	2
3	5	7

Note: - The magic square border sum is shown in Roman numerals.

In the magic square border puzzles the sum of the middle numbers in the horizontal border is equal to the sum of the middle numbers in the vertical borders. Call these sum of the middle numbers as 'S'.

In the square border puzzle A $4+8 = 7+5 = 12$

In the square border puzzle B $3+7 = 4+6 = 10$

In the square border puzzle C $5+3 = 2+6 = 8$

In the square border puzzle D $4+2 = 1+5 = 6$.

The arithmetic sum 'N' of the consecutive numbers starting from 1 to 8 making up the square border puzzle of the third order is 36. That is $[8(8+1)/2]=36$.

The corner border sums in the four solutions of the 3rd order magic square border puzzles are, in the square border puzzle A $1+2+3+6 = 12$

in the square border puzzle B $1+2+5+8 = 16$

in the square border puzzle C $1+4+7+8 = 20$

in the square border puzzle D $3+6+7+8 = 24$

Call these corner sums as 'C' numbers.

The magic border sums are 12, 13, 14 and 15. Call these numbers as 'M' numbers.

Now tabulate N, C and M numbers.

N	36	36	36	36
C	12	16	20	24
M	12	13	14	15

Draw another table

N+C	48	52	56	60
M	12	13	14	15

Now the sum of the consecutive numbers making up the square border 'N' with 'C' is equal to 4times 'M'. That is N+C=4M.

From this we can derive the fact that C = 4M – N.

Now tabulate S, C and M numbers

S	12	10	8	6
C	12	16	20	24
M	12	13	14	15

From this table we can derive S+C=2M. Combining with the N+C=4M we get S = N-2M.

The relationship between the magic square border sums and the magic square sum is $15 + 0 = 14 + 1 = 13 + 2 = 12 + 3$. In other words the cumulative deduction figures are simply the 15 components of the magic border sums.

Square border puzzle of first order is not possible and also the square border puzzle of second order is also not possible. So the square border puzzle starts from 3rd order onwards only.

Step by step approach for building magic square border of order 3 is as follows. Write the sequence of consecutive numbers starting from 1 onwards

For example 1 2 3 4 5 6 7 8;

Pick out the numbers which make the corner border sum (C=16) Here 1+2+5+8 = 16;

Fix the corner numbers (1, 2, 5 and 8) in diagonal wise order. That is

8		2
1		5

Note the adjacent corner number sum on each border outside the border of the square. That is adjacent sum of 5 and 2 is 7; 8 and 1 is 9; and so on.

10

8		2
1		5

6

Fix the complements of the adjacent corner sum numbers with respect to the magic border sum (M=13) For example 13-7 = 6; 13-9 = 4; and so on.

Now we get the magic border puzzle of order 3 with magic border sum of 13.

(13-10)

8	3	2
4		6
1	7	5

(13-6)

Magic square border puzzle of fourth order starting from 1 to 12 is as follows.

12	8	1	5
2			11
9			4
3	7	10	6

The sum is 26. Likewise we can extend the square border puzzles of higher orders.

A magic square border sum remains unchanged when rotated or reflected.

Tabulate the order number, consecutive numbers and the total number in the consecutive numbers as follows.

Order number	Consecutive numbers	Total number in the consecutive numbers
3	1,2,3,4,5,6,7,8	8 or 4(3-1)
4	1,2,3,4,5,6,7,8,9,10,11,12	12 or 4(4-1)
5	1,2,3,4,5,6,7,8,9,10,11,12,13,14,15,16	16 or 4(5-1)

From the table we can observe that if the order number is given, take its precursor and multiply it by 4 to get the total number of consecutive numbers.

Magic wheel puzzle

Magic wheels are the wheels having the peculiar property that the sum is constant regardless of adding the numbers diagonal-wise in any side. Magic wheel starting from 1 to 5 can be drawn as follows. Here 1 is the center of the magic wheel. The magic wheel sum is found by adding the numbers diagonal-wise in any side.

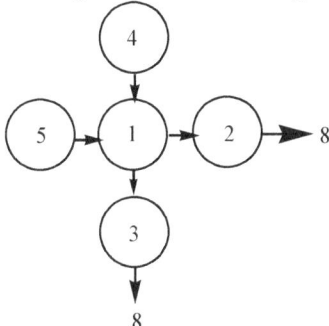

The order of the magic wheel is the number of integers present in a ray or spokes excluding the central number. In the above case it is a magic wheel of order 1 and since it has four rays it can also be called as four directional magic wheel of order one.

The magic wheels can further be developed by increasing the number of directions or rays to 6, 8, 12, 16, 24, 32 and so on.

The six directional magic wheel of order 1 can be formed in a similar way. Here the magic wheel sum is 10.

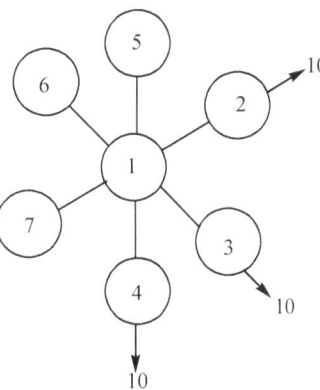

The eight directional magic wheel of order 1 is as follows. Here the magic wheel sum is12.

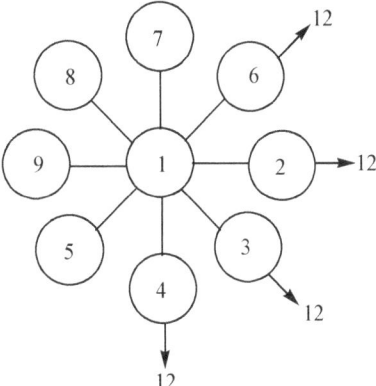

The eight directional magic wheel of order 1 can be drawn as follows.

222

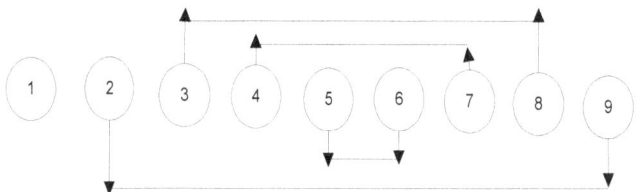

The eight directional magic wheel can be arranged in another way so that the magic wheel sum is 15 in all its diagonals.

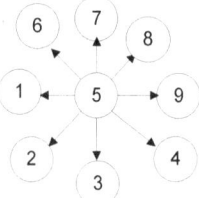

Except 4, we can arrange the numbers from 3 to 12 to express eight directional magic wheel of order one to have the magic wheel sum '20' in all sides.

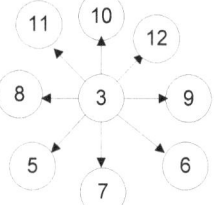

Ray sum is the sum of the integers in a ray of the magic wheel including the central number. The ray sum cannot be arranged equal in the first order magic wheel puzzles.

Magic wheel of order 2 with four directions is as follows.

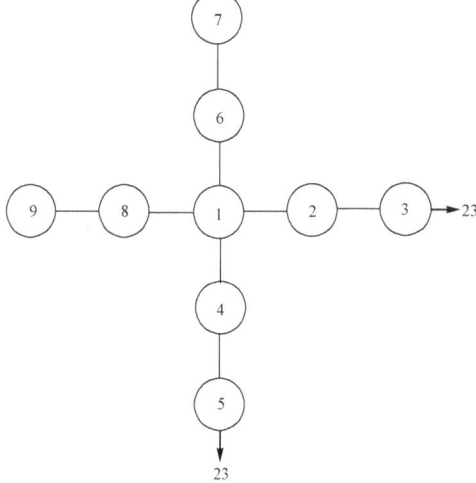

The magic wheel sum is 23. Here the ray sums are different.

The four directional magic wheel of order 2 can be arranged in such a way that the ray sums is also equal. In the following magic wheel puzzle, the ray sum is 12 and the wheel sum is 23.

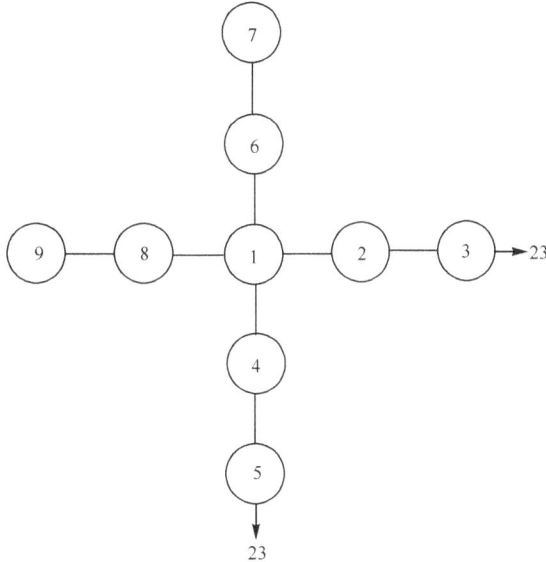

Eight directional magic wheel of order 2 is formed as follows.

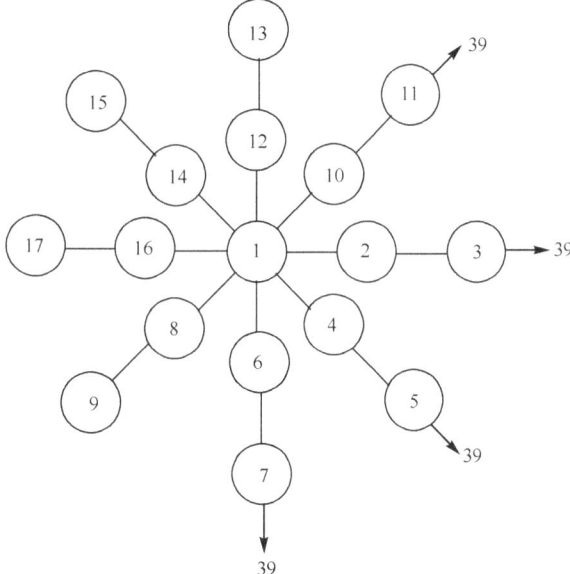

The 'n' order magic wheels will contain 'n' number of sub magic wheels. For example second order magic wheel will contain two sub magic wheels. One is the even numbered inner ring magic wheel and another one is the odd numbered outer ring magic wheel.

In the above case the inner ring sub magic wheel sum is 19 and the outer ring sub magic wheel sum is 21. Ray sum is the average of inner ring magic wheel sum and the outer ring magic wheel sum.

Inner ring sub magic wheel   Outer ring sub magic wheel

The even numbered sub magic wheel and the odd numbered magic sub wheel come alternately.

The summation of the inner ring sub magic wheel sum and the outer ring sub magic wheel when subtracted with central number will give the magic wheel sum.

In the above case 19 + 21 − 1 = 39. Or simply it is the sum of the numbers present in the magic wheel diagonal-wise in any side.

The steps for making eight directional 2nd order magic wheel puzzle is

Step 1   Step 2

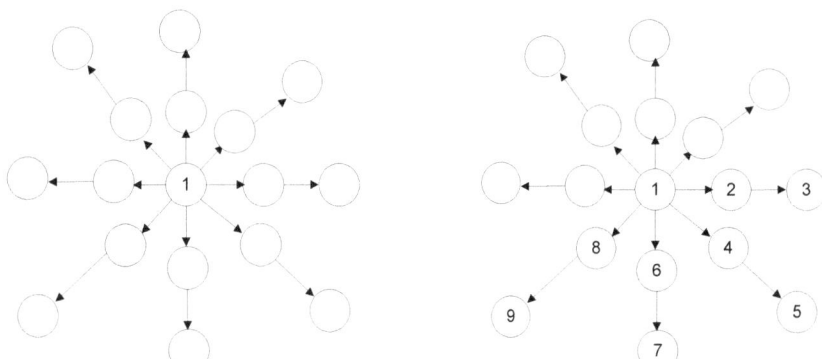

Step 3

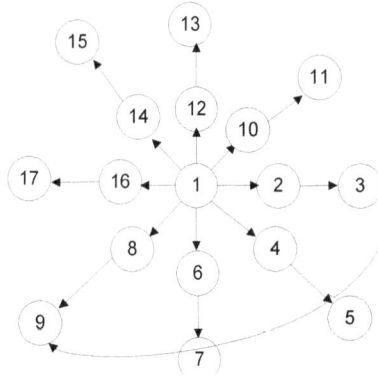

The magic wheel of order 2 can also be arranged so that the ray sums also become equal.

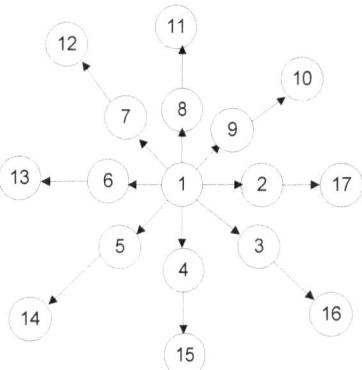

Another way of arranging the ray sum to equal is to position alternatively the even and odd numbers about the inner ring and the outer ring clockwise and anticlockwise respectively. In the following magic wheel puzzle the ray sum is 20 and the magic wheel sum is 39 in all sides.

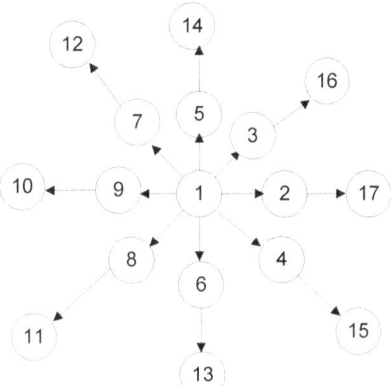

Magic star puzzle

The magic star can be formed in such a way that the sum of the numbers in every row and at the points is same as follows.

The six pointed magic star from 1 to 12 can be arranged in such a manner so that the sum of the numbers in every row and at the points are same. In the following six pointed magic star the sum is 26 in all sides and at the points (vertices).

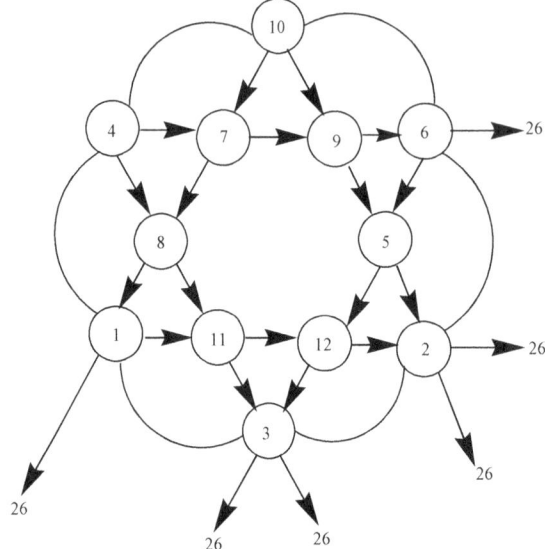

The sum of the numbers in all the directions as well as at the vertical points is 26.

The eight pointed magic star from 1 to 16 has the total of 34 in all sides except in the vertices.

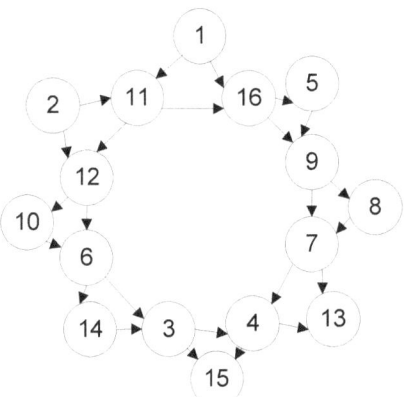

Magic hexagon: -

A magic hexagon can be formed using the numbers from 1 to 19, to get the magic hexagon sum 38 as follows.

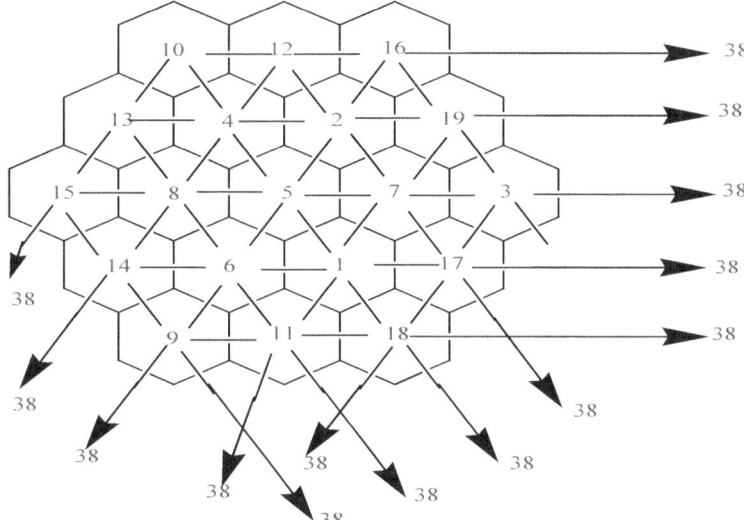

A magic hexagon of order 'n' is the number of hexagons arranged in the magic hexagon. For example in the above case it is a magic hexagon of order three.

The magic hexagon sum for the nth order magic hexagon is found by the following formula

$$\frac{9\left(n^4 - 2n^3 + 2n^2 - n\right) + 2}{2(2n-1)},$$

By applying the formula the magic constant sum sequence will be 1, 28/3, 38, 703/7, 1891/9, 4186/11, ... But this is an integer for only n=1(the trivial case of a single hexagon) and n=3.

Magic dumb bell: -

A magic dumb bell can be formed in such a way that the sum is same in all directions. For example a magic dumb bell is formed with prime numbers alone with the magic dumb bell sum 41 is as follows.

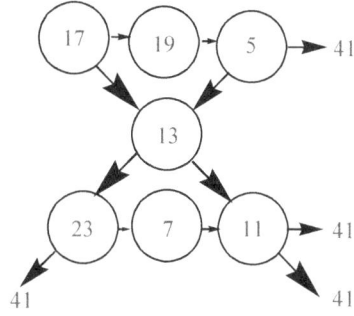

Magic circle: -

From 1 to 33 arranged on four concentric circles, with 9 at the center forms a magic circle as follows.

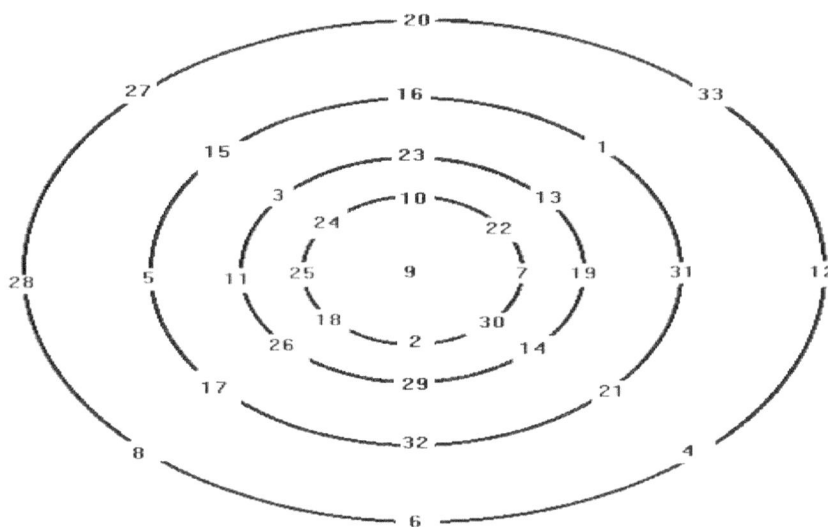

The sum of the numbers on four diameters is 147. That is the sum of 8 numbers plus 9 at the center is 147. For example 28+5+11+25+9+7+19+31+12=147.

The sum of eight radius without 9 is 69. For example 27 + 15 + 3 + 24 = 69.

The sum of all numbers on each circle (not including 9) = 2 × 69.

Fun with addition: -

Applying arithmetic addition only 28 ones can be used to make 1000, 23 twos can be used to make 1000, 16 fours can be used to make 1000, 20 fives can be used to make 1000 and 8 eights can be used to make 1000.

111	222	444	555	888
111	222	444	55	88
111	222	44	55	8
111	222	44	55	8
111	22	4	55	8
111	22	4	55	1000
111	22	4	55	
111	22	4	55	
111	22	4	55	
1	2	4	5	
1000	1000	1000	1000	

Likewise we can arrange pandigital numbers from 0 to 9 to get 100. For example 100= 3/7 + 16/28 + 49 + 50.

ANNEXURES

ANNEXURE 1:- GLOSSARY OF NUMBERS

Abundant number: - The number 'n' is abundant if the sum of all its positive divisors except itself is more than 'n'.

Alternating number: - A number is called as an alternating number, if in its representation odd and even digits come alternately.

Amicable number: - Two distinct positive integers (m, n) are said to be amicable when each integer is the sum of the proper divisors of the other.

Armstrong number: - If the addition of the cubes of each individual digit in a number is equal to the number itself, then it is called as Armstrong number.

Automorphic number: - If 'x' be a number with the property that x^n ends in the same digit as 'x' then it is an automorphic number.

Binary number: - It is the real number represented as the power of 2 and with the radix 2.

Bipolygonal number: - The general terminology for numbers whose dots form a lattice work of two polygons are called as bipolygonal numbers.

Cardinal number: - Counting of objects in terms of numbers like 1, 2, 3, 4… is called as cardinal number.

Carol prime number: - Prime numbers of the form $(2^n - 1)^2 - 2$ where n>2 are called as Carol primes.

Catalan number: - The numbers generated by the formula $C_n = (2n)!/(n! \times (n+1)!)$ where 'n' is a real number, are called as Catalan numbers.

Centered decagonal number: - A centered decagonal number is a centered figurate number that represents a decagon with a dot in the center and all other dots surrounding the center dot in a decagonal lattice.

Centered dodecagonal composite number: - The centered dodecagonal number which is also a composite number is called as centered dodecagonal composite number.

Centered dodecagonal number: - A centered dodecagonal number is a centered figurate number that represents a dodecagon with a dot in the center and all other dots surrounding the center dot in a dodecagonal lattice.

Centered dodecagonal prime number: - The centered dodecagonal number which is also a prime number is called as centered dodecagonal prime number.

Centered hendecagonal composite number: - The centered hendecagonal number which is also a composite number is called as centered hendecagonal number.

Centered hendecagonal number: - A centered hendecagonal number is a centered figurate number that represents a hendecagon with a dot in the center and all other dots surrounding the center dot in a hendecagonal lattice.

Centered hendecagonal prime number: - The centered hendecagonal number which is also a prime number is called as centered hendecagonal number.

Centered heptadecagonal composite number: - The centered heptadecagonal number which is also a composite number is called as centered heptadecagonal composite number.

Centered heptadecagonal number: - A centered heptadecagonal number is a centered figurate number that represents a heptadecagon with a dot in the center and all other dots surrounding the center dot in a heptadecagonal lattice.

Centered heptadecagonal prime number: - The centered heptadecagonal number which is also a prime number is called as centered heptadecagonal prime number.

Centered heptagonal composite number: -The centered heptagonal number which is also a composite number is called as centered heptagonal composite number.

Centered heptagonal number: -A centered heptagonal number is a centered figurate number that represents a heptagon with a dot in the center and all other dots surrounding the center dot in a heptagonal lattice.

Centered heptagonal prime number: -The centered heptagonal number which is also a prime number is called as centered heptagonal prime number.

Centered hexadecagonal composite number: -The centered hexadecagonal number which is also a composite number is called as centered hexadecagonal composite number.

Centered hexadecagonal number: - A centered hexadecagonal number is a centered figurate number that represents a hexadecagon with a dot in the center and all other dots surrounding the center dot in a hexadecagonal lattice.

Centered hexadecagonal prime number: -The centered hexadecagonal number which is also a prime number is called as centered hexadecagonal prime number.

Centered hexagonal composite number: -The centered hexagonal number which is also a composite number is called as centered hexagonal composite number.

Centered hexagonal number: -A centered hexagonal number, or hex number, is a centered figurate number that represents a hexagon with a dot in the center and all other dots surrounding the center dot in a hexagonal lattice.

Centered hexagonal prime number: -The centered hexagonal number which is also a prime number is called as centered hexagonal prime number.

Centered icosagonal composite number: -The centered icosagonal number which is also a composite number is called as centered icosagonal composite number.

Centered icosagonal number: - A centered icosagonal number is a centered figurate number that represents an icosagon with a dot in the center and all other dots surrounding the center dot in an icosagonal lattice.

Centered icosagonal prime number: -The centered icosagonal number which is also a prime number is called as centered icosagonal prime number.

Centered icosidigonal composite number: -Centered icosidigonal number which is also a composite number is called as centered icosidigonal composite number.

Centered icosidigonal number: - A centered icosidigonal number is a centered figurate number that represents an icosidigon with a dot in the center and all other dots surrounding the center dot in an icosidigonal lattice.

Centered icosidigonal prime number: -Centered icosidigonal number which is also a prime number is called as centered icosidigonal prime number.

Centered icosihenagonal composite number: -A centered icosihenagonal number which is also a composite number is called as centered icosihenagonal composite number.

Centered icosihenagonal number: - A centered icosihenagonal number is a centered figurate number that represents an icosihenagon with a dot in the center and all other dots surrounding the center dot in an icosihenagonal lattice.

Centered icosihenagonal prime number: -A centered icosihenagonal number which is also a prime number is called as centered icosihenagonal prime number.

Centered icosiheptagonal composite number: -The centered icosiheptagonal number which is also a composite number is called as centered icosiheptagonal composite number.

Centered icosiheptagonal number: - A centered icosiheptagonal number is a centered figurate number that represents an icosiheptagon with a dot in the center and all other dots surrounding the center dot in an icosiheptagonal lattice.

Centered icosiheptagonal prime number: -The centered icosiheptagonal number which is also a prime number is called as centered icosiheptagonal prime number.

Centered icosihexagonal composite number: -The centered icosihexagonal number which is also a composite number is called as centered icosihexagonal composite number.

Centered icosihexagonal number: - A centered icosihexagonal number is a centered figurate number that represents an icosihexagon with a dot in the center and all other dots surrounding the center dot in an icosihexagonal lattice.

Centered icosihexagonal prime number: -The centered icosihexagonal number which is also a prime number is called as centered icosihexagonal prime number.

Centered icosinonagonal number: - A centered icosinonagonal number is a centered figurate number that represents an icosinonagon with a dot in the center and all other dots surrounding the center dot in an icosinonagonal lattice.

Centered icosioctagonal composite number: -The centered icosioctagonal number which is also a composite number is called as centered icosioctagonal composite number.

Centered icosioctagonal number: - A centered icosioctagonal number is a centered figurate number that represents an icosioctagon with a dot in the center and all other dots surrounding the center dot in an icosioctagonal lattice.

Centered icosioctagonal prime number: -The centered icosioctagonal number which is also a prime number is called as centered icosioctagonal prime number.

Centered icosipentagonal composite number: -The centered icosipentagonal number which is also a composite number is called as centered icosipentagonal composite number.

Centered icosipentagonal number: - A centered icosipentagonal number is a centered figurate number that represents an icosipentagon with a dot in the center and all other dots surrounding the center dot in an icosipentagonal lattice.

Centered icosipentagonal number: - A centered icosipentagonal number is a centered figurate number that represents an icosipentagon with a dot in the center and all other dots surrounding the center dot in an icosipentagonal lattice.

Centered icosipentagonal prime number: -The centered icosipentagonal number which is also a prime number is called as centered icosipentagonal prime number.

Centered icositetragonal composite number: -The centered icositetragonal number which is also a composite number is called as centered icositetragonal composite number.

Centered icositetragonal number: - A centered icositetragonal number is a centered figurate number that represents an icositetragon with a dot in the center and all other dots surrounding the center dot in an icositetragonal lattice.

Centered icositetragonal prime number: -The centered icositetragonal number which is also a prime number is called as centered icositetragonal prime number.

Centered icositriacontagonal number: -It is a figurate number which can represent a centered triacontagon.

Centered icositrigonal composite number: -The centered icositrigonal number which is also a composite number is called as centered icositrigonal composite number.

Centered icositrigonal number: - A centered icositrigonal number is a centered figurate number that represents an icositrigon with a dot in the center and all other dots surrounding the center dot in an icositrigonal lattice.

Centered icositrigonal prime number: -The centered icositrigonal number which is also a prime number is called as centered icositrigonal prime number.

Centered nonadecagonal composite number: -A centered nonadecagonal number which is also a composite number is called as centered nonadecagonal composite number.

Centered nonadecagonal number: - A centered nonadecagonal number is a centered figurate number that represents a nonadecagon with a dot in the center and all other dots surrounding the center dot in a nonadecagonal lattice.

Centered nonadecagonal prime number: -A centered nonadecagonal number which is also a prime number is called as centered nonadecagonal prime number.

Centered nonagonal number: -A centered nonagonal number is a centered figurate number that represents a nonagon with a dot in the center and all other dots surrounding the center dot in a nonagonal lattice.

Centered octadecagonal composite number: -The centered octadecagonal number which is also a composite number is called as centered octadecagonal composite number.

Centered octadecagonal number: - A centered octadecagonal number is a centered figurate number that represents an octadecagon with a dot in the center and all other dots surrounding the center dot in an octadecagonal lattice.

Centered octadecagonal prime number: -The centered octadecagonal number which is also a prime number is called as centered octadecagonal prime number.

Centered octagonal number: -A centered octagonal number is a centered figurate number that represents an octagon with a dot in the center and all other dots surrounding the center dot in an octagonal lattice.

Centered pentadecagonal composite number: -The centered pentadecagonal number which is also a composite number is called as centered pentadecagonal composite number.

Centered pentadecagonal number: - A centered pentadecagonal number is a centered figurate number that represents a pentadecagon with a dot in the center and all other dots surrounding the center dot in a pentadecagonal lattice.

Centered pentadecagonal prime number: -The centered pentadecagonal number which is also a prime number is called as centered pentadecagonal prime number.

Centered pentagonal number: -A centered heptagonal number is a centered figurate number that represents a pentagon with a dot in the center and all other dots surrounding the center dot in a pentagonal lattice.

Centered square composite number: - A centered square number which is a composite number is called as centered square composite number.

Centered square number: -A centered square number is a centered figurate number that represents a square with a dot in the center and all other dots surrounding the center dot in a square lattice.

Centered square prime number: - A centered square number which is a prime number is called as centered square prime number.

Centered tetradecagonal composite number: -The centered tetradecagonal number which is also a composite number is called as centered tetradecagonal composite number.

Centered tetradecagonal number: - A centered tetradecagonal number is a centered figurate number that represents a tetradecagon with a dot in the center and all other dots surrounding the center dot in a tetradecagonal lattice.

Centered tetradecagonal prime number: -The centered tetradecagonal number which is also a prime number is called as centered tetradecagonal prime number.

Centered triacontagonal composite number: -A centered triacontagonal number which is also a composite number is called as centered triacontagonal composite number.

Centered triacontagonal number: - A centered triacontagonal number is a centered figurate number that represents a triacontagon with a dot in the center and all other dots surrounding the center dot in a triacontagonal lattice.

Centered triacontagonal prime number: -A centered triacontagonal number which is also a prime number is called as centered triacontagonal prime number.

Centered triangular composite number: - A centered triangular number which is a composite number is called as centered triangular composite number.

Centered triangular number: - A centered triangular number is a centered figurate number that represents a triangle with a dot in the center and all other dots surrounding the center dot in a triangular lattice.

Centered triangular prime number: - A centered triangular number which is a prime number is called as centered triangular prime number.

Centered tridecagonal composite number: -The centered tridecagonal number which is also a composite number is called as centered tridecagonal composite number.

Centered tridecagonal number: - A centered tridecagonal number is a centered figurate number that represents a dodecagon with a dot in the center and all other dots surrounding the center dot in a tridecagonal lattice.

Centered tridecagonal prime number: -The centered tridecagonal number which is also a prime number is called as centered tridecagonal prime number.

Circular prime number: - If cyclic numbers of a number remains prime then the number is called as circular prime number.

Climbing number or ascending number: -If the digits in a number are in an ascending order it means it is a climbing number.

Complementary number: - It is the number complement to the given decimal number with respect to the base 10.

Complex number:-It is a number consisting of a real and imaginary part. It can be written in the form $a + bi$, where a and b are real numbers, and i is the standard imaginary unit with the property $i^2 = -1$.

Composite ascending number: -The ascending number which is also a composite number is called as composite ascending number.

Composite deficient number: -The deficient number which is composite is called as composite deficient number.

Composite descending number: -The descending number which is also a composite number is called as composite descending number.

Composite happy number: -Happy number which is also a composite number is called as composite happy number.

Composite lucky number: -The lucky number which is also a composite number is called as composite lucky number.

Composite negative seesaw number: -Negative seesaw number which is composite is called as composite negative seesaw number.

Composite number or Divisible number: - All the real numbers except the prime numbers are called as composite numbers or divisible numbers.

Composite partition number: -The partition number which is also a composite number is called as composite partition number.

Composite Pascal number: -The composite number in the Pascal triangle is called as composite Pascal number.

Composite positive seesaw number: -Positive seesaw number which is composite is called as composite positive seesaw number.

Composite undulating number: -The undulating number which is composite in nature is called as composite undulating number.

Composite unhappy number: -Unhappy number which is also a composite number is called as composite unhappy number.

Concatenation number: - The number formed by placing the same number right to the original number is called as concatenation number.

Consecutive number or Following number: - It is the following number which presents successively in the given number sequence.

Coprime number: - Two numbers 'a' and 'b' are said to be co prime or relatively prime numbers if they share no common factor between them.

Cousin prime number: - Prime numbers that differ by 4 are called as cousin primes numbers.

Cube number: - The product of one number with the same number three times is called as the cube of the number.

Cullen number: -It is a number of the form $2^n n + 1$ where 'n' is a real number.

Cyclic number: - The numbers formed by constructing new numbers from the same digits by moving them cyclically are called as cyclic numbers.

Decagonal heptagon number: -The numbers whose dots form a lattice work of decagonal heptagons are called as decagonal heptagon numbers.

Decagonal hexagon number: -The numbers whose dots form a lattice work of decagonal hexagons are called as decagonal hexagon numbers.

Decagonal nonagon number: -The numbers whose dots form a lattice work of decagonal nonagons are called as decagonal nonagon numbers.

Decagonal number: - The numbers whose dots form a lattice work of decagons are called as decagonal numbers.

Decagonal octagon number: -The numbers whose dots form a lattice work of decagonal octagons are called as decagonal octagon numbers.

Decagonal pentagon number: -The numbers whose dots form a lattice work of decagonal pentagons are called as decagonal pentagon numbers or Pentagonal decagon numbers or Pentadecagonal numbers.

Decagonal pyramidal number: - The numbers whose dots form a lattice work of decagonal pyramids are called as decagonal pyramidal numbers.

Decagonal square number: - The numbers whose dots form a lattice work of decagonal squares are called as decagonal square numbers.

Decagonal triangular number: -The numbers whose dots form a lattice work of decagonal triangles are called as decagonal triangular numbers.

Decimal number: - It is the real number represented as the power of 10 with the radix 10.

Deficient number: -The number 'n' is deficient if the sum of all its positive divisors except itself is less than 'n'.

Demlo number: - If we add the first and last digit/s and we get the digit/s in between the numbers, then this type of phenomenon is called as demlo number.

Descending number: - If the digits in a number are in descending order then it is called as descending number.

Digitally balanced number: -

Divisible number: -If a number is divisible by another number/s without any remainder is called as divisible number.

Dodecagonal even number: -The dodecagonal number which is an even number is called as dodecagonal even number.

Dodecagonal odd number: -The dodecagonal number which is an odd number is called as dodecagonal odd number.

Dodecagonal or duodecagonal number: - The numbers whose dots form a lattice work of dodecagons are called as dodecagonal numbers.

Dodecagonal pyramidal number: - The numbers whose dots form a lattice work of dodecagonal pyramids are called as dodecagonal pyramidal numbers.

Dozenal number or Duodecimal number: - It is the real number represented as the power of 12 with the radix 12.

Dudeney number: - If the digital sum of a cube number is equal to the number itself or the cube root means it is called as Dudeney number.

Enneagonal pyramidal number: - The numbers whose dots form a lattice work of enneagonal pyramids are called as enneagonal pyramidal numbers.

Even alternating number: -Alternating number which is even is called as even alternating number.

Even Armstrong number: -Armstrong number which is even is called as even Armstrong number.

Even ascending number: -The ascending number which is an even number is called as an even ascending number.

Even centered dodecagonal number: -Centered dodecagonal number which is also an even number is called as even centered dodecagonal number.

Even centered hendecagonal number: -Centered hendecagonal number which is even is called as even centered hendecagonal number.

Even centered heptadecagonal number: -Centered heptadecagonal number which is even is called as even centered heptadecagonal number.

Even centered heptagonal number: -Centered heptagonal number which is even is called as even centered heptagonal number.

Even centered icosihenagonal number: -Centered icosihenagonal number which is even is called as even centered icosihenagonal number.

Even centered icosiheptagonal number: -Centered icosiheptagonal number which is even is called as even centered icosiheptagonal number.

Even centered icosinonagonal number: -Centered icosinonagonal number which is even is called as even centered icosinonagonal number.

Even centered icosipentagonal number: -Centered icosipentagonal number which is even is called as even centered icosipentagonal number.

Even centered icositrigonal number: -Centered icositrigonal number which is even is called as even centered icositrigonal number.

Even centered nonagonal number: - Centered nonagonal number which is even is called as even centered nonagonal number.

Even centered pentadecagonal number: -Centered pentadecagonal number which is even is called as even centered pentadecagonal number.

Even centered pentagonal number: - Centered pentagonal number which is even is called as even centered pentagonal number.

Even centered triangular number: - Centered triangular number which is even is called as even centered triangular number.

Even centered tridecagonal number: -Centered tridecagonal number which is even is called as even centered tridecagonal number.

Even composite number: - The even number which is also a composite number is termed as even composite number.

Even decagonal number: -The decagonal number which is even is called as even decagonal number.

Even decagonal pyramidal number: -Decagonal pyramidal number which is even is called as even decagonal pyramidal number.

Even deficient number: -The deficient number which is an even number is called as an even deficient number.

Even demlo number: -The demlo number which is even is called as even demlo number.

Even descending number: -The descending number which is an even number is called as an even descending number.

Even dodecagonal number: -Dodecagonal number which is even is called as even dodecagonal number.

Even dodecagonal pyramidal number: -Dodecagonal pyramidal number which is even is called as even dodecagonal pyramidal number.

Even enneagonal pyramidal number: -Enneagonal pyramidal number which is even is called as even enneagonal pyramidal number.

Even Fibonacci number: - Even Fibonacci number is the Fibonacci number which is even.

Even happy number: - The happy number which is also an even number is called as even happy number.

Even harshad number: -The harshad number which is even is called as even harshad number.

Even hendecagonal number: -Hendecagonal pyramidal number which is even is called as even hendecagonal pyramidal number.

Even hendecagonal pyramidal number: -Hendecagonal number which is even is called as even hendecagonal number.

Even heptadecagonal number: -Heptadecagonal number which is even is called as even heptadecagonal number.

Even heptagonal number: -Heptagonal number which is even is called as even heptagonal number.

Even heptagonal pyramidal number: -Heptagonal pyramidal number which is even is called as even heptagonal pyramidal number.

Even hexadecagonal number: -Hexadecagonal number which is even is called as even hexadecagonal number.

Even hexagonal number: -Hexagonal number which is even is called as even hexagonal number.

Even hexagonal pyramidal number: -Hexagonal pyramidal number which is even is called as even hexagonal pyramidal number.

Even icosagonal number: -Icosagonal number which is even is called as even icosagonal number.

Even icosahedral number: -Icosahedral number which is even is called as even icosahedral number.

Even icosidigonal number: -Icosidigonal number which is even is called as even icosidigonal number.

Even icosihenagonal number: -Icosihenagonal number which is even is called as even icosihenagonal number.

Even icosiheptagonal number: -Icosiheptagonal number which is even is called as even icosiheptagonal number.

Even icosihexagonal number: -Icosihexagonal number which is even is called as even icosihexagonal number.

Even icosinonagonal number: -Icosinonagonal number which is even is called as even icosinonagonal number.

Even icosioctagonal number: -Icosioctagonal number which is even is called as even icosioctagonal number.

Even icosipentagonal number: -Icosipentagonal number which is even is called as even icosipentagonal number.

Even icositetragonal number: -Icositetragonal number which is even is called as even icositetragonal number.

Even icositrigonal number: -Icositrigonal number which is even is called as even icositrigonal number.

Even Kaprekar number: -The Kaprekar number which is also an even number is called as even Kaprekar number.

Even negative seesaw number: -Negative seesaw number which is an even number is called as even negative seesaw number.

Even nonadecagonal number: -Nonadecagonal number which is even is called as even nonadecagonal number.

Even nonagonal heptagon number: -Nonagonal heptagon number which is even is called as even nonagonal heptagon number

Even nonagonal number: -Nonagonal number which is even is called as even nonagonal number.

Even number or Paired number: - If a real number is divisible by 2 or multiples of 2 means it is an even number.

Even octadecagonal number: -Octadecagonal number which is even is called as even octadecagonal number.

Even octagonal number: -The octagonal number which is even is called as even octagonal number.

Even octagonal pentagon number: -The octagonal pentagon number which is even is called as even octagonal pentagon number.

Even octagonal pyramidal number: -Octagonal pyramidal number which is even is called as even octagonal pyramidal number.

Even octahedral umber: -Octahedral number which is even is called as even octahedral number.

Even palindromic centered hendecagonal number: - The Palindromic centered hendecagonal number which is even is called as even palindromic centered hendecagonal number.

Even palindromic nonagonal number: -Palindromic nonagonal number which is even is called as even palindromic nonagonal number.

Even partition number: -The partition number which is also an even number is called as even partition number.

Even Pascal number: -The even number in the Pascal triangle is called as even Pascal number.

Even pentadecagonal number: -Pentadecagonal number which is even is called as even pentadecagonal number.

Even pentagonal number: -Pentagonal number which is even is called as even pentagonal number.

Even pentagonal pyramidal number: -Pentagonal pyramidal number which is even is called as even pentagonal pyramidal number.

Even pentagonal triangular number: -Pentagonal triangular number which is even is called as even pentagonal triangular number.

Even positive seesaw number: -Positive seesaw number which is an even number is called as even positive seesaw number.

Even power number: - The power number which is also an even number is called as even power number.

Even prime number: - Prime number which is also even number is called as an even prime number.

Even repdigit number: -Repdigit number which is even is called as even repdigit number.

Even smith number: -Smith number which is an even number is called as even smith number.

Even square pyramidal number: -Square pyramidal number which is even is called as even square pyramidal number.

Even square triangular number: -Square triangular number which is an even number is called as even square triangular number.

Even tetradecagonal number: -Tetradecagonal number which is even is called as even tetradecagonal number.

Even tetrahedral number: -Tetrahedral number which is even is called as even tetrahedral number.

Even triacontagonal number: -Triacontagonal number which is even is called as even triacontagonal number.

Even triangular number: -Triangular number which is even is called as even triangular number.

Even triangular pyramidal number: -Triangular pyramidal number which is even is called as even triangular pyramidal number.

Even tridecagonal number: -Tridecagonal number which is even is called as even tridecagonal number.

Even undulating number: -The undulating number which is even is called as even undulating number.

Even unhappy number: - The unhappy number which is also an even number is called as even unhappy number.

Evil number: - The number 'n' is evil number if it has an even number of 1's in its binary expansion.

Factorial prime number: - Prime numbers of the form $n! - 1$ or $n! + 1$ where 'n' is a real number are called as factorial prime numbers.

Fermat number: - Numbers of the form $2^{(2 \text{ to the power n})} + 1$, whether prime or composite are called Fermat numbers.

Fermat prime number: - Prime numbers of the form $2^{2^n} + 1$ are called as Fermat primes numbers.

Fibonacci number: - If a number falls in the Fibonacci sequence means it is a called as a Fibonacci number.

Fractional number:-If a number is not divided by another number perfectly means both the numerator and denominator are collectively called as fractional number.

Friedman number: - A Friedman number is an integer which, in a given base, is the result of an expression using all its own digits in combination with any of the four basic arithmetic operators (+, −, ×, ÷) and exponentiation.

Gaussian prime number: - Prime numbers of the form '4n+3' where 'n' is a whole number are called as Gaussian primes.

Happy number: - A number is said to be a happy number if we square its digits, and add them together, and then take the result and square its digits and add them together, and keep doing that over and over again, come down to the number 1.

Happy Pythagorean triplet number: -The happy numbers which are Pythagorean triplets also are called as happy Pythagorean triplets.

Harshad number: - It is defined as the number whose summation of the digits is one of the divisors of the number.

Hendecagonal composite number: -The hendecagonal number which is also a composite number is called as hendecagonal number.

Hendecagonal number: - The numbers whose dots form a lattice work of hendecagons are called as hendecagonal numbers.

Hendecagonal prime number: -The hendecagonal number which is also a prime number is called as hendecagonal prime number.

Hendecagonal pyramidal number: - The numbers whose dots form a lattice work of hendecagonal pyramids are called as hendecagonal pyramidal numbers.

Heptadecagonal composite number: -The heptadecagonal number which is also a composite number is called as heptadecagonal composite number.

Heptadecagonal number: - The numbers whose dots form a lattice work of heptadecagons are called as heptadecagonal numbers.

Heptadecagonal prime number: -The heptadecagonal number which is also a prime number is called as heptadecagonal prime number.

Heptagonal composite number: -The heptagonal number which is also a composite number is called as heptagonal composite number.

Heptagonal hexagon number: - The numbers whose dots form a lattice work of heptagonal hexagons are called as heptagonal hexagon numbers.

Heptagonal number: - The numbers whose dots form a lattice work of heptagons are called as heptagonal numbers.

Heptagonal pentagon number: -The numbers whose dots form a lattice work of heptagonal pentagons are called as heptagonal pentagon numbers.

Heptagonal prime number: -The heptagonal number which is also a prime number is called as heptagonal prime number.

Heptagonal pyramidal number: -The numbers whose dots form a lattice work of heptagonal pyramids are called as heptagonal pyramidal numbers.

Heptagonal square number: - The numbers whose dots form a lattice work of heptagonal squares are called as heptagonal square numbers.

Heptagonal triangular number: -The numbers whose dots form a lattice work of heptagonal triangles are called as heptagonal triangular numbers.

Hexadecagonal number: -The numbers whose dots form a lattice work of hexadecagons are called as hexadecagonal numbers.

Hexadecimal number: - It is the real number represented as the power of 16 with the radix 16.

Hexagonal number or cornered hexagonal number: - The numbers whose dots form a lattice work of hexagons are called as hexagonal numbers.

Hexagonal pentagon number: - The numbers whose dots form a lattice work of hexagonal pentagons are called as hexagonal pentagon numbers.

Hexagonal pyramidal number: -The numbers whose dots form a lattice work of hexagonal pyramids are called as hexagonal pyramidal numbers.

Hexagonal square number: -The numbers whose dots form a lattice work of hexagonal squares are called as hexagonal square numbers.

Hexagonal square triangular number: - The numbers whose dots form a lattice work of hexagon, square and triangular shape are called as hexagonal square triangular number.

Hexagonal tridecagonal triangular number: -The numbers whose dots form a lattice work of hexagon, tridecagon and triangular shape are called as hexagonal tridecagonal triangular number.

Icosagonal number: - The numbers whose dots form a lattice work of dodecagons are called as dodecagonal numbers.

Icosahedral number: - The numbers whose dots form a lattice work of icosahedrons are called as icosahedral numbers.

Icosidigonal composite number: -The icosidigonal number which is also a composite number is called as icosidigonal composite number.

Icosidigonal number: - The numbers whose dots form a lattice work of icosidigons are called as icosidigonal numbers.

Icosidigonal prime number: -The icosidigonal number which is also a prime number is called as icosidigonal prime number.

Icosihenagonal number: - The numbers whose dots form a lattice work of icosihenagons are called as icosihenagonal numbers.

Icosiheptagonal number: -The numbers whose dots form a lattice work of icosiheptagons are called as icosiheptagonal numbers.

Icosihexagonal number: -The numbers whose dots form a lattice work of icosihexagons are called as icosihexagonal numbers.

Icosinonagonal composite number: -The icosinonagonal number which is also a composite number is called as icosinonagonal composite number.

Icosinonagonal number: -The numbers whose dots form a lattice work of icosinonagons are called as icosinonagonal numbers.

Icosinonagonal prime number: -The icosinonagonal number which is also a prime number is called as icosinonagonal prime number.

Icosioctagonal number: -The numbers whose dots form a lattice work of icosioctagons are called as icosioctagonal numbers.

Icosipentagonal number: -The numbers whose dots form a lattice work of icosipentagons are called as icosipentagonal numbers.

Icositetragonal number: -The numbers whose dots form a lattice work of icositetragons are called as icositetragonal numbers.

Icositrigonal composite number: -The icositrigonal number which is also a composite number is called as icositrigonal composite number.

Icositrigonal number: - The numbers whose dots form a lattice work of icositrigons are called as icositrigonal numbers.

Icositrigonal prime number: -The icositrigonal number which is also a prime number is called as icositrigonal prime number.

Imaginary number: -It is the imaginary part of the complex number.

Impolite number: -It is a positive integer that cannot be written as the sum of two or more consecutive positive integers.

Indo Arabic numbers or Arabic numbers: - This is the widespread method of numbering system in the world inclusive of zero in it.

Inverse number: -The inverse of a number 'n' is 1/n where n≠0.

Irrational number: -The number which cannot be represented as p/q form where p and q are integers and q≠0 is an irrational number. For example $\sqrt{2}$, $\sqrt{3}$, π and e are some of the irrational number.

Isolated prime number: - Prime number p such that neither $p-2$ nor $p+2$ is an isolated prime number.

Kaprekar number: - When an 'n' digit number is squared and in the square number if the right hand 'n☐' digits are added to the left hand 'n☐☐ ☐' digits and is equal to the original number itself, it is called as Kaprekar number.

Kynea prime number: - Prime numbers of the form $(2^n+1)^2 - 2$ where 'n' is a whole number are called as Kynea prime number.

Leyland number: - A number which can be expressed as a sum of $x^y + y^x$ where x and y are integers greater than 1, is called as Leyland number.

Lucas number: - The Fibonacci rule of adding the latest two to get the next is kept, but here we start from 2 and 1 (in this order) instead of 0 and 1 for the (ordinary) Fibonacci numbers. The numbers obtained are called as Lucas numbers.

Lucas triangular number: - The Lucas numbers whose dots form a lattice work of triangles are called as Lucas triangular numbers.

Lucky number: - It is the natural number that is generated by keeping the first number in the first natural number series followed by keeping first and second numbers followed by keeping first, second and third numbers in the second series and so on.

Lychrel number: - It is a natural number which cannot form a palindrome through the iterative process of repeatedly reversing its base 10 digits and adding the resulting numbers.

Mersenne number: - A Mersenne number, named after Marin Mersenne, is a positive integer that is one less than a power of two. That is 2^n-1.

Mersenne prime number: - If the Mersenne number derived is a prime number then it is called as Mersenne prime number.

Multi polygonal number: - It is a collective terminology for representing bipolygonal numbers, tripolygonal numbers and so on.

Negative number: - Negative numbers are the reverse of real numbers. It ends in 0.

Negative palindromic seesaw number: - A negative seesaw number which is palindromic in nature is called as negative palindromic seesaw number.

Negative seesaw number: - If the left hand side of the digits in a number with respect to the middle digit/s is decreasing and again the right hand side of the number is increasing then the number is called as negative seesaw number.

Nice Friedman number: - A nice Friedman number is a Friedman number where the digits in the expression can be arranged to be in the same order as in the number itself.

Niven number: - It is any whole number that is divisible by the sum of its digits.

Nonadecagonal composite number: - The nonadecagonal number which is also a composite number is called as nonadecagonal composite number.

Nonadecagonal number: - The numbers whose dots form a lattice work of nonadecagons are called as nonadecagonal numbers.

Nonadecagonal prime number: - The nonadecagonal number which is also a prime number is called as nonadecagonal prime number.

Nonagonal heptagon number: - The numbers whose dots form a lattice work of nonagonal heptagons are called as nonagonal heptagon numbers.

Nonagonal hexagon number: - The numbers whose dots form a lattice work of nonagonal hexagons are called as nonagonal hexagon numbers.

Nonagonal number: - The numbers whose dots form a lattice work of nonagons are called as nonagonal numbers.

Nonagonal octagon number: - The numbers whose dots form a lattice work of nonagonal octagons are called as nonagonal octagon numbers.

Nonagonal pentagon number: - The numbers whose dots form a lattice work of nonagonal pentagons are called as nonagonal pentagon numbers.

Nonagonal square number: - The numbers whose dots form a lattice work of nonagonal squares are called as nonagonal square numbers.

Nonagonal triangular number: -The numbers whose dots form a lattice work of nonagonal triangles are called as nonagonal triangular numbers.

Normal or irregular undulating number: -In normal undulating number the digits are present randomly with ups and downs.

Octadecagonal number: - The numbers whose dots form a lattice work of octadecagons are called as octadecagonal numbers.

Octagonal heptagon number: -The numbers whose dots form a lattice work of octagonal heptagons are called as octagonal heptagon numbers.

Octagonal hexagon number: -The numbers whose dots form a lattice work of octagonal hexagons are called as octagonal hexagon numbers.

Octagonal number or cornered octagonal number: - The numbers whose dots form a lattice work of octagons are called as octagonal numbers.

Octagonal pentagon number: -The numbers whose dots form a lattice work of octagonal pentagons are called as octagonal pentagon numbers.

Octagonal pyramidal number: - The numbers whose dots form a lattice work of octagonal pyramids are called as octagonal pyramidal numbers.

Octagonal square number: - The numbers whose dots form a lattice work of octagonal squares are called as octagonal square numbers.

Octagonal triangular number: -The numbers whose dots form a lattice work of octagonal triangles are called as octagonal triangular numbers.

Octahedral number: - The numbers whose dots form a lattice work of octahedrons are called as octahedral numbers.

Octahedral palindromic number: -Octahedral palindromic number is an octahedral number that is palindromic in nature.

Octal number: - It is the real number represented as the power of 8 and with the radix 8.

Odd alternating number: -Alternating number which is odd is called as odd alternating number.

Odd Armstrong number: -Armstrong number which is odd is called as odd Armstrong number.

Odd ascending number: -The ascending number which is an odd number is called as an odd ascending number.

Odd centered dodecagonal number: -Centered dodecagonal number which is odd is called as odd centered dodecagonal number.

Odd centered hendecagonal number: -Centered hendecagonal number which is odd is called as odd centered hendecagonal number.

Odd centered heptadecagonal number: -Centered heptadecagonal number which is odd is called as odd centered heptadecagonal number.

Odd centered heptagonal number: -Centered heptagonal number which is odd is called as odd centered heptagonal number.

Odd centered icosihenagonal number: -Centered icosihenagonal number which is odd is called as odd centered icosihenagonal number.

Odd centered icosiheptagonal number: -Centered icosiheptagonal number which is odd is called as odd centered icosiheptagonal number.

Odd centered icosinonagonal number: -Centered icosinonagonal number which is odd is called as odd centered icosinonagonal number.

Odd centered icosipentagonal number: -Centered icosipentagonal number which is odd is called as odd centered icosipentagonal number.

Odd centered icositrigonal number: -Centered icositrigonal number which is odd is called as odd centered icositrigonal number

Odd centered nonagonal number: -Centered nonagonal number which is odd is called as odd centered nonagonal number.

Odd centered pentadecagonal number: -Centered pentadecagonal number which is odd is called as odd centered pentadecagonal number.

Odd centered pentagonal number: -Centered pentagonal number which is odd is called as odd centered pentagonal number.

Odd centered triangular number: -Centered triangular number which is odd is called as odd centered triangular number.

Odd centered tridecagonal number: -Centered tridecagonal number which is odd is called as odd centered tridecagonal number.

Odd composite number: - The odd number which is also a composite number is termed as odd composite number.

Odd decagonal number: -Decagonal number which is odd is called as odd decagonal number.

Odd deficient number: -The deficient number which is an odd number is called as an odd deficient number.

Odd demlo number: -The demlo number which is odd number is called as odd demlo number.

Odd descending number: -The descending number which is an odd number is called as an odd descending number.

Odd dodecagonal number: -Dodecagonal number which is odd is called as odd dodecagonal number.

Odd dodecagonal pyramidal number: -Dodecagonal pyramidal number which is odd is called as odd dodecagonal pyramidal number.

Odd enneagonal pyramidal number: -Enneagonal pyramidal number which is odd is called as odd enneagonal pyramidal number.

Odd Fibonacci number: - Fibonacci number which is odd is called as odd Fibonacci number.

Odd happy number: - The happy number which is also an odd number is called as odd happy number.

Odd harshad number: -The harshad number which is also an odd number is called as odd harshad number.

Odd hendecagonal number: -Hendecagonal number which is odd is called as odd hendecagonal number.

Odd hendecagonal pyramidal number: -Hendecagonal pyramidal number which is odd is called as odd hendecagonal pyramidal number.

Odd heptadecagonal number: -Heptadecagonal number which is odd is called as odd heptadecagonal number.

Odd heptagonal number: -Heptagonal number which is odd is called as odd heptagonal number.

Odd heptagonal pyramidal number: -Heptagonal pyramidal number which is odd is called as odd heptagonal pyramidal number.

Odd hexadecagonal number: -Hexadecagonal number which is odd is called as odd hexadecagonal number.

Odd hexagonal number: -Hexagonal number which is odd is called as odd hexagonal number.

Odd hexagonal pyramidal number: -Hexagonal pyramidal number which is odd is called as odd hexagonal pyramidal number.

Odd icosagonal number: -Icosagonal number which is odd is called as odd icosagonal number.

Odd icosahedral number: -Icosahedral number which is odd is called as odd icosahedral number.

Odd icosidigonal number: -Icosidigonal number which is odd is called as odd icosidigonal number.

Odd icosihenagonal number: -Icosihenagonal number which is odd is called as odd icosihenagonal number.

Odd icosiheptagonal number: -Icosiheptagonal number which is odd is called as odd icosiheptagonal number.

Odd icosihexagonal number: -Icosihexagonal number which is odd is called as odd icosihexagonal number.

Odd icosinonagonal number: -Icosinonagonal number which is odd is called as odd icosinonagonal number.

Odd icosioctagonal number: -Icosioctagonal number which is odd is called as odd icosioctagonal number.

Odd icosipentagonal number: -Icosipentagonal number which is odd is called as odd icosipentagonal number.

Odd icositetragonal number: -Icositetragonal number which is odd is called as odd icositetragonal number

Odd icositrigonal number: -Icositrigonal number which is odd is called as odd icositrigonal number.

Odd Kaprekar number: -The Kaprekar number which is also an odd number is called as odd Kaprekar number.

Odd negative seesaw number: -Negative seesaw number which is odd is called as odd negative seesaw number.

Odd nonadecagonal number: -Nonadecagonal number which is odd is called as nonadecagonal number.

Odd nonagonal heptagon number: -Nonagonal heptagon number which is odd is called as odd nonagonal heptagon number.

Odd nonagonal number: -Nonagonal number which is odd is called as odd nonagonal number.

Odd number or unpaired number: - If a real number is indivisible by 2 or multiples of 2, it is an odd number.

Odd octadecagonal number: -Octadecagonal number which is odd is called as octadecagonal number.

Odd octagonal number: -The octagonal number which is odd is called as odd octagonal number.

Odd octagonal pentagon number: -The octagonal pentagon number which is odd is called as odd octagonal pentagon number.

Odd octagonal pyramidal number: -Octagonal pyramidal number which is odd is called as odd octagonal pyramidal number.

Odd octahedral number: -Octahedral number which is odd is called as odd octahedral number.

Odd palindromic hendecagonal number: -Palindromic hendecagonal number which is odd is called as odd palindromic hendecagonal number.

Odd palindromic nonagonal number: -Palindromic nonagonal number which is odd is called as odd palindromic nonagonal number.

Odd partition number: -The partition number which is also an odd number is called as odd partition number.

Odd Pascal number: -The odd number in the Pascal triangle is called as odd Pascal number.

Odd pentadecagonal number: -Pentadecagonal number which is odd is called as odd pentadecagonal number.

Odd pentagonal number: -Pentagonal number which is odd is called as odd pentagonal number.

Odd pentagonal pyramidal number: -Pentagonal pyramidal number which is odd is called as odd pentagonal pyramidal number.

Odd pentagonal triangular number: -Pentagonal triangular number which is odd is called as odd pentagonal triangular number.

Odd positive seesaw number: -Positive seesaw number which is odd is called as odd positive seesaw number.

Odd power number: - The power number which is also an odd number is called as odd power number.

Odd prime number: -Prime number which is also odd number is called as an odd prime number.

Odd repdigit number: -Repdigit number which is odd is called as odd repdigit number.

Odd smith number: -Smith number which is an odd number is called as odd smith number.

Odd square pyramidal number: -Square pyramidal number which is odd is called as odd square pyramidal number.

Odd square triangular number: -Square triangular number which is an odd number is called as odd square triangular number.

Odd tetradecagonal number: -Tetradecagonal number which is odd is called as odd tetradecagonal number.

Odd tetrahedral number: -Tetrahedral number which is odd is called as odd tetrahedral number.

Odd triacontagonal number: -Triacontagonal number which is odd is called as odd triacontagonal number.

Odd triangular number: -Triangular pyramidal number which is odd is called as odd triangular pyramidal number.

Odd triangular pyramidal number: -Triangular number which is odd is called as odd triangular number.

Odd tridecagonal number: -Tridecagonal number which is odd is called as odd tridecagonal number.

Odd undulating number: -The undulating number which is odd is called as odd undulating number.

Odd unhappy number: - The unhappy number which is also an odd number is called as odd unhappy number.

Odious number: - The number 'n' is an odious number if it has an odd number of 1's in its binary expansion.

Opposite number: -

Ordinal number: - Counting of objects in terms of words like first, second, third… is called as ordinal number.

Palindrome number or symmetrical number: - When a number is written in the reverse order is equal to the original number itself, it is called as a palindrome number.

Palindromic centered decagonal number: - A centered decagonal number which is palindromic in nature.

Palindromic centered decagonal prime number: -A centered decagonal prime number which is palindromic in nature.

Palindromic centered dodecagonal number: -A centered dodecagonal number which is palindromic in nature.

Palindromic centered hendecagonal number: -A centered hendecagonal number which is palindromic in nature.

Palindromic centered heptagonal number: - A centered heptagonal number which is palindromic in nature.

Palindromic centered Nonagonal number: - A centered nonagonal number which is palindromic in nature.

Palindromic centered Octadecagonal number: -A centered Octadecagonal number which is palindromic in nature.

Palindromic centered Pentadecagonal number: -A centered Pentadecagonal number which is palindromic in nature.

Palindromic centered pentagonal number: - A centered pentagonal number which is palindromic in nature.

Palindromic centered square number: - A centered square number which is palindromic in nature.

Palindromic centered Tetradecagonal number: -A centered Tetradecagonal number which is palindromic in nature.

Palindromic centered Triangular number: -A centered triangular number which is palindromic in nature.

Palindromic composite number: -A composite number which is palindromic in nature is called as palindromic composite number.

Palindromic cube number: -A cubic number which is palindromic in nature is called as palindromic cube number.

Palindromic Cullen number: -A Cullen number which is palindromic in nature is called as palindromic Cullen number.

Palindromic deficient number: -The deficient number which is palindromic in nature is called as palindromic deficient number.

Palindromic demlo number: -The demlo number which is palindromic in nature is called as palindromic demlo number.

Palindromic even number: -An even number which is palindromic in nature is called as palindromic even number.

Palindromic factorial prime number: -The factorial prime which is palindromic in nature is called as palindromic factorial prime number.

Palindromic happy number: -The happy number which is palindromic in nature is called as palindromic happy number.

Palindromic hendecagonal number: -A hendecagonal number which is palindromic in nature.

Palindromic heptagonal number: -A heptagonal number which is palindromic in nature.

Palindromic hexagonal pyramidal number: -A hexagonal pyramidal number which is palindromic in nature.

Palindromic icosidigonal number: -An icosidigonal number which is palindromic in nature.

Palindromic isolated prime number: -The isolated prime number which is palindromic in nature is called as palindromic isolated prime number.

Palindromic Kaprekar number: -A Kaprekar number which is palindromic in nature is called as palindromic Kaprekar number.

Palindromic lucky number: -A lucky number which is palindromic in nature is called as palindromic lucky number.

Palindromic Nonagonal number: - A nonagonal number which is palindromic in nature.

Palindromic odd number: -An odd number which is palindromic in nature is called as palindromic odd number.

Palindromic partition number: -A partition number which is palindromic in nature is called as palindromic partition number.

Palindromic partition prime number: -A partition prime number which is palindromic in nature is called as palindromic partition prime number.

Palindromic Pascal number: -A palindromic number present in the Pascal triangle is called as palindromic Pascal triangle.

Palindromic pentagonal number: -The pentagonal number which is palindromic in nature is called as palindromic pentagonal number.

Palindromic prime number: -The prime number having the palindromic property is called as palindromic prime number.

Palindromic Pronic number: - A Pronic number which is palindromic in nature is called as palindromic pronic number.

Palindromic safe prime number: -A safe prime number which is palindromic in nature is called as palindromic safe prime number.

Palindromic seesaw number: -A seesaw number which is palindromic in nature is called as palindromic seesaw number.

Palindromic smith number: -The smith number which is palindromic in nature is called as palindromic smith number.

Palindromic square number: -If the square number is palindromic, it is called as palindromic square number.

Palindromic square pyramidal number: -A square pyramidal number which is palindromic in nature.

Palindromic star prime number: -The star prime number which is palindromic in nature is called as palindromic star prime number.

Palindromic Thabit prime number: -A Thabit prime number which is palindromic in nature.

Palindromic triacontagonal number: -A palindromic Triacontagonal number is a Triacontagonal number which is palindromic in nature.

Palindromic triangular number: -If the triangular number is palindromic, it is called as palindromic triangular number.

Palindromic tridecagonal number: -A Tridecagonal number which is palindromic in nature.

Palindromic undulating number: -An undulating number which is palindromic in nature is called as palindromic undulating number.

Palindromic Wagstaff prime number: -The Wagstaff prime number which is palindromic in nature is called as palindromic Wagstaff prime number.

Palindromic Wieferich prime number: -The Wieferich prime number which is palindromic in nature is called as palindromic Wieferich prime number.

Palindromic Woodall number: -The Woodall number which is palindromic in nature is called as palindromic Woodall number.

Pandigital number: - It is an integer that in a given base has among its significant digits each digit used in the base at least once.

Parasitic number: -When a number upon multiplication gives the resultant, where the unit digit of the multiplicand is shifted to first digit of the resultant followed by the same set of the digits as in the multiplicand.

Partition number: - The partition number is defined as the number of ways a given number can be written as a sum of positive integers.

Partition prime number: -Partition numbers which are also prime numbers are called as partition prime numbers.

Pascal number: - If a number falls in the Pascal sequence or if it is a member of Pascal triangle, it is a Pascal number.

Pentagonal bipyramidal number: -The numbers whose dots form a lattice work of pentagonal bipyramids are called as pentagonal bipyramidal numbers.

Pentagonal composite number: -The pentagonal number which is also a composite number is called as pentagonal composite number.

Pentagonal number: - The numbers whose dots form a lattice work of pentagons are called as pentagonal numbers.

Pentagonal prime number: -The pentagonal number which is also a prime number is called as pentagonal prime number.

Pentagonal pyramidal number: -The numbers whose dots form a lattice work of pentagonal pyramids are called as pentagonal pyramidal numbers.

Pentagonal square number: -The numbers whose dots form a lattice work of pentagonal squares are called as pentagonal square numbers.

Pentagonal triangular number: -The numbers whose dots form a lattice work of pentagonal triangles are called as pentagonal triangular numbers.

Pental number: -It is the real number represented as the power of 5 and with the radix 5.

Perfect number: - A perfect number is defined as the number whose summations of the factors (excluding the number itself) gives the same number.

Polite number: -It is a positive integer that can be written as the sum of two or more consecutive positive integers.

Polygonal number: - The numbers whose dots form a lattice work of polygons are called as polygonal numbers. This is the general terminology of all forming numbers from triangular number onwards.

Positive palindromic seesaw number: -A positive seesaw number which is palindromic in nature is called as positive palindromic seesaw number.

Positive seesaw number: - If the left hand side of the digits in a number with respect to the middle digit/s is increasing and again the right hand side of the number is decreasing, then the number is called as positive seesaw number.

Power or Exponent number: -It is a general terminology of a number whose product with the same number 'n' times where 'n' is called as power number or exponent.

Power Pascal number: -The power number in the Pascal triangle is called as power Pascal number.

Power triangular number: -The power number in the triangular number series is called as power triangular number.

Powerful number: - When the sum of individual digit of a number is raised to power is equal to the original number, then it is called as powerful number.

Preceding number or Precursor number: - It is the preceding number which presents prior to the number in the given number sequence.

Prime ascending number: -The ascending number which is also a prime number is called as prime ascending number.

Prime deficient number: -The deficient number which is prime is called as prime deficient number.

Prime descending number: -The descending number which is also a prime number is called as prime descending number.

Prime happy number: -The happy numbers which are also prime numbers are called as prime happy numbers.

Prime lucky number: - A prime lucky number is a lucky number that is prime number.

Prime negative seesaw number: -Negative seesaw number which is prime number is called as prime negative seesaw number.

Prime number or Indivisible number: - Prime number is a number which does not have a factor or divisor apart from 1 and the number itself.

Prime partition number: -The partition numbers which are also prime numbers are called as prime partition number.

Prime Pascal number: -The prime number in the Pascal triangle is called as prime Pascal number.

Prime positive seesaw number: -Positive seesaw number which is prime number is called as prime positive seesaw number.

Prime undulating number: -The undulating number which is prime number is called as prime undulating number.

Prime unhappy number: - If the unhappy number is ought to be a prime number, then it is called as prime unhappy number.

Probable number: -The possible number that can be formed using the given digits is called as probable number.

Pronic number: - It is a number which is the product of two consecutive integers, that is, $n(n+1)$ where 'n' is a real number.

Prorepunit number: -The palindromic number which can able to generate repunit number is called as prorepunit number.

Pyramidal number or Triangular pyramidal number: - The numbers whose dots form a lattice work of pyramids are called as pyramidal numbers.

Pythagorean triple number: - If x, y and z are real numbers and obey $x^2+y^2=z^2$ then x, y and z are called as Pythagorean triple numbers.

Ramanujan number: - Ramanujan number is defined as the number which can be represented as the summation of two numbers or with the powers in two different ways.

Rational number: - Rational numbers include negative numbers in the whole numbers system.

Real number or Natural number or Positive number: - Real numbers are the positive numbers. They start from 1 onwards.

Reciprocal number: - The number 1 divided by the rational number 'A' is called as the reciprocal number.

Rectangular number: - The numbers whose dots form a lattice work of rectangles are called as rectangular numbers.

Reflux number or Mirror image number: - This is the number obtained by the mirror image of the original number.

Reflux prime number: - Reflux prime means the reflux number of the prime numbers are also prime numbers.

Repdigit centered heptagonal number: - The centered heptagonal number which is repdigit in nature is called as repdigit centered heptagonal number.

Repdigit centered octadecagonal number: - The centered octadecagonal number which is repdigit in nature is called as repdigit centered octadecagonal number.

Repdigit happy number: - The happy number which is repdigit in nature is called as repdigit happy number.

Repdigit hendecagonal number: - The hendecagonal number which is repdigit in nature is called as repdigit hendecagonal number.

Repdigit lucky number: - The lucky number which is repdigit in nature is called as repdigit lucky number.

Repdigit nonagonal number: -The nonagonal number which is repdigit in nature is called as repdigit nonagonal number.

Repdigit number: -Numbers formed with repeated digits other than one is called as repdigit numbers.

Repunit centered hendecagonal number: -The centered hendecagonal number which is repunit in nature is called as repunit centered hendecagonal number.

Repunit centered nonagonal number: -The centered nonagonal number which is repunit in nature is called as repunit nonagonal number.

Repunit composite number: -Composite numbers which are in the repunit form are called as repunit composite numbers.

Repunit dodecagonal number: -The dodecagonal number which is repunit in nature is called as repunit dodecagonal number.

Repunit Fibonacci number: -The repunit number present in the Fibonacci sequence is called as repunit Fibonacci sequence.

Repunit hendecagonal number: -The hendecagonal number which is repunit in nature is called as repunit hendecagonal number.

Repunit hexagonal number: -The hexagonal number which is repunit in nature is called as repunit hexagonal number.

Repunit icosidigonal number: -The icosidigonal repunit number is an icosidigonal number that is repunit in nature.

Repunit nice Friedman number: -Repunit number which can be expressed as nice Friedman type is called as repunit nice Friedman number.

Repunit nonagonal number: -The nonagonal number which is palindromic in nature is called as repunit nonagonal number.

Repunit number: - Numbers formed with repeated digit of '1' only is called as repunit numbers.

Repunit partition number: -The repunit number which can be expressed as a partition number is called as repunit partition number.

Repunit Pascal number: -The repunit number which is present in a Pascal triangle is called as repunit Pascal number.

Repunit pentagonal number: -Pentagonal number which is repunit in nature is called as repunit pentagonal number.

Repunit prime number: -Prime numbers which are in the repunit form are called as repunit prime numbers.

Repunit triangular number: -Triangular number which is repunit in nature is called as repunit triangular number.

Rhombi cube octahedral number: - The numbers whose dots form a lattice work of rhombus as well as cube and octahedral are called as rhombic cube octahedral numbers.

Rhombic dodecahedral number: -The numbers whose dots form a lattice work of rhombus as well as dodecahedral are called as rhombic dodecahedral numbers.

Roman Friedman numbers: -The Roman numbers which can be expressed as Friedman numbers are called as Roman Friedman numbers.

Roman numbers: - Numbers starting from I onwards are called as Roman numbers. Here zero is not present.

Safe prime number: - A safe prime is a prime number of the form $2p + 1$, where p is also a prime.

Science notation number: -It is the method of representing all the numbers as the powers of 10.

Seesaw number: - If the left hand side of the digits in a number with respect to the middle digit/s is increasing/decreasing and again the right hand side of the number is decreasing/increasing then the number is called as seesaw number.

Smith number: - A composite number is called a Smith number if the sum of its digits equals the sum of all the digits appearing in its prime divisors.

Smooth or regular undulating number: -In smooth undulating number the digits are repeating uniformly.

Square bipyramidal number: - The numbers whose dots form a lattice work of square bipyramids are called as square bipyramidal numbers.

Square Cullen number: -The Cullen number which is also a square number is called as square Cullen number.

Square number: - The product of one number with the same number two times is called as the square of the number.

Square pyramidal number: - The numbers whose dots form a lattice work of square pyramids are called as square pyramidal numbers.

Square Pyramidal square Numbers: - The numbers whose dots form a lattice work of square pyramids as well as squares are called as square pyramidal square numbers.

Square Pyramidal tetrahedral Numbers: - The numbers whose dots form a lattice work of square pyramids as well as tetrahedrons are called as square pyramidal tetrahedral numbers.

Square Pyramidal triangular Numbers: - The numbers whose dots form a lattice work of square pyramids as well as triangles are called as square pyramidal triangular numbers.

Square triangular number: -The numbers whose dots form a lattice work of square triangles are called as square triangular numbers.

Squareful number: -A number is said to be "Squareful" if it contains at least one square in its factorization.

Star composite number: -A star number which is also a composite number is known as star composite number.

Star number:-A star number is a centered figurate number that represents a centered hexagram.

Star palindromic number: -A star palindromic number is a star number that is palindromic.

Star prime number: - A star prime number is a star number which is also a prime number.

Strobogrammatic number: - It is a number that, given a base and given a set of glyphs or styles, appears the same whether viewed normally or upside down.

Super even number: -If all the digits in a number is even, then it is called as super even number.

Super odd number: -If all the digits in a number is odd, then it is called as super odd number.

Ternary number: - It is the real number represented as the power of 3 and with the radix 3.

Tetradecagonal number: - The numbers whose dots form a lattice work of tetradecagons are called as tetradecagonal numbers.

Tetradic number: - A tetradic (or four-way) number is a number that remains unchanged when flipped back to front, mirrored up-down, or flipped up-down.

Tetrahedral number: - The numbers whose dots form a lattice work of tetrahedrons are called as tetrahedral numbers.

Tetrahedral palindrome number: -If the tetrahedral number is palindromic, it is called as tetrahedral palindrome number.

Tetrahedral square Numbers: - The numbers whose dots form a lattice work of tetrahedrons as well as squares are called as tetrahedral square numbers.

Tetrahedral triangular Numbers: - The numbers whose dots form a lattice work of tetrahedrons as well as triangles are called as tetrahedral triangular numbers.

Transcendental number: - It is a number (possibly a complex number) which is not algebraic, that is, it is not a root of a non-constant polynomial equation with rational coefficients. Other examples are $\prod$ and e.

Transpose number: - The numbers derived by interchanging the digits of the original number are called as transpose number.

Trapezium number: - The polite number representation starts with a number other than 1 is called as trapezium number

Triacontagonal number: -The numbers whose dots form a lattice work of triacontagons are called as triacontagonal numbers.

Triangular bipyramidal number: - The numbers whose dots form a lattice work of triangular bipyramids are called as triangular bipyramidal numbers.

Triangular number: - The numbers whose dots form a lattice work of triangles are called as triangular numbers.

Triangular square number: - The numbers whose dots form a lattice work of triangles as well as that of squares are called as triangular square numbers.

Tridecagonal composite number: -The tridecagonal number which is also a composite number is called as tridecagonal composite number.

Tridecagonal number: - The numbers whose dots form a lattice work of tridecagons are called as tridecagonal numbers.

Tridecagonal prime number: -The tridecagonal number which is also a prime number is called as tridecagonal prime number.

Trimorphic number: - A number 'n' is called as a trimorphic number if its n^3 ends with n.

Tripolygonal number: -The general terminology for numbers whose dots form a lattice work of three polygons are called as tripolygonal numbers.

Twin prime number: - Prime numbers that differ by 2 are called as twin prime number.

Undulating number: - Undulating numbers are numbers of the form abababab... in base 10.

Unhappy number: - A number is said to be a unhappy number if we square its digits, and add them together, and then take the result and square *its* digits and add them together, and keep doing that over and over again, not coming down to the number 1.

Vampire Friedman number: -All the vampire numbers are vampire Friedman numbers.

Vampire number: -The number 'n' is called a vampire number if there exists a factorization of n using n's digits.

Wagstaff prime number: -Prime number 'p' of the form p = $[(2^q + 1)/3]$ where q is another prime is a Wagstaff prime number.

Whole number: - Whole numbers include zero in the real number system. That is it starts from 0 onwards.

Wieferich prime number: -The prime number 'p' whose 'p^2' divides 'n^{p-1}' where n>2 is called as Wieferich prime number.

Wilson prime number: -The prime number 'p' whose p^2 divides $(p - 1)! + 1$ is called as Wilson prime number.

Woodall number: -It is the natural number of the form n · 2^n – 1 where 'n' starts from 1 onwards.

Woodall prime number: -Prime numbers of the form n · 2^n – 1 where n>2 are called as Woodall prime numbers.

Zeroless pandigital Friedman number: -Friedman numbers which are zeroless pandigital in nature are called as zeroless pandigital Friedman number.

Zeroless pandigital nice Friedman number: -Nice Friedman numbers which are zeroless pandigital in nature are called as zeroless pandigital nice Friedman number.

Zeroless pandigital number: - Pandigital number without zero in the given number system.

Annexure 2:- Comparison of mathematical properties

Comparison of mathematical properties

S.No	Properties	Arithmetic addition	Arithmetic subtraction	Arithmetic multiplication	Arithmetic division
1.	Additive property	Yes	No	No	No
2.	Subtractive property	No	Yes	No	No
3.	Multiplicative property	No	No	Yes	No
4.	Divisive property	No	No	No	Yes
5.	Determinative property	Yes	Yes	Yes	Yes
6.	Reflux property	No	No	No	No
7.	Palindrome property	No	No	No	No
8.	Symmetric property	Yes	No	Yes	No
9.	Transitive property	Yes	No	Yes	No
10.	Commutative property	Yes	No	Yes	No
11.	Anti commutative property	No	Yes	No	No
12.	Left associative property	Yes	No	Yes	No
13.	Right associative property	Yes	No	Yes	No
14.	Left cancellation property	No	No	Yes	Yes
15.	Right cancellation property	No	No	Yes	Yes
16.	Identity property	Yes	Yes	Yes	Yes
17.	Inverse property	Yes	Yes	No	No
18.	Left distributive property	Yes	Yes	No	No
19.	Right distributive property	Yes	Yes	No	No
20.	Closure property	Yes	Yes*	Yes	Yes
21.	Upset property	No	No	Yes	No
22.	Upset reflux property	No	No	No	No
23.	Opposite property	Yes	Yes	Yes	Yes
24.	Negative property	Yes	No	No	No

*The arithmetic subtraction of natural number and whole number are not having closure property but the other real numbers have closure property with respect to subtraction.

Comparison of mathematical properties (continued)

S.No	Properties	Powers	Mod	Ratio	Percentage	Average
1.	Additive property	Yes	Yes	Yes^f	Yes	Yes
2.	Subtractive property	Yes	Yes	Yes^g	Yes	No
3.	Multiplicative property	Yes	Yes	Yes	Yes	No
4.	Divisive property	Yes	Yes	Yes	Yes	No
5.	Determinative property	Yes	Yes	Yes	Yes	Yes
6.	Reflux property	No	No	Yes	No	No
7.	Palindrome property	No	No	No	No	No
8.	Symmetric property	No	Yes	Yes	No	No
9.	Transitive property	No	Yes	Yes	No	No
10.	Commutative property	No	Yes	Yes	No	No
11.	Anti commutative property	No	Yes^a	No	No	No
12.	Left associative property	No	Yes^b	Yes^h	No	No
13.	Right associative property	No	Yes^c	Yes^i	No	No
14.	Left cancellation property	Yes	Yes	Yes^j	No	No
15.	Right cancellation property	Yes	Yes	Yes^k	No	No
16.	Identity property	Yes(1^2)	Yes	No	No	No
17.	Inverse property	No	No	No	No	No
18.	Left distributive property	No	Yes^d	Yes^l	Yes	No
19.	Right distributive property	No	Yes^e	Yes^m	Yes	No
20.	Closure property	No	Yes	Yes^n	No	No
21.	Upset property	Yes*	No	No	No	No
22.	Upset reflux property	No	No	No	No	No
23.	Opposite property	Yes$^#$	No	No	No	No
24.	Negative property					

* Yes when powers are involved in multiplication operation only.
^{a} Yes for mod subtraction only.
^{B} Yes for mod addition and mod multiplication only.
^{c} Yes for mod addition and mod multiplication only.
^{d} Yes for mod addition and subtraction only.
^{e} Yes for mod addition and subtraction only.
^{f} Yes for addition of precedent part of ratio with another precedent part of the ratio.
^{g} Yes for subtraction of precedent part of ratio with another precedent part of the ratio.
^{h} Yes for ratio addition and multiplication only.
^{i} Yes for ratio addition and multiplication only.
^{j} Yes for multiplication and division only.
^{k} Yes for multiplication and division only.
^{l} Yes for multiplication and division only.
^{m} Yes for multiplication and division only.
^{n} Yes for multiplication and division only.
$^#$ The opposite property of powers is finding out the power roots.

Annexure 3:- Comparison of numbers

Number property	1	2	3	4	5	6	7	8	9	10
Abundant number	no	no	no	no	no	no	no	no	no	no
Alternating number	no	no	no	no	no	no	no	no	no	yes
Armstrong number	yes	no	no	no	no	no	no	no	no	no
Ascending number	no	no	no	no	no	no	no	no	no	no
Automorphic number	yes	no	no	no	yes	yes	no	no	no	no
Bipolygonal number	yes	no	no	no	no	no	no	no	yes	yes
Binary digital balanced number	no	no	yes	no	no	no	no	no	yes	yes
Catalan number	yes	yes	no	no	yes	no	no	no	no	no
Centered cube number	yes	no	no	no	no	no	no	no	no	no
Centered decagonal number	yes	no	no	no	no	no	no	no	no	no
Centered dodecagonal number	yes	no	no	no	no	no	no	no	no	no
Centered hendecagonal number	yes	no	no	no	no	no	no	no	no	no
Centered heptadecagonal number	yes	no	no	no	no	no	no	no	no	no
Centered heptagonal number	yes	no	no	no	no	no	no	yes	no	no
Centered hexadecagonal number	yes	no	no	no	no	no	no	no	no	no
Centered hexagonal number	yes	no	no	no	no	no	yes	no	no	no
Centered icosagonal number	yes	no	no	no	no	no	no	no	no	no
Centered icosidigonal number	yes	no	no	no	no	no	no	no	no	no
Centered icosihenagonal number	yes	no	no	no	no	no	no	no	no	no
Centered icosiheptagonal number	yes	no	no	no	no	no	no	no	no	no
Centered icosihexagonal number	yes	no	no	no	no	no	no	no	no	no
Centered icosinonagonal number	yes	no	no	no	no	no	no	no	no	no
Centered icosioctagonal number	yes	no	no	no	no	no	no	no	no	no
Centered icosipentagonal number	yes	no	no	no	no	no	no	no	no	no
Centered icositetragonal number	yes	no	no	no	no	no	no	no	no	no
Centered icositrigonal number	yes	no	no	no	no	no	no	no	no	no
Centered nonadecagonal number	yes	no	no	no	no	no	no	no	no	no
Centered nonagonal number	yes	no	no	no	no	no	no	no	no	yes
Centered octadecagonal number	yes	no	no	no	no	no	no	no	no	no
Centered octagonal number	yes	no	no	no	no	no	no	no	yes	no
Centered pentadecagonal number	yes	no	no	no	no	no	no	no	no	no
Centered pentagonal number	yes	no	no	no	no	yes	no	no	no	no
Centered square number	yes	no	no	no	yes	no	no	no	no	no
Centered tetradecagonal number	yes	no	no	no	no	no	no	no	no	no
Centered triacontagonal number	yes	no	no	no	no	no	no	no	no	no
Centered triangular number	yes	no	no	yes	no	no	no	no	no	yes
Centered tridecagonal number	yes	no	no	no	no	no	no	no	no	no
Composite number	no	no	no	yes	no	yes	no	yes	yes	yes
Concatenation number	no	no	no	no	no	no	no	no	no	no
Cube number	yes	no	no	no	no	no	no	yes	no	no
Cullen number	no	no	yes	no	no	no	no	no	yes	no
Decagonal number	yes	no	no	no	no	no	no	no	no	yes
Deficient number	no	no	no	yes	no	no	no	yes	yes	yes
Demlo number	no	no	no	no	no	no	no	no	no	no
Descending number	no	no	no	no	no	no	no	no	no	no
Dodecagonal number	yes	no	no	no	no	no	no	no	no	no

Number property	1	2	3	4	5	6	7	8	9	10
Dudeney number	yes	no	no	no	no	no	no	yes	no	no
Even number	no	yes	no	yes	no	yes	no	yes	no	yes
Evil number	no	no	yes	no	yes	yes	no	no	yes	yes
Fibonacci number	yes	yes	yes	no	yes	no	no	yes	no	no
Friedman number	no	no	no	no	no	no	no	no	no	no
Happy number	yes	no	no	no	no	no	yes	no	no	yes
Harshad number	no	no	no	no	no	no	no	no	no	no
Hendecagonal number	yes	no	no	no	no	no	no	no	no	no
Heptadecagonal number	yes	no	no	no	no	no	no	no	no	no
Heptagonal number	yes	no	no	no	no	no	yes	no	no	no
Hexadecagonal number	yes	no	no	no	no	no	no	no	no	no
Hexagonal number	yes	no	no	no	no	yes	no	no	no	no
Hogben number	yes	no	yes	no	no	no	yes	no	no	no
Icosagonal number	yes	no	no	no	no	no	no	no	no	no
Icosidigonal number	yes	no	no	no	no	no	no	no	no	no
Icosihenagonal number	yes	no	no	no	no	no	no	no	no	no
Icosiheptagonal number	yes	no	no	no	no	no	no	no	no	no
Icosihexagonal number	yes	no	no	no	no	no	no	no	no	no
Icosinonagonal number	yes	no	no	no	no	no	no	no	no	no
Icosioctagonal number	yes	no	no	no	no	no	no	no	no	no
Icosipentagonal number	yes	no	no	no	no	no	no	no	no	no
Icositetragonal number	yes	no	no	no	no	no	no	no	no	no
Icositrigonal number	yes	no	no	no	no	no	no	no	no	no
Impolite number	no	yes	no	yes	no	no	no	yes	no	no
Interprime number	no	no	no	yes	no	yes	no	no	yes	no
Kaprekar number	yes	no	no	no	no	no	no	no	yes	no
Leyland number	no	no	no	no	no	no	no	no	no	no
Lychrel number	no	no	no	no	no	no	no	no	no	no
Lucky number	yes	no	yes	no	no	no	yes	no	yes	no
Mersenne number	yes	no	yes	no	no	no	yes	no	no	no
Niven number	no	no	no	no	no	no	no	no	no	no
Nonadecagonal number	yes	no	no	no	no	no	no	no	no	no
Nonagonal number	yes	no	no	no	no	no	no	no	yes	no
Oblong number	no	yes	no	no	no	yes	no	no	no	no
Octadecagonal number	yes	no	no	no	no	no	no	no	no	no
Octagonal number	yes	no	no	no	no	no	no	yes	no	no
Odd number	yes	no	yes	no	yes	no	yes	no	yes	no
Odious number	yes	yes	no	yes	no	no	yes	yes	no	no
Palindrome number	no	no	no	no	no	no	no	no	no	no
Pandigital number	no	no	no	no	no	no	no	no	no	no
Partition number	yes	yes	yes	no	yes	no	yes	no	no	no
Pascal number	yes	yes	yes	yes	yes	yes	yes	yes	yes	yes
Pentadecagonal number	yes	no	no	no	no	no	no	no	no	no
Pentagonal number	yes	no	no	no	yes	no	no	no	no	no
Perfect number	no	no	no	no	no	yes	no	no	no	no
Polite number	no	no	yes	no	yes	yes	yes	no	yes	yes
Prime number	no	yes	yes	no	yes	no	yes	no	no	no

Number property	1	2	3	4	5	6	7	8	9	10
(Triangular) Pyramidal number	yes	no	no	yes	no	no	no	no	no	yes
First order Ramanujan number	no	no	no	no	yes	yes	yes	yes	yes	no
Rectangular number	yes	no	no	no	no	yes	no	yes	no	yes
Repdigit number	no	no	no	no	no	no	no	no	no	no
Repunit number	no	no	no	no	no	no	no	no	no	no
Seesaw number	no	no	no	no	no	no	no	no	no	no
Smith number	no	no	no	yes	no	no	no	no	no	no
Sphenic number	no	no	no	no	no	no	no	no	no	no
Square number	yes	no	no	yes	no	no	no	no	yes	no
Square pyramidal number	yes	no	no	no	yes	no	no	no	no	no
Super even number	no	no	no	no	no	no	no	no	no	no
Super odd number	no	no	no	no	no	no	no	no	no	no
Tetradecagonal number	yes	no	no	no	no	no	no	no	no	no
Tetrahedral number	yes	no	no	yes	no	no	no	no	no	yes
Triacontagonal number	yes	no	no	no	no	no	no	no	no	no
Triangular number	yes	no	yes	no	no	yes	no	no	no	yes
Tridecagonal number	yes	no	no	no	no	no	no	no	no	no
Trimorphic number	yes	no	no	yes	yes	yes	no	no	yes	no
Undulating number	no	no	no	no	no	no	no	no	no	No
Unhappy number	no	yes	yes	yes	yes	yes	no	yes	yes	no
Vampire number	no	no	no	no	no	no	no	no	no	no
Woodall number	yes	no	no	no	no	no	yes	no	no	no

Comparison of numbers (continued)

Number property	11	12	13	14	15	16	17	18	19	20
Abundant number	no	yes	no	no	no	no	no	yes	no	yes
Alternating number	no	yes	no	yes	no	yes	no	yes	no	no
Armstrong number	no	no	no	no	no	no	no	no	no	no
Ascending number	no	yes	yes	yes	yes	yes	yes	yes	yes	no
Automorphic number	no	no	no	no	no	no	no	no	no	no
Bipolygonal number	no	no	no	no	no	no	no	no	no	no
Catalan number	no	no	no	yes	no	no	no	no	no	no
Centered cube number	no	no	no	no	no	no	no	no	no	no
Centered decagonal number	yes	no	no	no	no	no	no	no	no	no
Centered dodecagonal number	no	no	yes	no	no	no	no	no	no	no
Centered hendecagonal number	no	yes	no	no	no	no	no	no	no	no
Centered heptadecagonal number	no	no	no	no	no	no	no	yes	no	no
Centered heptagonal number	no	no	no	no	no	no	no	no	no	no
Centered hexadecagonal number	no	no	no	no	no	no	yes	no	no	no
Centered hexagonal number	no	no	no	no	no	no	no	yes	no	no
Centered icosagonal number	no	no	no	no	no	no	no	no	no	no
Centered icosidigonal number	no	no	no	no	no	no	no	no	no	no
Centered icosihenagonal number	no	no	no	no	no	no	no	no	no	no
Centered icosiheptagonal number	no	no	no	no	no	no	no	no	no	no
Centered icosihexagonal number	no	no	no	no	no	no	no	no	no	no
Centered icosinonagonal number	no	no	no	no	no	no	no	no	no	no
Centered icosioctagonal number	no	no	no	no	no	no	no	no	no	no
Centered icosipentagonal number	no	no	no	no	no	no	no	no	no	no
Centered icositetragonal number	no	no	no	no	no	no	no	no	no	no
Centered icositrigonal number	no	no	no	no	no	no	no	no	no	no
Centered nonadecagonal number	no	no	no	no	no	no	no	no	no	yes
Centered nonagonal number	no	no	no	no	no	no	no	no	no	no
Centered octadecagonal number	no	no	no	no	no	no	no	no	yes	no
Centered octagonal number	no	no	no	no	no	no	no	no	no	no
Centered pentadecagonal number	no	no	no	no	no	yes	no	no	no	no
Centered pentagonal number	no	no	no	no	no	yes	no	no	no	no
Centered square number	no	no	yes	no	no	no	no	no	no	no
Centered tetradecagonal number	no	no	no	no	yes	no	no	no	no	no
Centered triacontagonal number	no	no	no	no	no	no	no	no	no	no
Centered triangular number	no	no	no	no	no	no	no	no	yes	no
Centered tridecagonal number	no	no	yes	no	no	no	no	no	no	no
Composite number	no	yes	no	yes	yes	yes	no	yes	no	yes
Concatenation number	yes	no	no	no	no	no	no	no	no	no
Cube number	no	no	no	no	no	no	no	no	no	no
Cullen number	no	no	no	no	no	no	no	no	no	no
Decagonal number	no	no	no	no	no	no	no	no	no	no
Deficient number	no	no	no	yes	yes	yes	no	no	no	no
Demlo number	no	no	no	no	no	no	no	no	no	no
Descending number	no	no	no	no	no	no	no	no	no	yes
Binary digital balanced number	no	yes	no	no	no	no	no	no	no	no
Dodecagonal number	no	yes	no	no	no	no	no	no	no	no

Number property	11	12	13	14	15	16	17	18	19	20
Dudeney number	no	no	no	no	no	no	yes	yes	no	no
Even number	no	yes	no	yes	no	yes	no	yes	no	yes
Evil number	no	yes	no	no	yes	no	yes	yes	no	yes
Fibonacci number	no	no	yes	no	no	no	no	no	no	no
Friedman number	no	no	no	no	no	no	no	no	no	no
Happy number	no	no	yes	no	no	no	no	no	yes	no
Harshad number	no	yes	no	no	no	no	no	no	no	yes
Hendecagonal number	yes	no	no	no	no	no	no	no	no	no
Heptadecagonal number	no	no	no	no	no	no	yes	no	no	no
Heptagonal number	no	no	no	no	no	no	no	yes	no	no
Hexadecagonal number	no	no	no	no	no	yes	no	no	no	no
Hexagonal number	no	no	no	no	yes	no	no	no	no	no
Hogben number	no	no	yes	no	no	no	no	no	no	no
Icosagonal number	no	no	no	no	no	no	no	no	no	yes
Icosidigonal number	no	no	no	no	no	no	no	no	no	no
Icosihenagonal number	no	no	no	no	no	no	no	no	no	no
Icosiheptagonal number	no	no	no	no	no	no	no	no	no	no
Icosihexagonal number	no	no	no	no	no	no	no	no	no	no
Icosinonagonal number	no	no	no	no	no	no	no	no	no	no
Icosioctagonal number	no	no	no	no	no	no	no	no	no	no
Icosipentagonal number	no	no	no	no	no	no	no	no	no	no
Icositetragonal number	no	no	no	no	no	no	no	no	no	no
Icositrigonal number	no	no	no	no	no	no	no	no	no	no
Impolite number	no	no	no	no	no	yes	no	no	no	no
Interprime number	no	yes	no	no	yes	no	no	yes	no	no
Kaprekar number	no	no	no	no	no	no	no	no	no	no
Leyland number	no	no	no	no	no	no	yes	no	no	no
Lucky number	no	no	yes	no	yes	no	no	no	no	no
Lychrel number	no	no	no	no	no	no	no	no	no	no
Mersenne number	no	no	no	no	no	no	yes	no	no	no
Niven number	no	yes	no	no	no	no	no	yes	no	yes
Nonadecagonal number	no	no	no	no	no	no	no	no	yes	no
Nonagonal number	no	no	no	no	no	no	no	no	no	no
Oblong number	no	yes	no	no	no	no	no	no	no	yes
Octadecagonal number	no	no	no	no	no	no	no	yes	no	no
Octagonal number	no	no	no	no	no	no	no	no	no	no
Odd number	yes	no	yes	no	yes	no	yes	no	yes	no
Odious number	yes	no	yes	yes	no	yes	no	no	yes	no
Palindrome number	yes	no	no	no	no	no	no	no	no	no
Pandigital number	no	no	no	no	no	no	no	no	no	no
Partition number	yes	no	no	no	yes	no	no	no	no	no
Pascal number	yes	yes	yes	yes	yes	yes	yes	yes	yes	yes
Pentadecagonal number	no	no	no	no	yes	no	no	no	no	no
Pentagonal number	no	yes	no	no	no	no	no	no	no	no
Perfect number	no	no	no	no	no	no	no	no	no	no
Polite number	yes	yes	yes	yes	yes	no	yes	yes	yes	yes
Prime number	yes	no	yes	no	no	no	yes	no	yes	no

Number property	11	12	13	14	15	16	17	18	19	20
(Triangular) Pyramidal number	no	no	no	no	no	no	no	no	no	yes
First order Ramanujan number	no	no	no	no	no	no	no	no	no	no
Rectangular number	no	yes	no	yes	yes	yes	no	yes	no	yes
Repdigit number	no	no	no	no	no	no	no	no	no	no
Repunit number	yes	no	no	no	no	no	no	no	no	no
Seesaw number	no	no	no	no	no	no	no	no	no	no
Smith number	no	no	no	no	no	no	no	no	no	no
Sphenic number	no	no	no	no	no	no	no	no	no	no
Square number	no	no	no	no	no	yes	no	no	no	no
Square pyramidal number	no	no	no	yes	no	no	no	no	no	no
Super even number	no	no	no	no	no	no	no	no	no	yes
Super odd number	yes	no	yes	no	yes	no	yes	no	yes	no
Tetradecagonal number	no	no	no	yes	no	no	no	no	no	no
Tetrahedral number	no	no	no	no	no	no	no	no	no	yes
Triacontagonal number	no	no	no	no	no	no	no	no	no	no
Triangular number	no	no	no	no	yes	no	no	no	no	no
Tridecagonal number	no	no	yes	no	no	no	no	no	no	no
Undulating number	no	no	no	no	no	no	no	no	no	no
Trimorphic number	no	no	no	no	no	no	no	no	no	no
Unhappy number	yes	yes	no	yes	yes	yes	yes	yes	no	yes
Vampire number	no	no	no	no	no	no	no	no	no	no
Woodall number	no	no	no	no	no	no	no	no	no	no

Comparison of numbers (continued)

Number property	21	22	23	24	25	26	27	28	29	30
Abundant number	no	no	no	yes	no	no	no	no	no	yes
Alternating number	yes	no	yes	no	yes	no	yes	no	yes	yes
Armstrong number	no	no	no	no	no	no	no	no	no	no
Ascending number	no	no	yes	yes	yes	yes	yes	yes	yes	no
Automorphic number	no	no	no	no	yes	no	no	no	no	no
Bipolygonal number	no	no	no	no	no	no	no	no	no	no
Catalan number	no	no	no	no	no	no	no	no	no	no
Centered decagonal number	no	no	no	no	no	no	no	no	no	no
Centered dodecagonal number	no	no	no	no	no	no	no	no	no	no
Centered hendecagonal number	no	no	no	no	no	no	no	no	no	no
Centered heptadecagonal number	no	no	no	no	no	no	no	no	no	no
Centered heptagonal number	no	yes	no	no	no	no	no	no	no	no
Centered hexadecagonal number	no	no	no	no	no	no	no	no	no	no
Centered hexagonal number	no	no	no	no	no	no	no	no	no	no
Centered icosagonal number	yes	no	no	no	no	no	no	no	no	no
Centered icosidigonal number	no	no	yes	no	no	no	no	no	no	no
Centered icosihenagonal number	no	yes	no	no	no	no	no	no	no	no
Centered icosiheptagonal number	no	no	no	no	no	no	no	yes	no	no
Centered icosihexagonal number	no	no	no	no	no	no	yes	no	no	no
Centered icosinonagonal number	no	no	no	no	no	no	no	no	no	yes
Centered icosioctagonal number	no	no	no	no	no	no	no	no	yes	no
Centered icosipentagonal number	no	no	no	no	no	yes	no	no	no	no
Centered icositetragonal number	no	no	no	no	yes	no	no	no	no	no
Centered icositrigonal number	no	no	no	yes	no	no	no	no	no	no
Centered nonadecagonal number	no	no	no	no	no	no	no	no	no	no
Centered nonagonal number	no	no	no	no	no	no	yes	no	no	no
Centered octadecagonal number	no	no	no	no	no	no	no	no	no	no
Centered octagonal number	no	no	no	no	yes	no	no	no	no	no
Centered pentadecagonal number	no	no	no	no	no	no	no	no	no	no
Centered pentagonal number	no	no	no	no	no	no	no	no	no	no
Centered square number	no	no	no	no	yes	no	no	no	no	no
Centered tetradecagonal number	no	no	no	no	no	no	no	no	no	no
Centered triacontagonal number	no	no	no	no	no	no	no	no	no	no
Centered triangular number	no	no	no	no	no	no	no	no	no	no
Centered tridecagonal number	no	no	no	no	no	no	no	no	no	no
Composite number	yes	yes	no	yes	yes	yes	yes	yes	no	yes
Concatenation number	no	yes	no	no	no	no	no	no	no	no
Cube number	no	no	no	no	no	no	yes	no	no	no
Cullen number	no	no	no	no	yes	no	no	no	no	no
Decagonal number	no	no	no	no	no	no	yes	no	no	no
Deficient number	yes	yes	no	no	yes	yes	yes	no	no	no
Demlo number	no	no	no	no	no	no	no	no	no	no
Descending number	yes	no	no	no	no	no	no	no	no	no
Binary digital balanced number	no	no	no	no	no	no	no	no	no	no
Dodecagonal number	no	no	no	no	no	no	no	no	no	no
Dudeney number	no	no	no	no	no	yes	no	no	no	no

Number property	21	22	23	24	25	26	27	28	29	30
Even number	no	yes	no	yes	no	yes	no	yes	no	yes
Evil number	no	no	yes	yes	no	no	yes	no	yes	yes
Fibonacci number	yes	no	no	no	no	no	no	no	no	no
Friedman number	no	no	no	no	yes	no	no	no	no	no
Happy number	no	no	yes	no	no	no	no	yes	no	no
Harshad number	no	no	no	yes	no	no	yes	no	no	yes
Hendecagonal number	no	no	no	no	no	no	no	no	no	yes
Heptadecagonal number	no	no	no	no	no	no	no	no	no	no
Heptagonal number	no	no	no	no	no	no	no	no	no	no
Hexadecagonal number	no	no	no	no	no	no	no	no	no	no
Hexagonal number	no	no	no	no	no	no	no	yes	no	no
Hogben number	yes	no	no	no	no	no	no	no	no	no
Icosagonal number	no	no	no	no	no	no	no	no	no	no
Icosidigonal number	no	yes	no	no	no	no	no	no	no	no
Icosihenagonal number	yes	no	no	no	no	no	no	no	no	no
Icosiheptagonal number	no	no	no	no	no	no	yes	no	no	no
Icosihexagonal number	no	no	no	no	no	yes	no	no	no	no
Icosinonagonal number	no	no	no	no	no	no	no	no	yes	no
Icosioctagonal number	no	no	no	no	no	no	no	yes	no	no
Icosipentagonal number	no	no	no	no	yes	no	no	no	no	no
Icositetragonal number	no	no	no	yes	no	no	no	no	no	no
Icositrigonal number	no	no	yes	no	no	no	no	no	no	no
Impolite number	no	no	no	no	no	no	no	no	no	no
Interprime number	yes	no	no	no	no	yes	no	no	no	yes
Kaprekar number	no	no	no	no	no	no	no	no	no	no
Leyland number	no	no	no	no	no	no	no	no	no	no
Lucky number	yes	no	no	no	yes	no	no	no	no	no
Lychrel number	no	no	no	no	no	no	no	no	no	no
Mersenne number	no	no	no	no	no	no	no	no	no	no
Niven number	yes	no	no	yes	no	no	yes	no	no	yes
Nonadecagonal number	no	no	no	no	no	no	no	no	no	no
Nonagonal number	no	no	no	yes	no	no	no	no	no	no
Oblong number	no	no	no	no	no	no	no	no	no	yes
Octadecagonal number	no	no	no	no	no	no	no	no	no	no
Octagonal number	yes	no	no	no	no	no	no	no	no	no
Odd number	yes	no	yes	no	yes	no	yes	no	yes	no
Odious number	yes	yes	no	no	yes	yes	no	yes	no	no
Palindrome number	no	yes	no	no	no	no	no	no	no	no
Pandigital number	no	no	no	no	no	no	no	no	no	no
Partition number	no	yes	no	no	no	no	no	no	no	yes
Pascal number	yes	yes	yes	yes	yes	yes	yes	yes	yes	yes
Pentadecagonal number	no	no	no	no	no	no	no	no	no	no
Pentagonal number	no	yes	no	no	no	no	no	no	no	no
Perfect number	no	no	no	no	no	no	no	yes	no	no
Polite number	yes	yes	yes	yes	yes	yes	yes	yes	yes	yes
Prime number	no	no	yes	no	no	no	no	no	yes	no
(Triangular) Pyramidal number	no	no	no	no	no	no	no	no	no	no

Number property	21	22	23	24	25	26	27	28	29	30
First order Ramanujan number	no	no	no	no	no	no	no	no	no	no
Rectangular number	yes	yes	no	yes	no	yes	yes	yes	no	yes
Repdigit number	no	yes	no	no	no	no	no	no	no	no
Repunit number	no	no	no	no	no	no	no	no	no	no
Seesaw number	no	no	no	no	no	no	no	no	no	no
Smith number	no	yes	no	no	no	no	yes	no	no	no
Sphenic number	no	no	no	no	no	no	no	no	no	yes
Square number	no	no	no	no	yes	no	no	no	no	no
Square pyramidal number	no	no	no	no	no	no	no	no	no	yes
Super even number	no	yes	no	yes	no	yes	no	yes	no	no
Super odd number	no	no	no	no	no	no	no	no	no	no
Tetradecagonal number	no	no	no	no	no	no	no	no	no	no
Tetrahedral number	no	no	no	no	no	no	no	no	no	no
Triacontagonal number	no	no	no	no	no	no	no	no	no	yes
Triangular number	yes	no	no	no	no	no	no	yes	no	no
Tridecagonal number	no	no	no	no	no	no	no	no	no	no
Trimorphic number	no	no	no	yes	yes	no	no	no	no	no
Undulating number	no	no	no	no	no	no	no	no	no	no
Unhappy number	yes	yes	no	yes	yes	yes	yes	no	yes	yes
Vampire number	no	no	no	no	no	no	no	no	no	no
Woodall number	no	no	yes	no	no	no	no	no	no	no

Comparison of numbers (continued)

Number property	31	32	33	34	35	36	37	38	39	40
Abundant number	no	no	no	no	no	yes	no	no	no	yes
Alternating number	no	yes	no	yes	no	yes	no	yes	no	no
Armstrong number	no	no	no	no	no	no	no	no	no	no
Ascending number	no	no	no	yes	yes	yes	yes	yes	yes	no
Automorphic number	no	no	no	no	no	no	no	no	no	no
Bipolygonal number	no	no	no	no	no	yes	no	no	no	no
Catalan number	no	no	no	no	no	no	no	no	no	no
Centered cube number	no	no	no	no	yes	no	no	no	no	no
Centered decagonal number	yes	no	no	no	no	no	no	no	no	no
Centered dodecagonal number	no	no	no	no	no	no	yes	no	no	no
Centered hendecagonal number	no	no	no	yes	no	no	no	no	no	no
Centered heptadecagonal number	no	no	no	no	no	no	no	no	no	no
Centered heptagonal number	no	no	no	no	no	no	no	no	no	no
Centered hexadecagonal number	no	no	no	no	no	no	no	no	no	no
Centered hexagonal number	no	no	no	no	no	no	yes	no	no	no
Centered icosagonal number	no	no	no	no	no	no	no	no	no	no
Centered icosidigonal number	no	no	no	no	no	no	no	no	no	no
Centered icosihenagonal number	no	no	no	no	no	no	no	no	no	no
Centered icosiheptagonal number	no	no	no	no	no	no	no	no	no	no
Centered icosihexagonal number	no	no	no	no	no	no	no	no	no	no
Centered icosinonagonal number	no	no	no	no	no	no	no	no	no	no
Centered icosioctagonal number	no	no	no	no	no	no	no	no	no	no
Centered icosipentagonal number	no	no	no	no	no	no	no	no	no	no
Centered icositetragonal number	no	no	no	no	no	no	no	no	no	no
Centered icositrigonal number	no	no	no	no	no	no	no	no	no	no
Centered nonadecagonal number	no	no	no	no	no	no	no	no	no	no
Centered nonagonal number	no	no	no	no	no	no	no	no	no	no
Centered octadecagonal number	no	no	no	no	no	no	no	no	no	no
Centered octagonal number	no	no	no	no	no	no	no	no	no	no
Centered pentadecagonal number	no	no	no	no	no	no	no	no	no	no
Centered pentagonal number	yes	no	no	no	no	no	no	no	no	no
Centered square number	no	no	no	no	no	no	no	no	no	no
Centered tetradecagonal number	no	no	no	no	no	no	no	no	no	no
Centered triacontagonal number	yes	no	no	no	no	no	no	no	no	no
Centered triangular number	yes	no	no	no	no	no	no	no	no	no
Centered tridecagonal number	no	no	no	no	no	yes	no	no	no	no
Composite number	no	yes	yes	yes	yes	yes	no	yes	no	yes
Concatenation number	no	no	yes	no	no	no	no	no	no	no
Cube number	no	no	no	no	no	no	no	no	no	no
Cullen number	no	no	no	no	no	no	no	no	no	no
Decagonal number	no	no	no	no	no	no	no	no	no	no
Deficient number	no	yes	yes	yes	yes	no	no	yes	yes	no
Demlo number	no	no	no	no	no	no	no	no	no	no
Descending number	yes	yes	no	no	no	no	no	no	no	no
Binary digital balanced number	no	no	no	no	yes	no	yes	yes	no	no
Dodecagonal number	no	no	yes	no	no	no	no	no	no	no

Number property	31	32	33	34	35	36	37	38	39	40
Dudeney number	no	no	no	no	no	no	no	no	no	no
Even number	no	yes	no	yes	no	yes	no	yes	no	yes
Evil number	no	no	yes	yes	no	yes	no	no	yes	yes
Fibonacci number	no	no	no	yes	no	no	no	no	no	no
Friedman number	no	no	no	no	no	no	no	no	no	no
Happy number	yes	yes	no	no	no	no	no	no	no	no
Harshad number	no	no	no	no	no	yes	no	no	no	yes
Hendecagonal number	no	no	no	no	no	no	no	no	no	no
Heptadecagonal number	no	no	no	no	no	no	no	no	no	no
Heptagonal number	no	no	no	yes	no	no	no	no	no	no
Hexadecagonal number	no	no	no	no	no	no	no	no	no	no
Hexagonal number	no	no	no	no	no	no	no	no	no	no
Hogben number	yes	no	no	no	no	no	no	no	no	no
Icosagonal number	no	no	no	no	no	no	no	no	no	no
Icosidigonal number	no	no	no	no	no	no	no	no	no	no
Icosihenagonal number	no	no	no	no	no	no	no	no	no	no
Icosiheptagonal number	no	no	no	no	no	no	no	no	no	no
Icosihexagonal number	no	no	no	no	no	no	no	no	no	no
Icosinonagonal number	no	no	no	no	no	no	no	no	no	no
Icosioctagonal number	no	no	no	no	no	no	no	no	no	no
Icosipentagonal number	no	no	no	no	no	no	no	no	no	no
Icositetragonal number	no	no	no	no	no	no	no	no	no	no
Icositrigonal number	no	no	no	no	no	no	no	no	no	no
Impolite number	no	yes	no	no	no	no	no	no	no	no
Interprime number	no	no	no	yes	no	no	no	no	yes	no
Kaprekar number	no	no	no	no	no	no	no	no	no	no
Leyland number	no	yes	no	no	no	no	no	no	no	no
Lucky number	yes	no	yes	no	no	no	yes	no	no	no
Lychrel number	no	no	no	no	no	no	no	no	no	no
Mersenne number	yes	no	no	no	no	no	no	no	no	no
Niven number	no	no	no	no	no	yes	no	no	no	yes
Nonadecagonal number	no	no	no	no	no	no	no	no	no	no
Nonagonal number	no	no	no	no	no	no	no	no	no	no
Oblong number	no	no	no	no	no	no	no	no	no	no
Octadecagonal number	no	no	no	no	no	no	no	no	no	no
Octagonal number	no	no	no	no	no	no	no	no	no	yes
Odd number	yes	no	yes	no	yes	no	yes	no	yes	no
Odious number	yes	yes	no	no	yes	no	yes	yes	no	no
Palindrome number	no	no	yes	no	no	no	no	no	no	no
Pandigital number	no	no	no	no	no	no	no	no	no	no
Partition number	no	no	no	no	no	no	no	no	no	no
Pascal number	yes	yes	yes	yes	yes	yes	yes	yes	yes	yes
Pentadecagonal number	no	no	no	no	no	no	no	no	no	no
Pentagonal number	no	no	no	no	yes	no	no	no	no	no
Perfect number	no	no	no	no	no	no	no	no	no	no
Polite number	yes	no	yes	yes	yes	yes	yes	yes	yes	yes
Prime number	yes	no	no	no	no	no	yes	no	no	no

Number property	31	32	33	34	35	36	37	38	39	40
(Triangular) Pyramidal number	no	no	no	no	yes	no	no	no	no	no
Ramanujan number	no	no	yes	no	no	no	no	no	no	no
Rectangular number	no	yes	yes	yes	yes	yes	no	yes	no	yes
Repdigit number	no	no	yes	no	no	no	no	no	no	no
Repunit number	no	no	no	no	no	no	no	no	no	no
Seesaw number	no	no	no	no	no	no	no	no	no	no
Smith number	no	no	no	no	no	no	no	no	no	no
Sphenic number	no	no	no	no	no	no	no	no	no	no
Square number	no	no	no	no	no	yes	no	no	no	no
Square pyramidal number	no	no	no	no	no	no	no	no	no	no
Super even number	no	no	no	no	no	no	no	no	no	yes
Super odd number	yes	no	yes	no	yes	no	yes	no	yes	no
Tetradecagonal number	no	no	no	no	no	no	no	no	yes	no
Tetrahedral number	no	no	no	no	yes	no	no	no	no	no
Triacontagonal number	no	no	no	no	no	no	no	no	no	no
Triangular number	no	no	no	no	no	yes	no	no	no	no
Tridecagonal number	no	no	no	no	no	yes	no	no	no	no
Trimorphic number	no	no	no	no	no	no	no	no	no	no
Undulating number	no	no	no	no	no	no	no	no	no	no
Unhappy number	no	no	yes	yes	yes	yes	yes	yes	yes	yes
Vampire number	no	no	no	no	no	no	no	no	no	no
Woodall number	no	no	no	no	no	no	no	no	no	no

Comparison of numbers (continued)

Number property	41	42	43	44	45	46	47	48	49	50
Abundant number	no	yes	no	no	no	no	no	yes	no	no
Alternating number	yes	no	yes	no	yes	no	yes	no	yes	no
Armstrong number	no	no	no	no	no	no	no	no	no	no
Ascending number	no	no	no	no	yes	yes	yes	yes	yes	no
Automorphic number	no	no	no	no	no	no	no	no	no	no
Bipolygonal number	no	no	no	no	no	no	no	no	no	no
Catalan number	no	yes	no	no	no	no	no	no	no	no
Centered cube number	no	no	no	no	no	no	no	no	no	no
Centered decagonal number	no	no	no	no	no	no	no	no	no	no
Centered dodecagonal number	no	no	no	no	no	no	no	no	no	no
Centered hendecagonal number	no	no	no	no	no	no	no	no	no	no
Centered heptadecagonal number	no	no	no	no	no	no	no	no	no	no
Centered heptagonal number	no	no	yes	no	no	no	no	no	no	no
Centered hexadecagonal number	no	no	no	no	no	no	no	no	yes	no
Centered hexagonal number	no	no	no	no	no	no	no	no	no	no
Centered icosagonal number	no	no	no	no	no	no	no	no	no	no
Centered icosidigonal number	no	no	no	no	no	no	no	no	no	no
Centered icosihenagonal number	no	no	no	no	no	no	no	no	no	no
Centered icosiheptagonal number	no	no	no	no	no	no	no	no	no	no
Centered icosihexagonal number	no	no	no	no	no	no	no	no	no	no
Centered icosinonagonal number	no	no	no	no	no	no	no	no	no	no
Centered icosioctagonal number	no	no	no	no	no	no	no	no	no	no
Centered icosipentagonal number	no	no	no	no	no	no	no	no	no	no
Centered icositetragonal number	no	no	no	no	no	no	no	no	no	no
Centered icositrigonal number	no	no	no	no	no	no	no	no	no	no
Centered nonadecagonal number	no	no	no	no	no	no	no	no	no	no
Centered nonagonal number	no	no	no	no	no	no	no	no	no	no
Centered octadecagonal number	no	no	no	no	no	no	no	no	no	no
Centered octagonal number	no	no	no	no	no	no	no	no	yes	no
Centered pentadecagonal number	no	no	no	no	no	yes	no	no	no	no
Centered pentagonal number	no	no	no	no	no	no	no	no	no	no
Centered square number	yes	no	no	no	no	no	no	no	no	no
Centered tetradecagonal number	no	no	yes	no	no	no	no	no	no	no
Centered triacontagonal number	no	no	no	no	no	no	no	no	no	no
Centered triangular number	no	no	no	no	no	yes	no	no	no	no
Centered tridecagonal number	no	no	no	no	no	no	no	no	no	no
Composite number	no	yes	no	yes	yes	yes	no	yes	yes	yes
Concatenation number	no	no	no	yes	no	no	no	no	no	no
Cube number	no	no	no	no	no	no	no	no	no	no
Cullen number	no	no	no	no	no	no	no	no	no	no
Decagonal number	no	no	no	no	no	no	no	no	no	no
Deficient number	no	no	no	yes	yes	yes	no	no	yes	yes
Demlo number	no	no	no	no	no	no	no	no	no	no
Descending number	yes	yes	yes	no	no	no	no	no	no	no
Binary digital balanced number	yes	yes	no	yes	no	no	no	no	yes	yes
Dodecagonal number	no	no	no	no	no	no	no	no	no	no

Number property	41	42	43	44	45	46	47	48	49	50
Dudeney number	no	no	no	no	no	no	no	no	no	no
Even number	no	yes	no	yes	no	yes	no	yes	no	yes
Evil number	no	no	yes	no	yes	yes	no	yes	no	no
Fibonacci number	no	no	no	no	no	no	no	no	no	no
Friedman number	no	no	no	no	no	no	no	no	no	no
Happy number	no	no	no	yes	no	no	no	no	yes	no
Harshad number	no	yes	no	no	yes	no	no	yes	no	yes
Hendecagonal number	no	no	no	no	no	no	no	no	no	no
Heptadecagonal number	no	no	no	no	no	no	no	yes	no	no
Heptagonal number	no	no	no	no	no	no	no	no	no	no
Hexadecagonal number	no	no	no	no	yes	no	no	no	no	no
Hexagonal number	no	no	no	no	yes	no	no	no	no	no
Hogben number	no	no	yes	no	no	no	no	no	no	no
Icosagonal number	no	no	no	no	no	no	no	no	no	no
Icosidigonal number	no	no	no	no	no	no	no	no	no	no
Icosihenagonal number	no	no	no	no	no	no	no	no	no	no
Icosiheptagonal number	no	no	no	no	no	no	no	no	no	no
Icosihexagonal number	no	no	no	no	no	no	no	no	no	no
Icosinonagonal number	no	no	no	no	no	no	no	no	no	no
Icosioctagonal number	no	no	no	no	no	no	no	no	no	no
Icosipentagonal number	no	no	no	no	no	no	no	no	no	no
Icositetragonal number	no	no	no	no	no	no	no	no	no	no
Icositrigonal number	no	no	no	no	no	no	no	no	no	no
Impolite number	no	no	no	no	no	no	no	no	no	no
Interprime number	no	yes	no	no	yes	no	no	no	no	yes
Kaprekar number	no	no	no	yes	no	no	no	no	no	no
Leyland number	no	no	no	no	no	no	no	no	no	no
Lucky number	no	no	yes	no	no	no	no	no	yes	no
Lychrel number	no	no	no	no	no	no	no	no	no	no
Mersenne number	no	no	no	no	no	no	no	no	no	no
Niven number	no	yes	no	no	yes	no	no	yes	no	yes
Nonadecagonal number	no	no	no	no	no	no	no	no	no	no
Nonagonal number	no	no	no	no	no	yes	no	no	no	no
Oblong number	no	yes	no	no	no	no	no	no	no	no
Octadecagonal number	no	no	no	no	no	no	no	no	no	no
Octagonal number	no	no	no	no	no	no	no	no	no	no
Odd number	yes	no	yes	no	yes	no	yes	no	yes	no
Odious number	yes	yes	no	yes	no	no	yes	no	yes	yes
Palindrome number	no	no	no	yes	no	no	no	no	no	no
Pandigital number	no	no	no	no	no	no	no	no	no	no
Pascal number	yes	yes	yes	yes	yes	yes	yes	yes	yes	yes
Partition number	no	yes	no	no	no	no	no	no	no	no
Pentadecagonal number	no	yes	no	no	no	no	no	no	no	no
Pentagonal number	no	no	no	no	no	no	no	no	no	no
Perfect number	no	no	no	no	no	no	no	no	no	no
Polite number	yes	yes	yes	yes	yes	yes	yes	yes	yes	yes
Prime number	no	no	no	no	no	no	yes	no	no	no

Number property	41	42	43	44	45	46	47	48	49	50
(Triangular) Pyramidal number	no	no	no	no	no	no	no	no	no	no
First order Ramanujan number	no	no	no	yes	no	no	no	no	no	no
Rectangular number	no	yes	no	yes	yes	yes	no	yes	yes	yes
Repdigit number	no	no	no	yes	no	no	no	no	no	no
Repunit number	no	no	no	no	no	no	no	no	no	no
Seesaw number	no	no	no	no	no	no	no	no	no	no
Smith number	no	no	no	no	no	no	no	no	no	no
Sphenic number	no	yes	no	no	no	no	no	no	no	no
Square number	no	no	no	no	no	no	no	no	yes	no
Square pyramidal number	no	no	no	no	no	no	no	no	no	no
Super even number	no	yes	no	yes	no	yes	no	yes	no	no
Super odd number	no	no	no	no	no	no	no	no	no	no
Tetradecagonal number	no	no	no	no	no	no	no	no	no	no
Tetrahedral number	no	no	no	no	no	no	no	no	no	no
Triacontagonal number	no	no	no	no	no	no	no	no	no	no
Triangular number	no	no	no	no	yes	no	no	no	no	no
Tridecagonal number	no	no	no	no	no	no	no	no	no	no
Trimorphic number	no	no	no	no	no	no	no	no	yes	no
Undulating number	no	no	no	no	no	no	no	no	no	no
Unhappy number	yes	yes	yes	no	yes	yes	yes	yes	no	yes
Vampire number	no	no	no	no	no	no	no	no	no	no
Woodall number	no	no	no	no	no	no	no	no	no	no

Note: (a) All triangular pyramidal numbers are tetrahedral numbers.

(b) 9 is a triangular square number and so a bipolygonal number.

(c) 10 is a triangular tetrahedral number and so a bipolygonal number.

(d) 36 is a triangular square number and so a bipolygonal number.

SOLVED EXAMPLES

1. What is the place value of the digit 5 in the number 5,726,631?
   Solution: Place value of 5 in 5726631 is ten lakh or million
2. Write the binary equivalent of 256
   Solution: 256/2 = 128 remainder is 0
   128/2 = 64 remainder is 0
   64/2 = 32 remainder is 0
   32/2 = 16 remainder is 0
   16/2 = 8 remainder is 0
   8/2 = 4 remainder is 0
   4/2 = 2 remainder is 0
   2/2 = 1 remainder is 0
   The binary equivalent of 256 is 100000000
3. The number which can divide 916916 is
   a. 101   c. 1001
   b. 11   d. 10001
   Solution: c. 1001
4. The alphanumeric system is
   a. Binary number system   c. octal number system
   b. Hexadecimal system   d. penta number system
   Solution: b. Hexadecimal system
5. Find out the sum of 51+52+53+...+98+99+100 by Gauss method of pairing of numbers.
   Solution

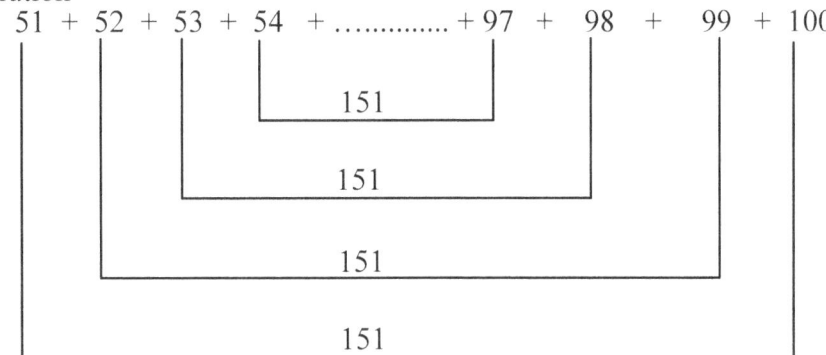

   The sum of consecutive natural numbers will be 151×25=3775.
6. Find out the missing in the series 3, 7, 9, 13, 15, 19, ?, 25...
   Solution: Missing number is 21
7. Find out the missing number in the series 7, 9, 13, 21, ?
   Solution: Missing number is 37
8. (153×109)+(82×153)-(153×91)=?
   Solution (153×109)+(82×153)-(153×91)=153(109+82-91)
   =153(100)
   =15300
9. If the addition sum of a real number with the same number is 20 and its multiplication product of the real number with the same real number is 100, find out the number?
   Solution: 10
10. Find the missing term in the following series
    81, 9, 64, 8, 49, 7, 36, 6, 25, ?

Solution: 5

11. If $\dfrac{32 \times 4 + \sqrt{x}}{36} = 4$, then find out the value of x

    Solution $(32 \times 4) + \sqrt{x} = 4 \times 36$

$$(32 \times 4) + \sqrt{x} = 144$$
$$\sqrt{x} = 144 - 128$$
$$\sqrt{x} = 16$$
$$x = 256$$

12. $1^3+2^3+3^3+4^3+5^3+6^3+7^3+8^3+9^3 = ?$

    Solution: The summation of the cubic numbers is correspondingly equal to the sum of the arithmetic progressive squared.

$$1^3+2^3+3^3+4^3+5^3+6^3+7^3+8^3+9^3 = (1+2+3+4+5+6+7+8+9)^2$$
$$1^3+2^3+3^3+4^3+5^3+6^3+7^3+8^3+9^3 = 45^2$$

13. Which is the largest among $(33)^3$, $(3+3)^3$, $(3\times3)^3$, 3^{33}

    Solution 3^{33}

14. Find out the missing number in the series 9, 81, ?, 6561…

Solution: 729

15. Convert 12% into ppm

    Solution: To convert % into ppm we have to multiply by 10,000. So we get 12,000ppm is the ppm amount of 12%.

16. Find the odd man out: 5, 7, 9, 17, 23, 37

    Solution: 9 is the answer. Other numbers are prime numbers.

17. If $(x+y) : (x-y) = 4 : 1$, then $(x^2+y^2) : (x^2-y^2) = ?$

    Solution $(x+y) : (x-y) = 4 : 1$

$$x+y = 1$$
$$x-y = 4$$

    Solving we get $x=5/2$ and $y=(-3/2)$

    So $(x^2+y^2) : (x^2-y^2) = (34/4) : (16/4)$

Cancelling the common terms we get $(x^2+y^2) : (x^2-y^2) = 17 : 8$

18. The arithmetic mean of ten numbers is -7. If 5 is added to every number then find out the new arithmetic mean

    Solution: The arithmetic mean can be added directly with 5 and so the new arithmetic mean will be -2.

19. Which is least fraction in the following $\dfrac{5}{16}, \dfrac{3}{8}, \dfrac{1}{4}, \dfrac{1}{2}$

    Solution Find the LCM of denominators of the above fractions

        LCM = 16

    Multiply both the numerator and denominator of the above fractions by 16, we get $\dfrac{5}{16}, \dfrac{6}{16}, \dfrac{4}{16}, \dfrac{8}{16}$

Comparing the numerators only we see $\dfrac{4}{16} = \dfrac{1}{4}$ is the least fraction.

20. Find the face value and place value of zero in the number 5021

    Solution: Face value is zero and place value is hundred.

21. What is the scientific notation number of 9845?

    Solution: 9.845×10^3

22. Find the missing in the following magic square

10	5	12
11	?	7
6	13	8

Solution: 9

23. Find the value of 70-30+10

Solution: 50

Exercise

1. What is the hexadecimal equivalent of the number 512
2. The number which can divide 353535 is
   a. 101   c. 1001
   b. 11   d. 10001
3. Find out the sum of 151+152+153+...+998+999+1000 by Gauss method of pairing of numbers.
4. If the addition sum of a real number with the same number is 30 and its multiplication product of the real number with the same real number is 225, find out the number?
5. Find out the value of $-(10)^2$ and (-10^2)
6. Find the value of $(2^3)^2$
7. Find the value of 2 to the power of 3 to the power of 2
8. Find the value of $\sqrt[4]{32} \times \sqrt[4]{8}$
9. Convert the surd number $\sqrt[8]{12}$ into index number or power number
10. The ratio of three numbers is 3:4:5 and the sum of their squares is 1250. Find out the sum of the three numbers?
11. The average weight of a group of 25 boys was calculated to be 30 kg. It was later found that one weight was misread as 25 kg instead of 35 kg. Then what is the correct average weight?
12. $48 \square 12 \times \left(\left(\frac{9}{8} of \frac{4}{3} \right) / \left(\frac{3}{4} of \frac{2}{3} \right) \right)$
13. Find the reciprocal of $\frac{-7}{13}$
14. What is the scientific notation number of 0.000482089?

Refer the next page number for the answer.

Answers

1. 200
2. a. 101
3. 489,175
4. 15
5. -100 and 100
6. 64
7. 512
8. 4
9. $(12)^{\frac{1}{8}}$
10. 60
11. 30.4 kg
12. 12
13. $\dfrac{-13}{7}$
14. 4.82089×10^{-4}

Subject index

Centered triangular number	Denominator
Centered triangular prime number	Descending number
Centered triacontagonal composite number	Determinative property
Centered triacontagonal prime number	Digital root (or) Digit sum
Centered tetradecagonal composite number	Digitally balanced number
Centered tetradecagonal prime number	Digitally imbalanced number
Centered tetrahedral number	Direct proportionality
Centered triacontagonal number	Dividend
Centered tridecagonal composite number	Divisible number
Centered tridecagonal number	Divisibility tests
Centered tridecagonal prime number	Divisive property
Chenshuwen's sequence	Divisor
Circular prime	Dodecagonal composite number
Climbing number	Dodecagonal number
Closure property	Dodecagonal prime number
Columnwise addition	Dodecagonal pyramidal number
Commutative property	Dozenal number or duodecimal number
Complementary number	Dozenal number system
Complex number	Dudeney number
Composite ascending number	**E**
Composite deficient number	Enneagonal pyramidal number
Composite descending number	Eratosthenes sieve
Composite Fibonacci number	Even Armstrong number
Composite happy number	Even centered dodecagonal number
Composite hexagonal pyramidal number	Even centered hendecagonal number
Composite lucky number	Even centered heptadecagonal number
Composite negative seesaw number	Even centered heptagonal number
Composite number	Even centered icosihenagonal number
Composite partition number	Even centered icosiheptagonal number
Composite Pascal number	Even centered icosinonagonal number
Composite positive seesaw number	Even centered icosipentagonal number
Composite undulating number	Even centered icositrigonal number
Composite unhappy number	Even centered nonagonal number
Concatenation number	Even centered pentadecagonal number
Consecutive number	Even centered pentagonal number
Coprime number	Even centered triangular number
Cousin prime	Even centered tridecagonal number
Cube number	Even decagonal number
Cube root	Even decagonal pyramidal number
Cullen number	Even deficient number
Cyclic number	Even demlo number
D	Even dodecagonal number
Decagonal number	Even dodecagonal pyramidal number
Decagonal pyramidal number	Even enneagonal pyramidal number
Decenary system	Even happy number
Decimal number	Even hendecagonal pyramidal number
Deficient number	Even heptadecagonal number
Demlo number	Even heptagonal number

Even heptagonal pyramidal number	Even triangular number
Even hexadecagonal number	Even triangular pyramidal number
Even hexagonal number	Even tridecagonal number
Even Kaprekar number	Even undulating number
Even unhappy number	**F**
Evil number	Face value
Exponential number see Power number	Factorial
Even ascending number	Factorization
Even composite number	Factorial number
Even descending number	Factorial prime number
Even harshad number	Fermat number
Even hexagonal pyramidal number	Fermat prime
Even icosagonal number	Fibonacci number
Even icosahedral number	Fibonacci series
Even icosidigonal number	Finite imperfect division
Even icosihenagonal number	Following number see Consecutive number
Even icosiheptagonal number	Fractions
Even icosihexagonal number	Friedman number
Even icosinonagonal number	**G**
Even icosioctagonal number	Gaussian prime number
Even icosipentagonal number	Geometric progression
Even icositetragonal number	Golden ratio
Even icositrigonal number	**H**
Even negative seesaw number	Happy number
Even nonadecagonal number	Harmonic progression
Even nonagonal heptagon number	Harshad number
Even nonagonal number	Hendecagonal number
Even number	Hendecagonal pyramidal number
Even octadecagonal number	Heptadecagonal composite number
Even octagonal number	Heptadecagonal number
Even octagonal pentagon number	Heptadecagonal prime number
Even octagonal pyramidal number	Heptagonal composite number
Even octahedral number	Heptagonal hexagon number
Even palindromic nonagonal number	Heptagonal number
Even partition number	Heptagonal pentagon number
Even Pascal number	Heptagonal prime number
Even pentadecagonal number	Heptagonal pyramidal number
Even pentagonal number	Heptagonal square number
Even pentagonal pyramidal number	Heptagonal triangular number
Even pentagonal triangular number	Heteromecic number see Oblong number
Even positive seesaw number	Hexadecagonal number
Even repunit number	Hexadecimal number system
Even smith number	Hexagonal number
Even square pyramidal number	Hexagonal pentagon number
Even square triangular number	Hexagonal pyramidal number
Even tetradecagonal number	Hexagonal pyramidal palindromic number
Even tetrahedral number	Hexagonal square number
Even triacontagonal number	Hexagonal square triangular number

Odd icosihexagonal number	Odd octagonal pyramidal number
Odd icosinonagonal number	Odd octahedral number
Odd icosioctagonal number	Odd palindromic nonagonal number
Odd icosipentagonal number	Odd Pascal number
Odd icositetragonal number	Odd pentadecagonal number
Odd icositrigonal number	Odd pentagonal number
Odd nonadecagonal number	Odd positive seesaw number
Odd nonagonal heptagon number	Oneness
Odd nonagonal number	Opposite property
Octahedral palindrome number	Ordinal number
Octal number	Odd ascending number
Octal number system	Odd descending number
Odd Armstrong number	Odd happy number
Odd centered dodecagonal number	Odd Kaprekar number
Odd centered hendecagonal number	Odd negative seesaw number
Odd centered heptadecagonal number	Odd partition number
Odd centered heptagonal number	Odd pentagonal pyramidal number
Odd centered icosihenagonal number	Odd pentagonal triangular number
Odd centered icosiheptagonal number	Odd prime number
Odd centered icosinonagonal number	Odd repunit number
Odd centered icosipentagonal number	Odd smith number
Odd centered icositrigonal number	Odd square pyramidal number
Odd centered nonagonal number	Odd square triangular number
Odd centered pentadecagonal number	Odd tetradecagonal number
Odd centered pentagonal number	Odd tetrahedral number
Odd centered triangular number	Odd triacontagonal number
Odd centered tridecagonal number	Odd triangular number
Odd composite number	Odd triangular pyramidal number
Odd decagonal number	Odd tridecagonal number
Odd decagonal pyramidal number	Odd undulating number
Odd deficient number	Odd unhappy number
Odd demlo number	Odious number
Odd dodecagonal number	**P**
Odd dodecagonal pyramidal number	Palindrome number
Odd enneagonal pyramidal number	Palindrome number magic square
Odd harshad number	Palindromic seesaw number
Odd hendecagonal pyramidal number	Palindrome property
Odd heptadecagonal number	Palindromic centered decagonal number
Odd heptagonal number	Palindromic centered decagonal prime number
Odd heptagonal pyramidal number	Palindromic centered dodecagonal number
Odd hexadecagonal number	Palindromic centered hendecagonal number
Odd hexagonal number	Palindromic centered heptagonal number
Odd hexagonal pyramidal number	Palindromic Woodall number
Odd icosagonal number	Palindromic Cullen number
Odd number	Palindromic happy number
Odd octadecagonal number	Pandigital number
Odd octagonal number	Parasitic number
Odd octagonal pentagon number	Partition number

Partition prime number	Prime descending number
Pascal number	Prime Fibonacci number
Pascal prime	Prime happy number
Pascal series	Prime hexagonal pyramidal number
Pascal triangle	Prime negative seesaw number
Pentadecagonal number	Prime number
Pentagonal number	Prime partition number
Pentagonal pyramidal number	Prime Pascal number
Pentagonal square number	Prime positive seesaw number
Pentagonal triangular number	Prime triplets
Palindromic centered nonagonal number	Prime unhappy number
Palindromic centered octadecagonal number	Probable number
Palindromic centered pentadecagonal number	Prorepdigit number
Palindromic centered pentagonal number	Prorepunit number
Palindromic centered tetradecagonal number	Pythagorean triples
Palindromic centered triangular number	Pental number
Palindromic composite number	Pental number system
Palindromic cube number	Percentages
Palindromic deficient number	Perfect division
Palindromic demlo number	Perfect number
Palindromic even number	Place value
Palindromic factorial prime number	Polite number
Palindromic hendecagonal number	Positive number
Palindromic heptagonal number	Positive palindromic seesaw number
Palindromic icosidigonal number	Positive seesaw number
Palindromic isolated prime number	Power number
Palindromic Kaprekar number	Power Pascal number
Palindromic lucky number	Power triangular number
Palindromic nonagonal number	Powerful number
Palindromic odd number	Prime ascending number
Palindromic partition number	Prime lucky number
Palindromic partition prime number	Prime number
Palindromic pentagonal number	Prime quadruplets
Palindromic prime number	Prime undulating number
Palindromic pronic number	Pental number
Palindromic safe prime number	Pental number system
Palindromic smith number	**Q**
Palindromic square number	Quotient
Palindromic star prime number	**R**
Palindromic Thabit prime number	Rachnisky sequence
Palindromic triacontagonal number	Radicand
Palindromic triangular number	Radix
Palindromic tridecagonal number	Ramanujan number
Palindromic undulating number	Ratio
Palindromic Wagstaff prime number	Rational number
Palindromic Wieferich prime number	Real number see Natural number
Precursor number	Reciprocal property
Prime deficient number	Reciprocal number

Rectangular number	Strobogrammatic number
Recurring infinite division	Subtractive property
Reflux number	Surd
Reflux prime number	Symmetric property
Reflux property	**T**
Remainder theorem	Ternary number
Repdigit centered heptagonal number	Ternary number system
Repdigit centered octadecagonal number	Tetradecagonal number
Repdigit lucky number	Tetradic number
Repdigit nonagonal number	Tetrahedral number
Repunit centered nonagonal number	Tetrahedral palindrome number
Repunit composite number	Tetrahedral square number
Repunit dodecagonal number	Tetrahedral triangular number
Repunit hexagonal number	Trailing zero
Repunit nice Friedman number	Transcendental number
Repunit nonagonal number	Transitive property
Repunit number	Transpose number
Repunit partition number	Trapezium number
Repunit prime number	Triacontagonal number
Repunit triangular number	Triangular number
Repdigit number	Triangular pyramidal number
Repunit Fibonacci number	Tridecagonal composite number
Repdigit happy number	Tridecagonal number
Repdigit hendecagonal number	Tridecagonal prime number
Repunit Pascal number	Tripolygonal number
Rhombus number see Square number	Twin prime number
Right cancellation property	**U**
Right distributive property	Undulating number
Roman Friedman number	Unhappy number
Roman number	Unit number
S	Upset property
Safe prime number	Upset reflux property
Scientific notation number	**V**
Seesaw number	Vampire Friedman number see Vampire number
Smith number	Vampire number
Square Cullen number	**W**
Square number	Wagstaff prime number
Square pyramidal number	Weird property
Square pyramidal square number	Whole number
Square pyramidal tetrahedral number	Wieferich prime
Square pyramidal triangular number	Wilson prime number
Square root	Woodall number
Square triangular number	Woodall prime number
Squareful number	**Z**
Star composite number	Zero
Star number	Zeroless pandigital Friedman number
Star palindrome number	Zeroless pandigital nice Friedman number
Star prime number	Zeroless pandigital number

REFERENCES

1. Ya. Perelman: Mathematics can be fun, Mir publishers, Moscow (1985).
2. J. T. Glover: Vedic mathematics for schools Book 1, 2 & 3, Motilal Bansaridass Publishers, Delhi (1999).
3. H.K.Gupta: Vedic mathematics, bpi India private limited.
4. T. A. Ramasubban: Speed arithmetic, Efficient prints, Chennai (1999).
5. H.C.Khare: Issues in Vedic mathematics, Rashtriya Veda vidya Pratishthan.
6. Kenneth R. Williams: Discover Vedic mathematics, Motilal Bansaridass Publishers, Delhi.
7. Jagadhguru Sri Bharti Krishna Maharaj: Vedic mathematics, Motilal Bansaridass Publishers, Delhi.
8. Jayanth V. Narlikar: Fun and fundamentals of mathematics, University press, Delhi (2001).
9. Amit Garg: Fun with numbers, Pustak Mahal, Delhi, (1999).
10. D.Jaganmohan Rao: Mathematics quizmaster, Neelkamal publications private limited, New Delhi, (2003).
11. Vandana Sighal: Vedic mathematics for all ages, Motilal Bansaridass Publishers, Delhi, (2007).
12. D.Jaganmohan Rao: Sensations in mathematics, Neelkamal publications private limited, New Delhi, (2007).
13. Dr. S.K.Kapoor: Vedic mathematics skills, Lotus press, New Delhi (2006).
14. Dhaval Bathia: Vedic mathematics, Jaico publishing house, Mumbai.
15. G.Polya: How to solve it, prentice hall of India private limited, New Delhi.
16. K. Venkataraman: Think without ink, TWI foundation, Bangalore.
17. Shailesh Shirali: A primer on number sequences, University press.
18. Shailesh Shirali: A primer on divisibility, University press.
19. Shailesh Shirali: Adventures in problem solving, University press.
20. T.S. Bhanu Murthy: A modern introduction to Indian mathematics, Wiley Eastern limited, New Delhi (1995).
21. Shakuntala Devi: The book of numbers, Orient paperbacks, New Delhi (1984).
22. Ravi Narula: Brain Teasers, Jaico publishing house, Mumbai (2000).
23. Shakuntala Devi: The joy of numbers, Orient paperbacks, New Delhi (2000).
24. Shakuntala Devi: Puzzles to puzzle you, Orient paperbacks, New Delhi (2000).
25. Shakuntala Devi: More puzzles to puzzle you, 40th printing, Orient paperbacks, New Delhi (2006).
26. Dr.S.K.Kapoor: Learn and Teach Vedic Mathematics, Lotus press, New Delhi (2005).
27. Robert D. Carmichael: The Theory of numbers, MJP publishers, Chennai (2008).
28. Debra Anne Ross: Master Math Basic math and pre algebra, Master mind books, Bangalore (2002).
29. Bibhutibhushan Datta: History of Hindu mathematics, Bharatiya kala prakashan, Delhi (2004).

30. Varun Shastri: Academic dictionary of mathematics, ISHA books, Delhi (2005).

31. S. Haridas: Vedic Mathematics Part I, Bharatiya Vidhya Bhavan, Mumbai (2000).

32. Hans Rademacher, The Enjoyment of mathematics, Princeton university press, New Jersey, USA (1957).

33. L.M.Singvi: The Natural Calculator, Motilal Bansaridass Publishers, Delhi (1991).

Features

➢ Important arithmetic terms and definitions
➢ Lucid explanation of terminologies
➢ Glossary of arithmetic numbers
➢ Comparative study of mathematical properties
➢ Understanding explanation of basic arithmetic
➢ A companion guide for preparing competitive examinations
➢ Clearly an amazing arithmetic book for further studies

About the author

Mr. R. Sridharan M.Sc., B.Ed., M.Phil. is working as an Assistant professor, having 5 years of teaching experience to the B.E. and B.Tech. students. Prior to that he had 14 years of industrial experience in chemical companies. Though the core subject of the author is Chemistry, he is interested in the areas of Mathematics especially in Arithmetic.